The Political Economy of Environmental Policy:
an Australian Introduction

The Political Economy of Environmental Policy:

An Australian Introduction

K. J. Walker

UNSW PRESS

For Mary

Published in Australia by
UNIVERSITY OF NEW SOUTH WALES PRESS
Kensington NSW Australia 2052
Telephone (02) 398 8900
Fax (02) 398 3408

© K. J. Walker, 1994
First published 1994

This book is copyright. Apart from any fair dealing for the purposes of private study, research, criticism or review, as permitted under the Copyright Act, no part may be reproduced by any process without written permission from the publisher.

National Library of Australia
Cataloguing-in-Publication entry:

Walker, K.J.(Kenneth James), 1940 -
   The Political economy of environmental policy: an
   Australian introduction

   Bibliography.
   Includes index

   ISBN 0 86840 070 X.

   1. Environmental policy — Australia . 2. Environmental policy — Murray River Region (N.S.W.–S. Aust.). 3. Environmental policy — New South Wales — Darling River Region. 4. Australia — Politics and Government. I. Title.

333.70994

Formatted in Quark Xpress, Goudy 11/12pt
Printed by Southwood Press, Marrickville

Available in North America through:
ISBS Inc
Portland Oregon 97213-3644
Tel: (503) 287 3093
Fax: (503) 280 8832

Contents

Preface	vii
Acknowledgements	xi
Introduction	1

PART 1 Environmental Problems — 7

1	Toxic Waste Pollution	9
2	Humans and Their Impact on the Environment	34
3	Ecology and Politics	51

PART 2 The Murray–Darling System — 70

4	Salinity in the Murray–Darling River System	71
5	Irrigation: A History of Conflict	87
6	Gridlock and Landslide	105
7	Taking Stock	126

PART 3 A Dash of Theory — 139

8	Scarcity, Competition, and Collective Goods	141
9	Strategic Games	151
10	Social Choice	169
11	Making Decisions	185

PART 4: Political Economy and Public Policy	**202**
12 The State in Environmental Management	204
13 The Modern State	217
14 Development and Environment in Australia	232
15 The Political Economy of 'Development' in Australia	247

PART 5 Environmental Policy in Australia	**261**
16 Making Policy	263
17 Evaluating Environmental Policy	282
18 The Challenge of Environmental Policy	303

Notes	317
Bibliography	321
Index	331

PREFACE

There are lots of books about 'our environmental crisis', with copious details of the problems. There is a growing literature on the management of Australia's natural resources. There are numerous studies of the making of public policy: fiscal, development, social welfare, infrastructure, and so on. But there are very few about environmental policy, systematically linking the environmental and ecological issues to their social, economic, and political expressions.

That is one purpose of this book. Put very simply, environmental problems are almost always about common goods, and frequently about the stresses placed on them by scarcity, congestion and competition. They are difficult to deal with by mechanisms such as the market, and intractable subjects for bargaining and horse-trading. The collective nature of environmental problems, together with the fact that they are fundamental, means that they inevitably entail making choices, and that those choices are interdependent: collective or social choices. A very considerable literature exists on the processes of social choice, and its implications permeate the social and political institutions of human societies. These confront government with serious challenges, the full impli-

Preface

cations of which have not yet been grasped. It is important to sketch out these issues and explore their ramifications.

Environmental policy is, in a sense, the cement that binds many of these seemingly disparate issues. But policy cannot be understood without some critical understanding of the various underlying problems, physical, biological, ecological, social and economic. A full synthesis would be a massive and probably impossible task; further, as argued in the penultimate chapter of this book, it may not be an especially worthwhile one, given the urgency and variability of many of the issues. But for understanding many complex contemporary problems, a sort of pragmatic interdisciplinarity is a necessary minimum, and no book on environmental problems should avoid it. Consequently, many of this book's chapters attempt to bring together considerations from a wide range of disciplines, and show why and how they interrelate.

Furthermore, environmental issues — in particular, the question of the long-term sustainability of human civilisation — raise some of the thorniest and most enduring questions of political philosophy, albeit with a novel twist. To explore these problems in any depth would require a further book or books; but the philosophical questions remain in the background and should not be forgotten. Occasionally they will be touched upon here.

In addition, this book approaches environmental problems in a radical way, attempting to link them closely to day-to-day, bread-and-butter issues. In doing so, it hopes to advance the debate on environmental policy substantially. For unless environmental concerns are woven far more directly and explicitly into public policy, in Australia as elsewhere, important and fundamental issues will be systematically misinterpreted and neglected.

Taking an environment-oriented policy perspective requires attention to issues neglected by texts emerging from disciplines such as geography, biology or the earth sciences. Concerns to which they, quite properly, pay much attention must be pushed into the background. This book is therefore complementary to, rather than competitive with, more orthodox textbooks; it does not attempt to reproduce their content, though it necessarily makes links with it. Readers wishing to know more about the formation of public policy, for example, would do well to consult the text by Davis, Wanna, Warhurst, and Weller (*Public Policy in Australia*); and for political economy, works by Bell and Head. Mercer (*A Question of Balance*) and Cocks (*Use With Care*) both offer issue-oriented reviews of major Australian environmental problems, though their decision process models do not draw on the latest knowledge. Smith's Open Learning collection, *The Unique Continent*, can be recommended as a basic introduction, linking biological and ecological themes to climatic, geological, anthropological and practical perspectives.

SOURCES OF INFORMATION

This book is intended mainly for the lay reader or the student encountering environmental policy as an area of study for the first time. To improve readability, it departs from the normal academic style of referencing and sourcing of material. All sources consulted in the preparation of this book will be found in the Bibliography. Within each chapter, references in the form of endnotes have been kept to the minimum: generally, only quotations and points of substantial importance have been given explicit page references. Students are cautioned that following the same technique in their essay work may not maximise their marks!

Additionally, each chapter has an *annotated bibliography* at the end of the chapter, under the heading 'Further Exploration'. This suggests some useful readings bearing on the topics discussed in the chapter, and attempts briefly to indicate their relevance and importance. In general, much of the information and many of the ideas presented in the chapter will derive from these sources or be directly relevant to them.

In selecting sources, *readability* has been an important criterion; in some cases readings which are accessible and enjoyable have been preferred to more recent, but less accessible material. Furthermore, each chapter has been written to bring out the connections with these sources, as well as to suggest issues they may have ignored, or problems which link apparently disparate areas.

Some important *typographical conventions* have been adopted throughout this book. Firstly, new ideas and terms are presented in **bold** type on their first appearance. These are either essential to understanding the material being presented, or recur later in the book. Often they are associated with a definition. Secondly, *italics* are used for emphasis, as well as for the titles of books and journals, according to established convention. Thirdly, the proper names of specific institutions are given in capitals (the State of New South Wales; the Whitlam Labor Government) whereas the same terms applied in a general context are not capitalised (the state as a form of social organisation, government in general). Fourthly, acronyms — abbreviations of the names of organisations, etc., made by using the initials — are given in capitals, thus: CSIRO, not Csiro.

Preface

WEIGHTS AND MEASURES

Since decimalisation and metrication, the citing of weights, measures, and currency have created continual difficulties, especially for those of us brought up in the 'real', non-metric or decimal systems.

There is little use in converting monetary values of the late 19th Century, expressed in pounds (£) into modern dismal guernsey, because inflation renders the comparison meaningless. Pre-decimal currency amounts are therefore given in their original units; interested readers anxious to know the purchasing power may find it useful to compare the amount with the then minimum wages.

Similarly, most weights and measures have been given in the original format. This can cause confusion where US, Imperial, and metric dimensions are all used. In direct quotes from US sources in this book, a 'ton' means a 'short' ton of 2,000 lbs; pints and gallons are 4/5ths of their Imperial size. Thus a 55-gallon (US) drum is the same as a 44-gallon (Imperial) one. In the body of the text, US dimensions have generally been converted to Imperial, except for some metric examples. These are clearly marked, either by established conventional usage — 'ton' for Imperial ton, 'tonne' for the metric ton — or by terminology. Few readers will experience difficulty in distinguishing litres from pints, kilometres from miles.

Acknowledgements

Intellectual endeavours cannot be separated from the histories of institutions, given that intellectuals have to eat, and are typically employed by universities or other similar foundations. Had the dream of an integrated, truly interdisciplinary Environmental Studies programme at Griffith University materialised, this book might have been written as much as ten years earlier; and it might well have been a joint effort. True interdisciplinarity has been frustrated by the conservatism of Australian academia and, especially since the Dawkins 'reforms' to tertiary eduction, remains a dream.

As it is, this book's belated appearance owes much to the encouragement and advice of a handful of good friends and colleagues. Chief among these must be named Reg Henry, for continuing support and wise counsel. Steve Bell and Ian Lowe have had significant, even subliminal, influence. To the late Phil Tighe, as well as to Ros Taplin and Tim Doyle, PhD supervisees turned good colleagues, a great deal is also owed. Mark Carden collaborated on an early draft of what became Chs. 1-5, intended for a different purpose, and has been a constant source of ideas and inspiration. He has been a major catalyst in crystallising a series of broad orientations into a far more structured and coherent intel-

Acknowledgements

lectual orientation. At longer range are debts to Ross Annels, Bruce Davis, John Formby, Peter Hay, Bob Goodin, Aynsley Kellow, Roger Lawrey, Les Milbrath, Hugh Ward and John Young for ideas, encouragement and interest. Needless to remark, none of them bears any blame for the final result.

The author's wife, Mary Heath, gave invaluable assistance in editing and compiling the index. Robin Brown drew Figures 1.4, 1.5, 4.2, 4.3, 4.4, and 14.1 . Sharon Wong drew Figures 4.1, 6.1, 14.2 and 14.3. The maps of Kingston appearing in Ch. 1 are Dudley Nott's handiwork.

Figures 4.3, 4.4 and 6.1 incorporate Crown Copyright material which is reproduced by permission of the Government Printer of the State of Victoria. It was first published by the Law Printer, PO Box 292, South Melbourne, Vic. 3205. The figures appearing in this book are not an official copy of Crown Copyright material, and the State of Victoria accepts no responsibility for their accuracy. Figures 14.2 and 14.3 are modified from three maps appearing in D.W. Meinig's *On the Margins of the Good Earth*. The copyright material contained in them is reproduced by permission of Rand McNally Inc. Figure 1.5 incorporates copyright material from Levine, A., *Love Canal*, and Hall, R.H., 'Poisoning the Lower Great Lakes: the failure of U.S. Environmental Legislation', while Figure 14.1 is drawn from Figure 1.1, Distribution of arid region climates' in Dregne, H.E., *Desertification of Arid Lands;* each is reproduced by kind permission of the authors. Figure 4.3 was extensively redrawn using material from ASTEC, *Environmental Research in Australia: Case Studies*, Figure 4, p. 94 and Jakeman, A.J., et. al., *Water Resource Management in the River Murray*, Figure 2.4, p. 47, used by permission of Commonwealth Information Services and the authors respectively.

INTRODUCTION

Nowadays, environmental problems are global in scope and increasingly frequent. Daily, new disasters are reported in the press or appear as graphic pictures on television. Lakes are drying up, soil is eroding, the atmosphere is heating up, floods are becoming worse and more frequent, people are dying from starvation, toxic contamination, military mayhem, and so on. Yet there is very little understanding of why these problems occur, or how, let alone how to fix them. Environmental policy as a serious area of study is still in its infancy, and few as yet appreciate its enormous importance.

In June 1992, the United Nations Commission for Environment and Development (UNCED) held a major international conference, the 'Earth Summit', at Rio de Janiero in Brazil. They had hoped to finalise a series of substantial agreements on checking the growth of greenhouse gases in the atmosphere, protection of biodiversity, and assistance to the poorer countries in achieving better lives for their people while minimising environmental impacts. These aims were not realised, due in part to 'dog-in-the-manger' behaviour by the world's most powerful and profligate nation, the USA. What the conference did bring out, however, was the fact that in many areas the ecological condition of the world is

Introduction

deteriorating rapidly, threatening the survival of millions. Experts drew attention to declining productivity of previously fertile land, and shortages of water. One claimed that 'safe drinking water is becoming a chemical curiosity in some areas of the world' as a consequence of pollution.[1]

Rio was the political tip of an enormous iceberg, the base of which lies in ecology. While alarmist notions of a global 'environmental crisis' in which the oceans suddenly become stagnant with rotting marine life and atmospheric oxygen declines catastrophically have not (so far) come true, the fact remains that environmental degradation is occurring at a rapidity and on a scale which are quite simply unprecedented. One example may suffice.

The world is experiencing the most massive extinction of species it has ever seen. Both animals and plants are vanishing at a rate of two species per hour.[2] This is a rate far greater than those associated with the five known episodes of mass extinction in the past, such as the loss of the dinosaurs at the end of the Cretaceous, 65 million years ago.[3]

But the causes of all previous extinction episodes have been natural. Stephen Jay Gould identifies some eight hypothesised causes in two major groups: intra- and extraterrestrial. The former include climate change due to continental drift, and collapse of food chains; the latter the much-discussed meteor impact hypothesis. Scientific argument about the merits of these hypotheses still rages.[4] By contrast, the present problem is unambiguously due to a single species, Homo sapiens, which is systematically wiping out not only its competitors but many of the species on which its own future depends. At the same time, it is also interfering seriously with fundamental biological processes on a global scale. The fact of this interference is far more important than the symptoms, of which species extinction is only one. There is of course no master plan for this destruction. It is the sum of the impacts of the most selfish, ruthless and intelligent species the planet has ever seen, single-mindedly pursuing its own short-term advantage with the aid of ever more powerful technologies.

It is a truism that the planet is finite. More important is the fact that most human technologies depend on 'key' resources, which are themselves scarce and represent limiting factors. These may include space, clean air and water, and good arable land, as well as crucial feedstocks for production. Even abundant resources are resistant to exploitation, and thus resources in forms suitable for human use are always relatively scarce.

Furthermore, if populations of humans are large in relation to resources, **congestion** effects will occur: access to desired resources will not always be possible, due not only to shortages, but also to the unavailability of particular resources at desired times: scheduling problems (compare the demand for central-city parking space at 3 am with that at 3 pm, for example). And congestion gives rise to **competition,** which despite its propagandists, is not always desirable. Competition can and frequently does lead to **conflict,** which can be highly destructive. Nobody, for example, would wish to encourage the genocidal forms of competition for land adopted in Croatia and Bosnia during 1991–3.

Congestion, and consequently competition, will be worsened if pollution occurs. Pollution is often defined — carelessly — as the adverse environmental effects of human activity. But such effects are generally perceived as adverse only when they reach human notice and result in the degradation of some amenity, whether it be utilitarian or ësthetic. The perception of pollution is part of the **feedback** from the natural environment to humans. The significance of pollution is that it increases scarcity, and hence congestion and conflict. Loss of unspoiled wilderness, contamination of farmland, fouling of streams, all sharpen competition for the remaining usable resources — sometimes with disastrous effects on those in turn.

These are all social problems, and they are the reason why environmental crises are also political ones. Just as economics is about the allocation of scarce resources, so politics is for settling disputes about ways of increasing, maintaining or decreasing those resources, as well as about the broad strategies for distributing and allocating them. All human societies, even those without governments, perform such functions. Occasionally, but with disappointing rarity, political systems look ahead in an attempt to plan for their long-term welfare. But the pressures of the short term, among them survival in competition with other humans, often lead to short-sighted and ecologically irrational behaviour. Worse, human societies are often poorly equipped for the challenges that face them. The available techniques for making choices are uncertain. Most ordinary humans typically lack understanding of their predicament, leading to an unwillingness to contemplate threats which may well be invisible. Information systems such as press and media have inbuilt, profoundly conservative distortions, frequently trivialising and ignoring important issues. All these factors lead to an institutionalisation of this short-sightedness.

Environmental problems therefore increase the already considerable stresses on the capacities of human societies for **coopera-**

3

Introduction

tion and **coordination**. Worse, they critically strain their capacity for making **collective choices**.

There is nothing particularly novel about this. All human societies, by the very nature of human adaptation, face the **ecological problem**: the question of providing food, shelter, clothing and other desirables for populations which by definition exceed the 'natural' carrying capacity for a specific environment. What is novel about the contemporary crisis is simply that it is global, and it thus presents an urgent and irreversible threat to continued human existence. It also presents some major dilemmas: for example, should human numbers be maximised, even at the cost of falling living standards and serious risks to other species, or should they be limited, in order to secure both a more commodious living standard for humans and the possible survival of other species? Clearly, the policy implications of these courses will be radically different; but either will pose serious dilemmas in implementation.

In order to understand environmental policy — or the lack of it — in modern times, it is necessary to examine actual policy-making processes, their impact on human society, and their impact on its behaviour.

This is particularly true because the available stock of ideas — theories — about the relationship between ecosystems and human social activity is inadequate. Knowledge of ecosystem characteristics themselves is incomplete and, worse, inconclusive. Knowledge of the effects on human societies — even of such effects as toxicity of pollutants — is radically incomplete. This is not altogether surprising; most scientific advances necessarily occur in conditions of uncertainty, without a supporting framework of established theory. Discoveries come first; explanations later. This can and does lead to a fixation with explaining the past at the expense of understanding the future, which has long dogged the social sciences. But diagnosis and prediction are as necessary as explanation; understanding ongoing **processes** of change and applying that understanding are essential first steps in changing the world. Where problems which reflect on policy questions are concerned, policy evaluation often has to take place in advance of explanation, diagnosis or prediction, not least because of poor resources in all the relevant fields of knowledge. It has long been appreciated that social scientists working on environmental questions not only have to contend with **scientific uncertainty** but also with the need to give quick — though unavoidably imprecise — answers without foreclosing important options.[5] But, in the absence of received wisdom about the human–environment interface, the only way to understand both the problems and the ramifications is to examine actual environmental problems and follow their policy implications through.

PLAN OF THE BOOK

This book is divided into five Parts. In Part I, some of the most obvious environmental problems are introduced, by way of two case studies of industrial pollution, one Australian, the other North American. The novelty and complexity of such problems is stressed, as are the inadequacies of government responses, and the effects of incomplete data and scientific uncertainty. This leads to a broader discussion of the impact of technology on the natural environment, introducing a sketch of human adaptation and its dependence on language, sociality and technology. The implications for ecosystemic integrity of a species which necessarily modifies its environment to survive are suggested.

Part II seeks to explain how physical, climatic and ecological constraints can affect human exploitation of a region. It consists of an extended case study of salinity in the Murray–Darling Basin in Australia. The nature of the environment, and the constraints it imposes on development, is first described, followed by the history of the exploitation of the Basin, and the intercolonial disputes to which it gave rise. Examination of the history and environmental impact of irrigation, and especially the growth of salination, precede an exploration of political and administrative responses, in particular the formation of the River Murray Commission and its later supplanting by the Murray–Darling Basin Ministerial Council. The suitability of these responses is assessed.

Part III explains the difficulties explored in the preceding sections by introducing some elementary theoretical problems affecting environmental policy, many of which, sadly, are not widely understood or appreciated. The collective nature of most environmental goods is spelt out, and the deleterious effects of congestion and competition explored. The situation logic of collective goods provision is briefly sketched, making use of the theory of strategic games. This is followed by a review of some of the simpler paradoxes and problems of social choice, especially those associated with Utilitarian 'head-counting' techniques such as voting. Finally, a critical overview of modern theories of decision-making completes this section.

While social choice processes and decision theory explain many problems, further difficulties are created by social structure and institutional rigidity. Part IV reviews the political economy of state societies, and the powerful constraints it imposes on policy. The rǒle of the state since earliest times is explored in the light of the tensions between statecraft and environmental management. The disruptive global impact of the great European expansion of the last 500 years, and of the Industrial Revolution, is sketched. Australia's place in the global colonial economy is related to the manner of its settlement and exploitation, and the ecological consequences. The political economy of modern Australia is briefly reviewed, with special emphasis on the myth of 'development' and its implications for environmental policy.

Part V, the final section, sketches the processes and constraints affecting the making of environmental policy in Australia, especially within government. It points to the domination of the policy process by networks of 'insiders', and the consequent conservatism and bias. It critically reviews techniques for making environmental policies, such as cost-benefit analysis (CBA) and environmental impact analysis (EIA), concluding that both are vulnerable to systematic bias and to fragmentation of overall policy aims. The final chapter explores the challenges posed by environmental issues to 'business as usual', both at a global level, showing how many of the problems explored in the preceding chapters have not been understood or addressed by existing political systems, and how they pose significant challenges to future well-being unless they are resolved in a timely fashion.

Part 1
Environmental Problems

Most people associate the word 'environment' with pollution. Pollution, defined as undesirable feedback from the natural environment, is frequently the most visible and spectacular of the myriad human environmental impacts. Few Australians would be aware of the massive rate of extinction of native flora and fauna since 1788; but most have heard of incidents such as Melbourne's Coode Island fire of 1991, or the persistent pollution of Bondi beach by raw sewage.

Starting with pollution, Part I explores several characteristics of environmental problems, from the human impact on the natural environment to ways of arguing about political — and especially environmental — issues. In Chapter 1, toxic waste pollution problems challenge the capacity of government to recognise, anticipate and deal with issues of a scientific/technical nature. Chapter 2 examines the characteristic human mode of adaptation, showing how its reliance on technology creates ongoing environmental problems. Chapter 3 notes that day-to-day politics is often far removed from urgent environmental problems, attributing its lack of engagement to preoccupation with other issues of more immediate importance to politicians. It shows how problems are often *displaced*, rather than solved, and how arguments about environmental questions can be very dubious.

CHAPTER 1

TOXIC WASTE POLLUTION

Pollution, as one of the most visible human environmental impacts, often causes concern to ordinary people. Toxic wastes frequently generate alarm and political tension. The cases described here raise all the major issues reviewed in the Introduction. All share six characteristics:

1. In none of the cases was the problem clearly defined.
2. Scientific research was inadequate to resolve the problems.
3. Political response was slow.
4. Affected citizens were unhappy with the political system's response.
5. Doubts and fears about the wastes were never finally allayed.
6. Longer-term and geographically remote consequences were not acknowledged.

TOXIC WASTE AT KINGSTON

A number of recent toxic waste contamination scares have resulted from residential development on former industrial sites. In 1987, at Kingston, a working-class suburb on the outskirts of Brisbane in Queensland, toxic acid ooze began surfacing beneath some houses. A serious controversy erupted over the safety of housing in the vicinity of Mount Taylor.

Background
Gold was discovered at Mt. Taylor in 1885, and small-scale mining went on until 1913, four shafts being sunk. Between 1913 and 1932, the mine was worked only sporadically. From 1932 a new management exploited the mine, supplementing underground mining by the development of two open cuts, a large one at the top of the hill, and a smaller one to the north. The mine closed, its gold exhausted, in 1955.

Gold was extracted from Mt. Taylor's low grade ore on site, initially by crushing the ore and separating the metal by amalgamation with mercury, after which the amalgam was 'heated in order to drive off the mercury, and then smelted with sodium carbonate and borax glass to yield gold bullion'.[1] Gold was also recovered from tailings, by using potassium cyanide to dissolve it, after which it was precipitated by being passed over zinc shavings. From 1932 to 1955, the more modern cyanide process was used, dramatically improving recoveries of gold, especially after 1938. Cyanide compounds, now a major waste product of the refining process, accumulated in two tailings dams near the base of the hill. Two spillages were reported to the Queensland Government, in 1936 and 1951, and more probably occurred.

After the mine's closure, the site remained under the control of South Queensland Mines until the company was wound up in 1982. From 1955 to 1967, the smaller, more northerly open cut was leased to the Mobil oil company for disposal of an acidic sludge resulting from the reprocessing of lubricating oil; this included sulphuric acid and heavy metals such as lead. Other firms may also have dumped waste there, including such things as battery acid.[2] This small open cut later became known as the 'oil pit'.

Other waste materials, including mining machinery, were dumped into the original mine shaft and open cut on the hill top, which became a municipal waste dump; more than 30,000 cubic metres were dumped between 1968 and 1973. The site was unfenced and unsupervised; no records were kept. However, witnesses have claimed that at least 12,300 litres of polychlorinated biphenyls (PCBs) and 400 containers of poisonous gas and acid were dumped into the open cut and mine shaft. The exact contaminants and their quantities remain unknown. Only drilling could discover what was deposited; that would be costly and possibly more dangerous than simply 'capping' the site.

Toxic Waste Pollution

Fig 1.1
Mt Taylor Gold Mine at closure in 1960.

Chapter 1

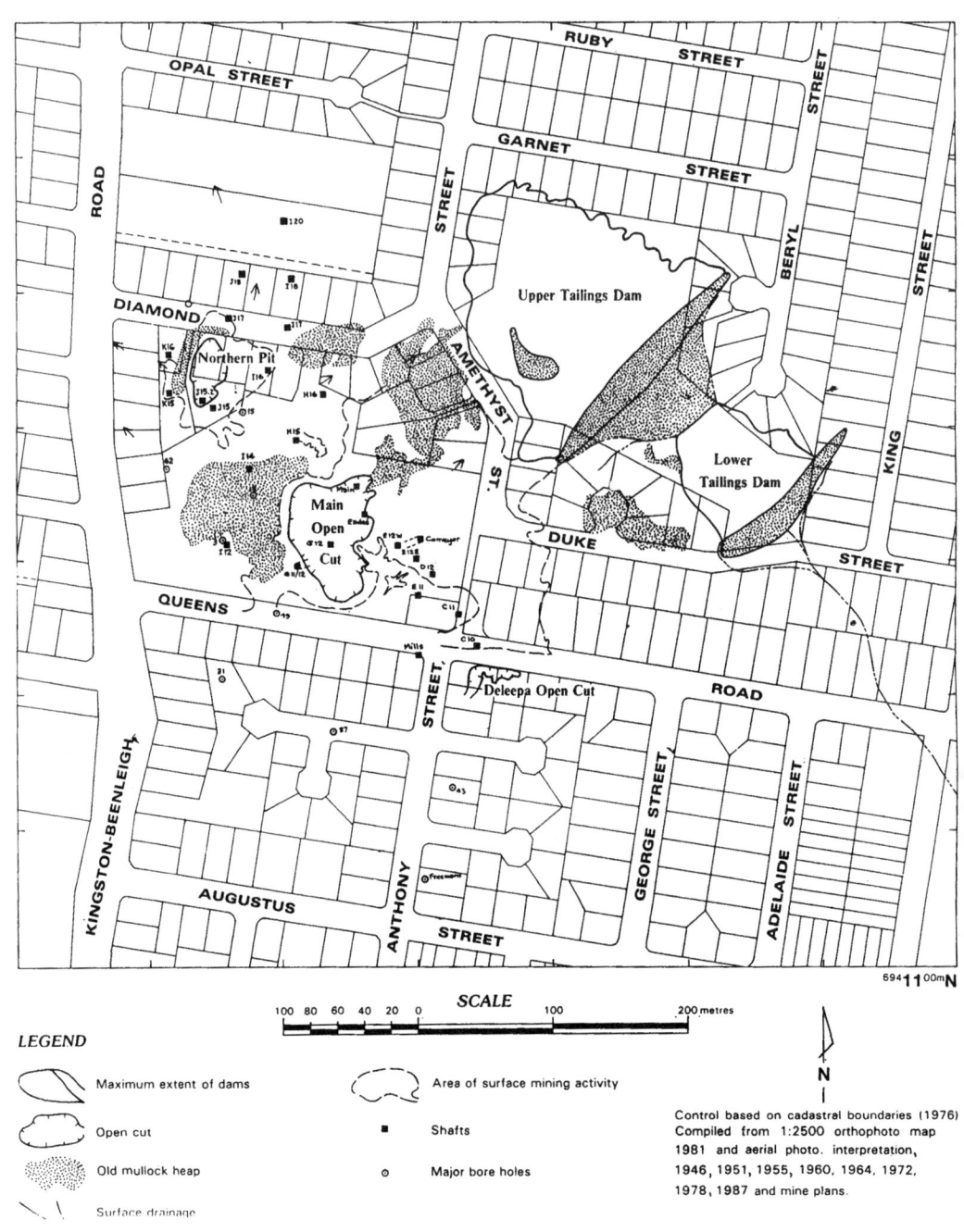

Fig 1.2
Waste sites at Mt Taylor.

BOX 1.1 PCBs AND THE DISPOSAL PROBLEM

Polychlorinated biphenyls (PCBS) are a family of chemicals, widely used throughout industry since 1930 in transformers and capacitors, as hydraulic and heat transfer fluids, as plasticisers, and in paints, varnishes, adhesives, inks, lubricants and sealants. Many of these uses are dissipative — i.e., small quantities are dispersed irrecoverably throughout the environment — but some applications use large quantities (for example, transformer coolants).

PCBs are lethal to some fish at very small concentrations (parts per billion, or less). They are less toxic to mammals and birds, though one case of serious poisoning occurred in Japan when in 1968 rice bran oil was contaminated by leaking coolant, poisoning 1000 people.

However, PCBs are considerably more persistent in the environment than DDT and its derivatives. In consequence, dangers from bioaccumulation over long periods loom large. PCBs are also intractable wastes, needing high temperature incineration or chemical treatment to render them safe.

Except for electrical equipment — capacitors and transformers — the use of PCBs is being phased out. Even so, PCB wastes are accumulating in Australia at the rate of 1000 tonnes per annum; until very recently, incineration seemed the only option for their disposal, but vigorous opposition from affected communities has so far aborted every proposal. The disposal problem remains serious, because of the persistence of the material

---

Some land in the area was subdivided in 1955. In 1968 the 'oil pit' was closed because nearby residents complained of odours and smoke from fires which occasionally erupted in it. Shortly afterwards, on the application of the owner of South Queensland Mines, the area was zoned for residential development by Albert Shire. Urbanisation since closure of the mine and dump has been rapid, and by the mid-1980s the area was extensively built up.

Approval was granted in 1979 for a shopping centre and a bowling alley at the top of the hill over the old mine and open cut. Houses were built on Diamond Street, above the 'oil pit'. Below, to the northeast, the main tailings dams were filled and made into a public park, where children played. They also played in the mine shafts near the top of the hill, where unprotected entrances and pools of water were irresistible lures. It was only after media pres-

Chapter 1

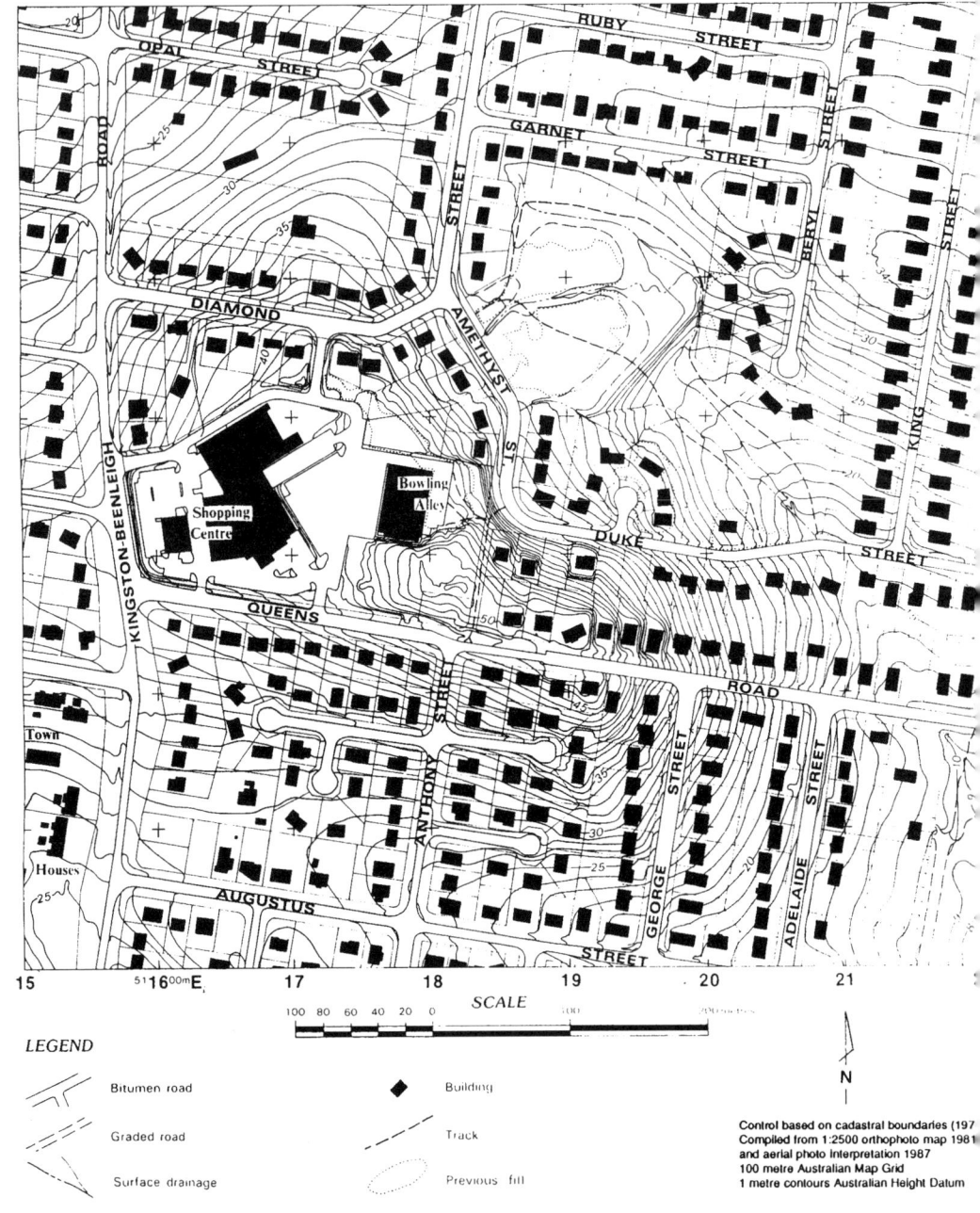

Fig 1.3
Mt Taylor area, Kingston (1987). Extent of commercial and residential sub-divisions.

sure in 1986 that the shafts were finally sealed.

In 1987, after waste in the oil pit had oozed to the surface under the houses built over it, serious public concern was aroused. Subsequently, some of the affected houses were evacuated and removed. Adverse health effects were reported from as far away as King Street, down the hill and across the creek beyond the tailings dams. Complaints included anecdotes about dogs dying or losing their hair; hands that blistered after short bouts of gardening in allegedly polluted soil; constant headaches and other sicknesses. One theory about the King Street problems, never tested, is that they may be the outcome of random, illegal dumping when the area was bushland.

Residents became very concerned, but were unable to get satisfactory answers to their questions. Scientific testing was left up to individual families, who found it difficult to get reliable results. Consequently, many residents began to study the side effects of chemicals but, without adequate scientific training, concluded that all the effects they were reading about were necessarily present at Kingston.

Liability
Liability for damages at Kingston, let alone responsibility for a clean-up, was difficult, if not impossible, to establish. South Queensland Mines finished operations in 1955 and was wound up in 1982. It was no longer around — although its former owner was — to be made accountable for cleaning up the toxic wastes from its activities. The oil waste had been dumped with government approval, and so the company that dumped it could not be held liable. Those who illegally deposited toxic waste in the mine could not be traced and prosecuted. In any case, claims at law for damages must establish that the waste deposited by a particular party is directly causing ill health to some person. Where the party cannot be traced, and the exact nature and source of the pollutant are uncertain, this is nearly impossible. Moves to amend and tighten the law still rely on liability for clean-up costs, which is only weakly deterrent in its effect, rather than on prevention. Furthermore, some proposals (legislation varies from State to State, and is still evolving) would impose some of the liability for clean-up costs on the residents.

The local authority primarily responsible for the mess, Albert Shire, no longer controls the area. Its successor, Logan City, excised from Albert Shire in 1979, was poorly endowed financially, and was concerned at the unknown and potentially escalating

CHAPTER 1

costs of an open-ended commitment to a clean-up. The Queensland Government did not wish to admit responsibility, although key Departments approved the rezoning of Mount Taylor and had supposedly advised the Albert Shire. By refusing to admit that a problem existed, the State Government could avoid or at least delay committing resources to cleaning it up; it was fearful that it would set a precedent by acknowledging its responsibility in the matter. Mount Taylor is not the only toxic waste site in the increasingly crowded Brisbane region.

Political Response
Consequently, government responses to the problem were slow and haphazard, undermining public confidence. In 1987 Logan City Council commissioned environmental consultants Sinclair Knight and Partners to assess the extent of hazardous waste contamination at Diamond Street, and prepare a management plan. Their report recommended removal of affected dwellings on Diamond Street, and excavation and removal of the sludge, at an estimated cost of $700,000. This firm had little experience in environmental matters; indeed some of its previous work had been publicly criticised by the Tasmanian Fisheries Department. Its handling of the Kingston problem came under attack by experts as unsystematic and unscientific. A programme of test drilling had assumed that the risk and impact on health declined uniformly with the distance from the centre of the 'oil pit', ignoring the possible significance of the major open cut and the tailings dams. Attempts made to trace waste movement on the surface ignored possible underground pathways. Most of the complaints were coming from an area to the north and east of the site, which is not surprising, given the lie of the land and the location of the tailings dams. But the drilling and the health surveys included some areas which were almost certainly not affected; not surprisingly, the results were inconclusive. Some areas — for example, King Street — from which persistent complaints emerged were not tested.

In 1988 the State Government created a Task Force to investigate the problem and advise both it and local government. The person appointed to head this group was a sociologist specialising in the effects of stress on community health. The Task Force's statistical surveys purported to show no more ill-health than would be present in any working-class population, disproportionately exposed to toxic substances in the workplace.

After the 1987 scare, five familes were relocated from Diamond

Street, and three houses removed. The plan to remove the sludge from the 'oil pit' and dump it elsewhere remained in abeyance, due to the difficulties of finding a suitable dumping site.

It was not until the Labor Party won the 1989 Queensland general election and formed a new State Government that the problem was tackled decisively. The Brisbane firm Envirotest was commissioned to report on the earlier investigations as well as to carry out its own. The tests it conducted and commissioned identified a number of hazardous substances at the site including sulphuric acid, petroleum sludge, polyaromatic hydrocarbons (PAHs), and residues of lead, zinc, copper, mercury, arsenic, chlorobenzenes and cyanide. The tests also identified 'excessive' contamination of groundwater with metals. Blood and urine tests made in 1990 indicated that some residents had high levels of arsenic.

In 1990 the State Government opted to cap the area with clay, in order to minimise upward migration of hazardous materials, and to remove a further 26 houses on Diamond and Amethyst Sts. This was judged less hazardous than attempting to remove the waste and dump it elsewhere. The land was to be used subsequently as a public park, with regular monitoring for pollutants. The cost of this programme to the State Government was $8,000,000. But some issues, such as the problems reported by the residents of King Street, were not resolved.

The Failure to Plan
In the 25 years to 1990, the population of the south-east Queensland region increased by 206% and the area of urbanisation in Brisbane grew by about 300%. Such rapid growth creates a need for planning, in order to minimise land-use conflicts, ensure that corridors are reserved for transport purposes, and provide for needs such as parkland and open space, even before environmental impacts are considered.

Queensland legislated in 1971 to provide for environmental assessment by local councils. But the procedures proved cumbersome and impracticable, due to lack of resources and expertise, and the planning mechanisms quickly fell into desuetude. After some years the legislation was quietly repealed.

This experiment commenced after the major 1968 subdivision at Mt. Taylor, which was not subject to land-use controls. But the use of the land for housing was approved despite the fact that many local politicians, and the government agencies responsible, were aware of the site's history. The effect of this carelessness was

Chapter 1

to throw a significant part of the burden onto the local residents, rather than those who should have exercised responsibility.

Since the Goss Government's election in 1989, some moves have been made towards more effective planning legislation, and a Green Paper issued in 1991 discussed measures for ensuring that contaminated land is properly identified and recorded, and that those responsible bear the costs of restoration.

Too Little, Too Late
Too little attention was paid to Kingston residents when they first complained of problems. Scientific investigation was tardy. No legal framework existed for assigning responsibility. Governments were reluctant to assume it, both because of relative incompetence and of concern about financial implications. The election of the Goss Government saw a partial change of policy in 1990; the solution adopted, though recommended by a reputable firm, remained a 'cover-up' — in the literal sense — and knowledge of the extent and nature of contamination remains very incomplete.

Kingston was an avoidable problem. By the time the major subdivisions took place, the dangers of toxic waste were already well known. It was also relatively minor. By the 1970s, a far more serious problem had already created a major scandal in the USA.

CHEMICAL WASTE POLLUTION AT NIAGARA FALLS

The Great Lakes area of North America straddles the border between the USA and Canada. It is economically rich; the bulk of the old-established industry in both countries is within easy reach. Water transport played a major part in the development of a thriving steel industry: ores from the Iron Range at the west end of Lake Superior are shipped to steelworks at Chicago, Detroit, Toledo, Cleveland, Erie, and Buffalo on the US side. Many ancillary industries have grown up nearby. Across the border, Canada's heavy industry is found predominantly in the Province of Ontario, bordering Lakes Erie, Superior, and Ontario. Toronto, a complex the size of Melbourne or Sydney, is the largest Canadian city adjacent to the Lakes. Lower down the St. Lawrence River, which drains Lake Ontario, are the Canadian cities of Montreal and Quebec.

The Niagara River connects Lake Erie with Lake Ontario, the lowest of the Great Lakes. The Niagara Falls mark the edge of the hard rock of the Niagara Escarpment, which runs along the southern end of Lake Ontario. The river, forming the border between Canada and the USA, flows through a deep gorge below the Falls.

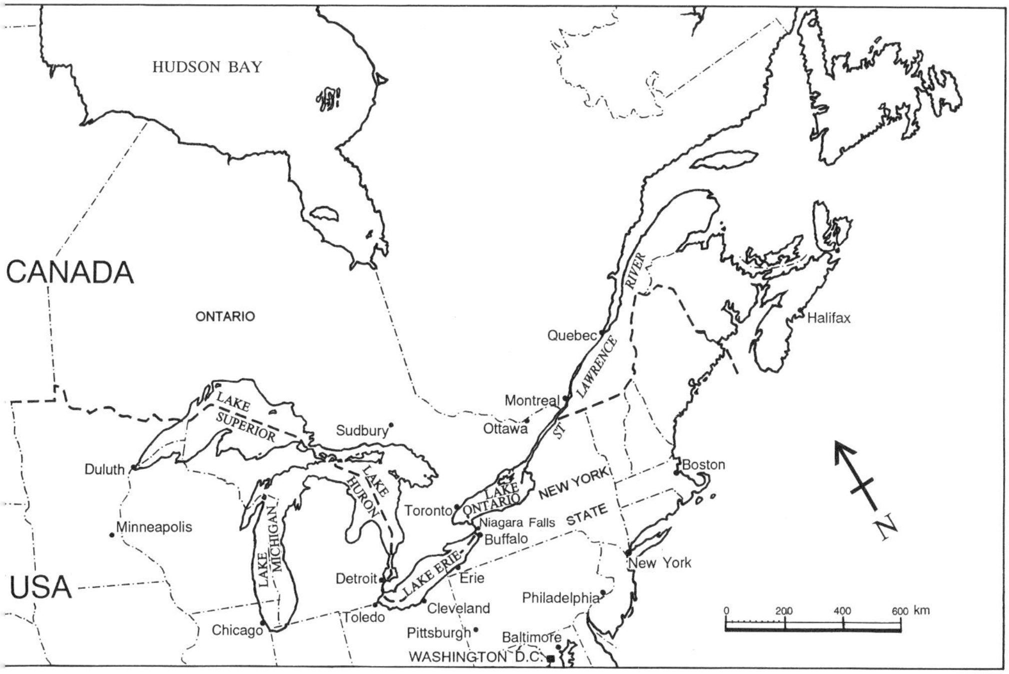

Toxic Waste Pollution

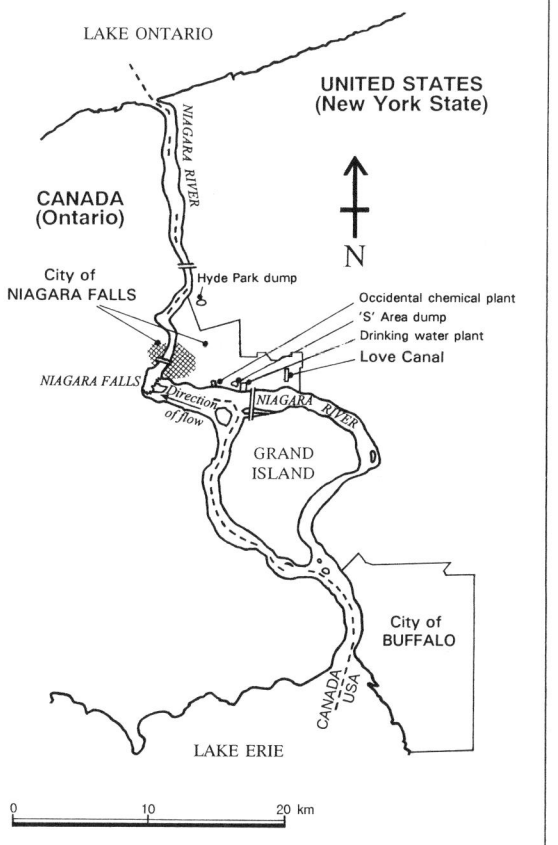

Fig 1.4
The Great Lakes region.

Fig. 1.5
Niagara Falls and Love Canal.
(Sources: Levine, p.8; Hall p. 119)

Chapter 1

The US and Canadian cities of Niagara Falls are popular tourist destinations. Niagara Falls, USA (population: 217,000), is governed by a City Council and is part of Niagara County. A school board under the authority of the City is responsible for primary and secondary education.

From the 1920s, the City of Niagara Falls adopted an active policy of attracting industry by supplying cheap hydro-electricity. This resulted in the second largest concentration of chemical industries in the USA. Their wastes add substantially to those generated by daily life in the cities. The waste disposal problem has grown progressively more acute over time; there are now 215 chemical waste dumps in the Niagara Falls area, containing 8,000,000 tons of chemicals. All are simply pits excavated in the earth and then covered over. Only three have management plans.

Love Canal

The Love Canal housing estate, a suburb of Niagara Falls, was built on top of one such dump. Its occupants were exposed to a number of chemicals which leaked to the surface, causing many adverse health effects.

History

In 1894, one William Love launched a project for a navigable canal connecting the upper and lower Niagara rivers, combined with hydro-electric power generation. But after construction had commenced, an economic slump resulted in the withdrawal of Love's financial backing, and a large excavation, which became known as 'Love Canal', was left uncompleted.

In 1942 the Hooker Chemical company leased Love Canal from its owners, the county power company. It drained the excavation, lined it with clay, and dumped at least 82 different compounds totalling nearly 20,000 tons of chemicals into the excavation. In 1947 Hooker purchased Love Canal and the adjacent land.

After the dump closed it was buried under a layer of clay and earth. In 1953 Hooker sold part of the site to the Niagara Falls Board of Education for one dollar. The Board was aware that the area contained dumped chemical wastes and agreed to accept full legal responsibility, indemnifying Hooker against liability for damages should any incident involving the chemicals occur. In doing so, it ignored its own lawyer's advice that chemical tests should be conducted before concluding the agreement.

At that time, with an urgent need to expand schooling to meet the post-war 'baby boom', school sites were urgently sought. By the 1960s, problems were evident: land was subsiding, releasing toxic chemicals, people complained of chemical smells, and children were burned in a long series of minor accidents. Although government bodies were notified on a number of occasions, no action was taken and residents were assured there was no problem.

In 1976 a noxious black ooze began to emerge in the basements of houses. Residents initially responded by pumping this material into the local sewer. Chemical analysis revealed that it contained a variety of dangerous substances, and had originated from the Hooker dump. In response to the problem the County Health Department threatened to fine residents who continued to pump the material into the sewer. Subsequently, it reassured residents that the sludge was not toxic; then warned them not to allow children to come into contact with it!

During 1976, both the City and the New York State Department

Box 1.2 Noxious Substances at Love Canal

The wastes at Love Canal were mostly contained in 44 (imperial) gallon drums, though some liquid wastes were poured directly into the excavation, which had been lined with clay. Among the substances deposited were:

Chemical	Effect on humans
Benzene	leukemia and anæmia
chloroform	carcinogen affecting nervous, respiratory and gastrointestinal systems
lindane	causes convulsions and extra production of white blood cells
trichloroethylene	carcinogen; attacks nervous system, genes and liver
methylene chloride	chronic respiratory distress and death
trichlorophenol	contains dioxin, a powerful carcinogen

The sludge appearing in the residents' basements, upon analysis, proved to be 'a horrendous brew of 15 organic chemicals, including three chlorinated hydrocarbons that are toxic by inhalation, ingestion and skin absorption.'[3]

of Environmental Conservation carried out surveys of water movement at the site, but neither acted. The city drew up plans for a project, costing $400,000, to contain the toxic chemicals at Love Canal, but rejected the plan as too expensive after spending $20,000 to research it. By the end of 1977 the clean-up bill was estimated at $50 million, far beyond the resources of School Board, City or County.

Health Effects
A New York State Department of Health (NYDOH) survey in 1978 found that 95% of the houses in the 'inner ring' surrounding the school site were contaminated; the US Environmental Protection Authority (EPA) identified health hazards from toxic vapours in their basements. Statistical evidence was accumulating that the health of residents was severely affected; cancers, skin diseases, miscarriages, birth defects, and nervous disorders were more frequent than expected. Despite this, the City and the County in particular were reluctant to act and issued contradictory statements, angering residents.

The residents compiled health data suggesting that the waste was migrating along underground channels, or swales, actually the beds of creeks filled in during urban development. The worst affected cases were those who lived near them. The NYDOH refused to accept this hypothesis, even though its own studies initially gave support to it. No attempt was made to verify or disprove it *scientifically* until the test drilling of the 1980 investigation, which appeared to disprove it. Had the swales hypothesis been accepted, it would have been necessary to evacuate several hundred extra families.

Health Emergencies
During 1978 considerable buck-passing took place between the governments concerned. The New York State Government ordered Niagara County to 'clean up' the canal area; the County in turn passed on the order to the City, which refused on the grounds that it was the County's responsibility, and that it (the City) could be legally liable if it intervened. It also feared it would be unable to raise enough money. The County's Health Commissioner allegedly failed to act on complaints from as early as 1970, and the County, also concerned about finance, refused to investigate the resulting charges, attempting to lay the blame on the US Government.

In August 1978, the State Health Commissioner declared a health emergency and urged pregnant women and young children under

Box 1.3 Short and Long Term Toxicity; Synergistic Effects

The reason why the Love Canal problem aroused such serious concern is because there is an important distinction between the short and long term toxicity of environmental agents. Most short-term effects, except where individuals are exceptionally sensitive, are caused by massive exposures to pollutants, often far exceeding levels likely to be found even in industrial environments. Examples include farmers overcome by pesticides while spraying crops, or spray-painters inhaling lethal doses of solvents. Short-term exposure to many chemicals, provided that it is in low doses and is not prolonged, frequently causes no detectable ill-effects.

Prolonged exposure is different, and highly controversial. Even measuring the exposure is fraught with difficulty. It is nearly impossible to control for exposures to other chemicals over a number of years, and the quantities of the pollutants present in individual environments are very small, often being measured in parts per billion. They are difficult and sometimes impossible to detect. Consequently it is extremely difficult to gather or to interpret data about long-term hazards from pollutants. Even if the presence of pollutants can be demonstrated, assessing the risk they pose is very hard. Most of the data is statistical, and consequently inferential; and results are frequently hotly contested.

Nonetheless, it is thought that long-term exposure to many chemical pollutants, even in minute quantities, does pose a potential health hazard. DDT and its breakdown products, for example, are known to accumulate in fatty tissues and are reputed to attack the liver. Certain chemicals are thought to cause potentially serious genetic damage. The chemical industry and government face the risk that citizens exposed to chemicals at concentrations once thought to be safe may later sue for damages, which can be costly. But scientific uncertainty means that firms can claim that no danger exists, or even try to conceal the extent of known dangers; and that attempts to prove damage, especially if they come to court, may prove costly and inconclusive. This has meant that workers in industries which expose them to such dangers have frequently found it impossible to establish a case, even when scientific opinion regards it as well founded. Only recently have courts — in Australia as elsewhere — recognised that 'black lung' in coal miners or mesothelemia, a very virulent form of cancer, in asbestos

Chapter 1

> workers, are employment-related.
>
> And there are synergistic effects: 'witches' brews' such as the basement sludge at Love Canal may have impacts which none of the contaminants has by itself. Even less is known about the toxicity of such combinations, especially over the long term, than about specific, individual substances. At Love Canal, claims by a scientist specialising in cancer, that illnesses were being caused by chemicals migrating along the swales, fuelled direct controversy.
>
> Government did not react at Love Canal until incontrovertible evidence of short-term effects could no longer be ignored. There seems to have been little awareness of the long-term risks and even less attempt to deal with them.

two to evacuate the area immediately. Some 200 families were moved. Shortly afterwards, the President of the USA declared the area a national disaster site and made available funds for relocation.

A second emergency was declared in May 1980, permitting the relocation of a further 710 families with federal funds. Meantime, the State Government was proposing to 'revitalise' the area by allowing people to reoccupy abandoned houses. The plan coincided with the release of medical studies suggesting that residents had suffered both chromosomal and nerve damage. A Bill for 'revitalisation' failed to pass the State legislature.

Eventually 227 houses were removed, clearing the 'inner ring' of houses immediately adjacent to the dump site. Costs to government, including those for the EPA's health survey, were of the order of $30 million, of which $17 million were borne by New York State, $1.3 million by the City of Niagara Falls, and $9 million by the US Government. Total private costs, for relocation, medical treatment, and litigation, among other items, are unknown.

Scientific Investigation

Several scientific studies were carried out. They were uncoordinated, difficult to interpret, and mutually contradictory. The New York State Department of Health carried out a study in 1979, after the first evacuation, suggesting that the area suffered an unusually high rate of miscarriages, congenital birth defects and babies with low birth weights; but evidence of a direct connection with chemical exposure was lacking. The US EPA and Department of Justice launched an independent survey, in ignorance of the DOH work, which seemed to show chromosomal damage in Love Canal residents; this time the results were inconclusive due to a lack of controlled comparisons with a 'normal' population. Two panels concluded that these results were unreliable; three independent geneticists — one engaged by Hooker Chemical — then declared them valid. This study triggered the second Presidential declaration of emergency, and an extensive US EPA investigation.

Not only was the inconclusiveness of the scientific investigations striking, but what information was available was poorly communicated. The scientists involved made little attempt to render their data intelligible to the citizens and were reluctant to attempt any assessment of the public health implications. So great was the hostility this engendered that in summer 1980 the residents' association announced a boycott of all future government health studies.

In August 1980, after the second evacuation, the EPA's comprehensive scientific survey of the site got under way. It was claimed at the time to be the most comprehensive scientific investigation of its kind. Nearly 7000 samples, collected from 174 bores and various other locations, were analysed for more than 150 substances. Some dioxin was found in sediments in storm drains and sewers, directly traceable to Love Canal. Leakage from the dump was found to be confined to the immediate area of the 'inner ring,' where permeable sands and unusually wet conditions had led to the seepages of 1976; impermeable geological structures meant that no contamination of bedrock nor of the surface outside the immediate vicinity of the dump had occurred. The 'swales' hypothesis was not confirmed.

The EPA's environmental monitoring report, issued in 1982, recommended the complete removal of the 'inner ring' houses and the permanent sealing of the area. It endorsed the project, already under way, to prevent superficial groundwater movement by digging deep trenches around the 'inner ring' and filling them with impermeable clay. It declared the rest of the site habitable, and played down health risks.

Though authoritative, the EPA report was controversial. Technical objections concerning the study's design, sampling techniques and handling of statistical data were raised. Opponents cited scientific uncertainty, the problems of detecting trace contaminants and the lack of knowledge of the long-term effects of low-level exposure as reasons for distrusting its recommendations. They rejected the view that unless material danger could be shown, no action should be taken, urging instead that if a reasonable apprehension of danger existed, there was cause for concern. This view had significant influence on the legal argument, as well as the scientific.

CHAPTER 1

Evaluation

The Love Canal problem was never properly resolved. Citizens were left in doubt about past and future effects on their health. While the EPA study made some informed judgements about the direction, rate of flow, composition and possible effects of subsurface toxic waste, controversy continued. Plans to 'revitalise' the area remained suspended.

There were several reasons for this striking lack of resolution. Scientific evaluation was belated, commencing in earnest only after the two declarations of emergency. Citizens were kept in the dark; there was poor dissemination of information about the nature and possible impact of the toxic waste. Scientific uncertainty was the consequence of a political reluctance to invest in knowledge.

Government response was slow. Local government in particular, but also the State of New York, were unwilling to become involved for fear that the cost of the clean-up would fall on them. Even the Federal Government in Washington was reluctant, but its vastly greater resources meant that once it did become involved, it could act more decisively. Prolonged political pressure by the Homeowners Association, backed by medical evidence, was the major cause of action.

Love Canal is an example of a problem which was technologically novel, and for which existing institutional arrangements were inadequate. In the 1920s, and even as late as the 1950s, very few people in government understood the dangers associated with the wastes from modern chemical industries; though there is little doubt that the firms themselves did, at least in part. This 'time lag' phenomenon, where popular understanding of a technology follows only long after its widespread introduction, is a common characteristic of environmental issues. The existing institutional arrangements — the city, county, state, and even the federal government — lacked the ability to cope promptly and effectively with an unforeseen problem.

THE HYDE PARK AND 'S' AREA DUMPS: THE FAILURE OF US LAW

The 215 chemical waste dumps at Niagara pose a direct threat to the water supplies of 6,000,000 people in communities bordering Lake Ontario, 4,500,000 of them in Canada. This transboundary element is common in environmental problems. Nor is the problem simply a public health matter. Toxic substances have already had adverse effects on wildlife as well as people.

Even before Love Canal became controversial, similar problems had surfaced at the 102nd St. dump, also a Hooker repository. Exposed wastes, including chlorates and phosphorous, burned children playing at the dump much as they did at Love Canal. Two other cases, involving the Hyde Park and 'S' Area dumps in Niagara Falls, illustrate the policy problems.

THE HYDE PARK DUMP

In 1970, reproductive failures in herring gull communities in Lake Ontario were traced to dumped chemicals. The main culprit was identified as dioxin.

Studies by Canadian and New York State officials showed that dioxin and other chemical wastes were steadily seeping into the Niagara River, or being dumped directly into the river by US plants. The Canadian Department of Environment successfully pressured the US authorities for legislation to stop direct dumping. Dioxin levels in herring gulls fell for a number of years, but levelled off in 1980, still at unacceptably high levels.

The studies had also detected some seepage through the face of the Niagara Gorge. The main source was thought to be the 80,000 ton Hyde Park dump, containing one ton of dioxin; the largest concentration in the world. This dump drains into a local creek called Bloody Run, but, more important than surface runoff, chemicals were also thought to be seeping into the rock below, to reappear at the gorge face.

The Legal Process

The US Environmental Protection Authority (EPA) reacted by developing a policy of prosecuting the major polluters, but demanded only containment, rather than removal and destruction of the wastes. It devoted only a small team to the project and ignored evidence that its preferred solution would not eliminate seepage. The case went to court in 1982, but the resulting decision did little to eliminate the problem.

The Hyde Park court case demonstrated two failings in the US legal process. Firstly, Judge Curtin of the US District Court refused to admit citizens groups, including some from Canada, to **standing** in the Court. This meant that they could present evidence but could not challenge the Court's decisions or appeal them.

Secondly, much scientific evidence was ignored by the Court. The biological effects of dioxin wastes were either not admitted or rejected as irrelevant. No account was taken of the biomagnification of toxic wastes in food chains. Other ecological evidence was

Box 1.4 Properties of Dioxin

Dioxin is the common name for 2,3,7,8-tetrachlorodibenzo-p-dioxin, or TCDD, one of a family of similar compounds. It is formed in at least three manufacturing processes: the bleaching of pulp in paper production using the chlorine process; as a by-product and contaminant in the production of the herbicide 2,4,5T; and in the manufacture of pentachlorophenol, a wood preservative. It can also be formed by incineration of 'chlorinated organic substances' and is consequently found in gases and ash from waste incinerators.

2,4,5T, reputedly contaminated with dioxin, was one of the constituents of Agent Orange, used extensively as a defoliant in the Vietnam war, and implicated in cancers and birth deformities both in Vietnam and in the USA, the latter as a result of soldier exposure during combat. Though controversy continues, dioxin is regarded by many authoritative sources as highly toxic, carcinogenic, and teratogenic (i.e., likely to cause birth deformities).

In July 1976 a serious accident involving dioxin occurred at Seveso, near Milan, in Northern Italy. Unreacted feedstocks for the manufacture of 2,4,5T were stored over the weekend in a reactor at a plant owned by Industrie Chimiche Meda Societa Anonima (ICMESA), a subsidiary at one remove of the Swiss conglomerate Hoffman-LaRoche. A cloud of vapour containing dioxin (among other contaminants) was released from the plant, settling on 700 acres (285 ha) of land nearby. ICMESA did not fully inform the local residents of the nature of the contamination until nine days later; evacuation of some 700 residents did not take place for two and a half weeks. There was considerable mortality among domestic animals, especially in the three days immediately following the accident. The reaction by local emergency service agencies was a shambles. Eventually, some 267 acres (108 ha) were sealed off and quarantined, later being subjected to drastic decontamination measures; another 665 acres (270 ha) were sealed to the public, though the residents were allowed to continue living there. Numerous cases of chloracne, a skin complaint like, but more severe than, normal acne persisted for a long period after the accident; evidence of carcinogenic or teratogenic effects is lacking, due in part to lack of control data for statistical analysis. Government compensation, in Italy's corrupt and incompetent political system, did not reach those worst affected, and enforcement of quarantine measures was incomplete.

ignored, and no notice was taken of the fact that a 'clean-up' of the wastes might become impossible if not attempted soon.

The Court uncritically accepted evidence from the Occidental (formerly Hooker) Chemical Company's own geologists, who claimed that the wastes could not seep into the underlying rocks. On the basis of this 'evidence' the Court negotiated a cheap 'clean-up' scheme with the company. Shallow retaining walls were constructed around the dump, to prevent surface runoff, but nothing was done about wastes that had seeped deeper into the rocks. The cost of removing this deeper waste would have been far greater.

No attempt was made to seek confirmation of theories that wastes from the Hyde Park dump were seeping into permeable rock and emerging at the face of the Niagara Gorge, although a programme of test drilling would have been feasible. Once again, scientific uncertainty led to a decision which was at best uninformed.

THE 'S' AREA DUMP

The 'S' Area dump is located 200 yards (180 metres) from the Niagara City water treatment plant. 70,000 tons of chemicals are dumped on a landfill only four feet (1.2 metres) above river level. Noxious odours are frequently present at the water supply plant, and tests have detected 22 chemicals in city water. Both State and County officials have expressed concern at the risk of contamination.

The city water supply draws water from the river through a tunnel running below water level, through bedrock. In 1978, while investigating a blockage, divers discovered chemical wastes in the tunnel, very similar to those leaking through the face of the gorge, and thought to come from the Hyde Park dump. This evidence could be interpreted to mean that chemicals had contaminated the bedrock underlying Niagara Falls as a whole; at the very least it meant that waste from the 'S' Area dump had broken through.

Prosecution followed only after the EPA and the State Government applied pressure. In 1982 the same District Court judge handed down a judgement very similar to that on Hyde Park.

The total cost of Curtin's decisions to the company was $15,000,000 for Hyde Park, and $35,000,000 for the 'S' Area dump, spread over 35 years; a little over 1% of the company's predicted sales for the same period, at $35,000,000,000. The cost to the taxpayer was several hundreds of thousands of dollars. The cost of removal and destruction of the chemicals at 'S' Area alone might have reached as much as $US100,000,000. The settlement between the Court and the company blocked any further appeals by citizen

CHAPTER 1

groups, and makes further serious environmental damage inevitable — at incalculable cost.

The Failure of Politics and Law

Hall argues that the fundamental flaw lay in the legal system's failure to address the real problems of citizen health and welfare. Instead, the process had favoured the interests of major companies, and had ignored important scientific evidence, as well as the risks inherent in refusing to deal with problems in a timely fashion. If toxic waste was indeed leaching into the bedrock below Niagara Falls, the longer the delay in cleaning up, and the more serious the problem would become. If left long enough, no remedy would be feasible. Thus the legal system was ignoring not only the need for reliable scientific data, but also the inherent urgency and irreversibility of the problems.

Despite considerable legislative activity from 1965 to 1990, US laws have not been adequate to prevent serious damage to the Great Lakes. Nor has there been sufficient political will, especially in the USA, to ensure that effective legislation was passed. Even the existing laws were not fully enforced by the agencies charged with administering them. The US EPA estimated during the 1980s that there were 60,000 'major' sources of toxic wastes in the USA, producing 90,000,000,000 pounds annually, of which 90% would be disposed of improperly, one quarter of it within the Great Lakes region. The courts and the political system were wide open to pressure from large companies, with vested interests in neglecting the problem. The results are often unsatisfactory compromises, with permitted pollution levels being set, not by scientific criteria, but by bargaining between lawyers.[4]

Liability for toxic waste was tightened by further US legislation in 1980, prompted by the events of Love Canal. In addition to imposing increased liability on previous owners of sites, a 'superfund' was created by imposing a levy on chemical manufacturers to provide resources for future clean-ups. However, the legislation has been inadequate to the scale of the problem; contaminants are still being produced faster than they are cleaned up.

OVERVIEW: SLOW RESPONSE, POOR UNDERSTANDING

Thus, both at Kingston and at Niagara Falls, political systems created a problem through neglect or ignorance. They then responded sluggishly to the needs of those affected, neglecting necessary scientific information-gathering and failing to appreciate the magnitude of the problem. Such inadequate responses are common in many cases in Australia and abroad; for example, in 1991, a major fire and explosions occured at the antiquated, poorly-maintained Coode Island chemical storage facility close to the centre of Melbourne. The Victorian Government, after a hasty study, decided to move the storage facility to Point Wilson, near Geelong, home of a rare and endangered species of parrot. Further controversy promptly erupted, underlining the increasingly intense competition between various land uses.

Yet the firms and individuals disposing of the toxic wastes at Niagara and Kingston knew of the likely consequences. Their actions suggest that their primary concern was to minimise their own costs, for example by choosing to dump wastes rather than neutralise them by processing. They also sought to evade legal responsibility for the consequences of their actions, most notably in the case of the indemnity signed by the Niagara Falls School Board. This suggests that firms will ignore the consequences of their actions — however serious — if the political and legal system will let them. In other words, individual goodwill cannot be relied upon to prevent environmental disasters. In the absence of controls, knowledgeable actors — such as private firms — may still behave in antisocial ways.

In a recent review of legislation in the USA and Australia, Johnsen concluded that in neither country was the response adequate to prevent future contamination or to deal effectively with what exists.

These examples also demonstrate that unlike some other political issues — such as rates of taxation, for example — environmental problems cannot be resolved by political debate alone. Scientific knowledge must be procured and brought to bear; without it, no policy can be evolved. But the citizens involved must also have a say, for they are often closely aware of the problems, which cannot be resolved simply by bringing in 'experts' to make decisions for them. Decision-makers must have access to all the relevant information (and preferably minimise the amount of *irrelevant* information reaching them) before they can operate. Even then, there are serious questions of equity and morality to be considered.

Niagara Falls and Kingston are not unique. In recent years there have been several major chemical spills in Queensland alone, most of which could have been prevented or contained had laws been more effective.

And other features of the cases are typical. Where large populations crowd into small areas, and especially where there is indus-

Chapter 1

try, problems of congestion will arise. This leads to competition between various uses: waste dumps versus schools and houses, for example. Because of Niagara Falls' location on an international boundary, **transboundary** and **riparian** problems emerged: the Canadians, downstream, have to put up with discharges of waste from the USA, but have no way of retaliating in kind. In fact, repeated complaints have achieved far too little. The relationship between Canada and the USA is **asymmetrical** in this respect. Unsurprisingly, riparian issues have generated a good deal of litigation. They remain a vast, particularly difficult area of international law.

Political systems frequently fail to respond effectively to environmental problems, for numerous reasons, including:

- ignorance of critically important ecological principles;
- failure to assimilate scientific knowledge and apply it;
- institutional inadequacies in the existing political, economic and legal systems; and
- lack of foresight and political will where large-scale problems for the future are concerned.

These are recurring themes in political ecology. The first and second directly address biological and ecological aspects; the remaining two, social and political.

Further Exploration

There is no definitive, readily accessible account of the Kingston case, though Tamvakis' Honours thesis (Griffith University, 1990) dealt with it. The 1990 report to the Goss Government by the firm Envirotest (though unpublished) is also useful; it has a potted history of the problem, a review of the measures that were undertaken to rectify it, and a series of recommendations. An article by J. Phillips, 'Sludge II — the sequel?' appearing in the Brisbane *Courier-Mail* early in 1989, briefly summarised events. The ABC broadcast a documentary *Poisonous Neighbours*, on 3 April 1989, and the Environmental Studies School at Griffith University produced a teaching video, *Out of Sight, Out of Mind*, in 1993.

The weaknesses of Queensland's environmental legislation were pilloried in 1973 by the (Brisbane) Public Interest Research Group's book *Legalised Pollution*. Interestingly, the last years of the Bjelke-Petersen Government were marked by a series of controversies over the disposal of toxic and radioactive wastes, as well as four spectacular chemical fires, three of them in Brisbane. The Goss

Government, elected in 1989, made some changes to environmental legislation; a review of their efforts is in a special issue of the *Legal Service Bulletin* for April 1991; see especially Keim and Richards. Johnsen's review of Australian and US legislation focuses mainly on New South Wales.

There is a brief account of Love Canal in Mokhiber and Shen; Levine's book *Love Canal: Science, Politics, and People* is longer and more comprehensive. For details of the Hyde Park and 'S' Area issues, see the article by Hall, 'Poisoning the Lower Great Lakes: the Failure of U.S. Environmental Legislation'. For a broader, popular, and well-organised account of the environmental problems of the Great Lakes in general, see Ashworth's *The Late, Great Lakes*; Chapter XIV is especially useful in placing the problems at Niagara in perspective.

It is always dangerous for a lay person to recommend any source on a technical issue. However, the information in this chapter derives from several such. Freedman's *Environmental Ecology* is useful and accessible, enumerating a number of the stress processes affecting modern ecologies, with due attention to toxic materials. More specialised is Zakrzewski's *Principles of Environmental Toxicology*, which, as it is a 1991 publication of the American Chemical Society, may be considered authoritative and up to date. Connell and Miller, in *Chemistry and Ecotoxicology of Pollution*, offer a comprehensive and reasonably accessible, though rather technical, overview. Whiteside's *The Pendulum and the Toxic Cloud* is a popular account of the Seveso disaster.

For Australia, Allan Gilpin sketches the dangers of hazardous chemicals in his *Environment Policy in Australia*, at pp. 207–15. Garry J. Smith, in his recent book *Toxic Cities*, deals at length with Australian toxic waste problems, especially in Chapters 4 and 5, briefly mentioning Kingston. Whitelock's *A Dirty Story* dates to 1971. Ben Selinger's admirable *Chemistry in the Market Place* is an extremely useful source of data on the properties of many common household and industrial chemicals.

CHAPTER 2

HUMANS AND THEIR IMPACT ON THE ENVIRONMENT

Attempting to deal with environmental problems without understanding the *processes* — both natural and human — which cause them is rather like looking at a still photograph and trying to puzzle out what actions are depicted. A more appropriate tool would be a moving picture, in which the subject can be seen changing over time.

Understanding Environmental Problems
Several kinds of processes underlie environmental problems. Some are **simply physical**. The movement of toxic waste in groundwater, for example, is a physical process. So is the solar radiation on which terrestrial life depends. Other processes are *chemical*: the impact of pollutants on air and water, for example, or the **synergistic** effects caused by the interaction of compounds. These may be beneficial, as when potential toxins are bound and metamorphosed in rock or soil; or harmful, as when the mixing of different toxic compounds leads to a 'witches brew' more deadly than any of the constituents. Other processes are *biological* or *ecological*. For example, toxins such as heavy metals may be taken up by algae or bacteria, concentrated by filter feeders such as shell-

fish, and finally eaten by humans who then display symptoms. This process of **biomagnification** is common in nature. Birds of prey, for example, highly dependent on fish or other species already 'high' up the food chain, were among the first victims of persistent pesticides such as DDT, which caused them to lay thin-shelled eggs which broke in the nest.

Typically, problems in environmental policy require a clear understanding of the physical, chemical, biological and ecological processes underlying them. But they are policy problems precisely because of the human element. The toxic waste problems at Niagara Falls and Kingston were side effects of the economic process of production. The decisions to build schools and houses were political; they were made by the School Board at Niagara Falls and the Albert Shire Council in Kingston. Thus an understanding of social, economic and political processes is essential to link environmental issues to humans and their activities. *It is the discovery and elucidation of this linkage — the human–environment interface — which makes the study of environmental policy uniquely difficult and uniquely fascinating.*

The human **mode of adaptation**, and its impact on the physical and biological environment, constitute the first key.

HUMAN ADAPTATION

Humans differ crucially from other animals in the way they modify their environment. Inadvertent environmental modification is not uniquely human. Predators can eat out their prey and die; herbivores can exhaust their supply of vegetation. But humans deliberately modify their environments; more extensively than any other species.

Specialisations

Other species adapt by *physical specialisation*, developing characteristics which fit them for specific environments. Some are very obvious: polar bears have thick furry coats to protect them from the cold; honeyeaters have long thin beaks for sucking up nectar; fish have fins for swimming; tigers have claws and teeth for killing and munching up on prey; kangaroos have huge hind legs for hopping; and so on. Less obvious adaptations are things like abilities to digest specific foods; special sensory capabilities like the bats' sonar navigational ability; or the cane toad's poisonous secretions, which are a passive protection against predators. Such adaptations come at a price: individual members of the species are tied, often

Chapter 2

very closely, to specific habitats and lifestyles. They are, in effect, *programmed*; the necessary behavioural information is encoded on their genes, as DNA. The behaviour patterns of many species — such as insects — are largely fixed. But some species, mammals in particular, supplement this genetically transmitted information by **learning** additional patterns of behaviour; but they can only do so by example, which requires direct contact between generations, and their basic behaviour patterns are not greatly modified. Thus, in general, non-human species can adapt to new habitats only by *genetic* change: the information encoded on their DNA must change, over generations. This is basic to Darwinian natural selection: only genetic information can be passed from generation to generation, not 'acquired' information, or experience.

Humans are far more generalised, specialising for intelligence and sociality rather than a specific environment or lifestyle. An important human characteristic is *plasticity*: the ability to adapt to a wide range of situations. As individuals, this makes humans more adaptable, but also far more vulnerable. They compensate by **technology** and **cooperation**. Technology can be defined simply as 'tools and know-how'. Humans make a whole series of the teeth and claws they lack, to be taken up and discarded at will. Tool-making and -using are not themselves uniquely human; many species will use materials to hand, and some will actually make tools. Apes, for example, will select branches and strip them of leaves to use in extraction of termites from their nests. But the human use of tools is far more elaborate: a quantitative difference so large that it has a very significant qualitative impact. It also differs in the elaboration of *knowledge* (or 'know-how') about the tools on an unprecedented scale. Without knowledge, tools must remain very primitive, and the elaboration of complexes of inter-related tools (called 'cultures' by archæologists) is unimaginable.

Human cultures depend on cooperation and **specialisation**. For efficiency, tasks are 'batched' — i.e., a whole lot of one thing is done at once, and the result kept for later use — or divided among specialists; both entailing cooperation or at least coordination. This requires **communication**; and the human solution to this is **open language**, the capacity for which is itself a genetic endowment unique to humans.

Extrasomatic Information Transfer

Many species use sets of conventional call signs. But only open language can *express novel ideas*. The key is **syntax**: languages have a structure of rules by which the specialised parts — nouns, verbs, adjectives, conjunctions — can be combined in very different ways. Any person knowing a particular language can 'decode' any statement and understand it *whether or not they have heard it before*. Furthermore, language permits **abstraction**: the capacity to think about things remote in space or time, or even things which have not happened or could never happen (fiction and speculation, for example). It also makes possible remote communication: it is possible to read the words or view the pictures of people long dead, or to send messages to people spatially remote from the sender. This gives rise to things such as books, letters, and modern communications technologies. Even simple 'do-it-yourself' books combine the characteristics of learning, remote communication, and novelty.

This powerful capacity to generate and communicate knowledge, practical and abstract, underlies the exceptionally rapid adaptive capability humans enjoy. It makes possible the **extrasomatic** — non-genetic — transmission of adaptive information. This differs from genetically-based information transfer via DNA in two crucial ways. Firstly, the genetic process is discrete: two gametes fuse at a particular moment, and the behavioural information encoded in their DNA is passed on to the new organism. By contrast, extrasomatic information transfer is a diffuse process, occurring over many years, and the important, adaptively relevant information is mixed up with much irrelevant data or 'noise'. Although DNA does appear to store some redundant data, the problem of identifying what information it carries is primarily one of identifying the *sites* on the chain at which it is stored: in principle, a relatively simple problem. By contrast, it is quite hard to sort out the adaptively essential information in any human cultural complex, because of the vast amount of 'noise'. There is no 'site' in a culture at which the adaptively relevant information is guaranteed to be stored. Secondly, language, by conferring the capability to master abstract learning, rather than simply to learn by example, makes possible the elaboration of highly complex analytical knowledge.

Human adaptive capability, therefore, is Lamarckian, not Darwinian: humans do not have to wait for genetic change, because they can pass on 'acquired' characteristics — in the form of knowledge — within a very short time. This difference is one of the many reasons why laws and assumptions valid for the study of animal species cannot be applied directly to an understanding of humans.

Dependence on extrasomatic information transfer, however, has drawbacks. For example, a human society could lose its store of

CHAPTER 2

information if a catastrophe were to strike. There is no nuclear physics encoded in human genes: only the capacity to develop it and learn it. When great civilisations collapse, they take with them large bodies of knowledge and skills. This has happened many times; Weiskel lists ancient Mesopotamia, Phoenicia, Palestine, Greece and Rome, but there are numerous others in Asia, Africa, and the Americas.[1]

Nonetheless, the great flexibility of language and culture undoubtedly confers a critically important adaptive advantage on large, slow-breeding animals such as humans. But it means that humans are essentially niche-expanding animals, perpetually seeking to increase the carrying capacity of particular ecological niches, or to find new ones.

Population and Technology
Through specialisation, non-human species have colonised a vast range of habitats, in a process known as adaptive radiation. In a striking parallel, human cultures have spread to a remarkable range of environments. Humans now live all over the globe, from the arctic to the tropics, exploiting everything from seals to bananas.

Human population growth has been **exponential**. The *rate* of growth has remained roughly constant for long periods so that the numbers added to each population grow constantly. Just as with compound interest, the 'base number' grows larger each year. This has meant that the 'doubling time' — the time taken for each population to double in numbers — has persistently shortened; in the late twentieth century it is now approximately 35 years.

Such an explosive growth of human populations suggests that human societies have always used increases in productivity made possible by technological innovation to increase their populations, rather than to improve their standard of living. Spectacular increases in population followed the two major technological revolutions: the invention of agriculture about 10,000 years ago, and the Industrial Revolution, which began in earnest about 1750. Both, by sharply increasing output, made possible greatly increased populations. The massive improvement in public health and hygiene which accompanied the Industrial Revolution further boosted population.

Since the Industrial Revolution, seemingly endless rises in standards of living and a growing cornucopia of consumer goods have fed the belief, deeply embedded in popular perceptions and major political philosophies, that 'Progress' is inevitable, beneficent, and desirable. But the basis of the Industrial Revolution was its exploitation of fossil resources, especially fuels, at a rate far greater than they can be replaced. This sparked a rise in productivity so massive that substantial increases in standard of living for the richer countries over the last 100 years have accompanied a growth in population that remains very rapid in historical terms. This point is very important: the Industrial Revolution did not bring about any reduction in population growth rates: in fact they increased, stimulated by the substantial improvements in material welfare. But material progress through enhanced consumption is essentially a one-off: it cannot be repeated due to depletion of easily-accessible resources, and alternatives in energy and materials have to be found before fossil reserves run critically low. Even if this were done, massive population growth still underlies many contemporary environmental problems, which may be insoluble unless global populations are stabilised and even reduced. A further consequence of industrialisation is that the modification of the natural environment by human activity is qualitatively far greater than for any other species; a fact that, again, underlies many problems.

The 'Limits to Growth' school of thought stresses the exponential growth of world population, and the impossibility of maintaining it on a finite planet. Limits are pervasive: non-renewable resources can be depleted, renewable ones can be exploited faster than they can regenerate, and the impact of wastes from human economic activity — pollution — can overwhelm the capability of natural ecosystems to render them safe. These problems are all aggravated by technology.

Catton has further pointed out the dependence of the richer nations on 'ghost' resources and acreage, on someone else's territory. This system creates inequalities such as the export of protein from poor countries, to be replaced by far less nutritious grains in the form of 'food aid'. But 'ghost' acreage and resources can also lead to ecological imbalances, such as overpopulation and falling regional self-sufficiency, with increased risks of overloading the local ecosystem, through waste disposal and other impacts of an elevated population of humans. The importers become vulnerable to disruptions of their supply lines or sources of supply, while the exporters often suffer from depletion.

THE HUMAN ENVIRONMENTAL IMPACT

Like any other animal, humans withdraw the necessities of life from the environment and add wastes to it. Technology increases the impact of these withdrawals and additions, increasing the probability of massive environmental modification; while some outcomes are desirable, many are unintended and inimical. Human environmental modification underlies a wide spectrum of problems, including pollution, land degradation, overpopulation, poor health, and resource depletion. Technologically-based adaptation means that human populations are invariably — even in the most 'primitive' societies — above the 'natural' level at which they would stabilise in the (hypothetical) case that there was no technology. Thus the ecological problem for human societies is to find resources for their support without disruption of the supporting ecosystem. The conspicuous failures of the past suggest that this is not always easy.

Early, 'primitive' human societies adapted slowly, by trial and error, and the successful ones achieved stable equilibria, sometimes lasting millennia. But more elaborate technologies exploit an ever-growing list of natural resources, and their environmental impact has increased by leaps and bounds. Modern technology has been changing at breakneck pace in the 250 years since the Industrial Revolution began. The latest technologies — such as genetic engineering or nuclear energy — pose many novel problems. They have also worsened old problems — such as waste management and deforestation — so that it is now hard to cope with them using existing management methods and institutions.

One cause is that technology *insulates* people from their environment. Clothing and shelter protect and separate humans from harsh or unwanted environmental factors; agricultural and industrial techniques provide food and material surpluses, overcoming periodic and seasonal shortages that might otherwise make life difficult or hazardous. The greater the technological capacity to modify the environment, the more humans become unaware of the ecosystems on which they depend for materials and waste disposal. The **feedback loop** linking society and environment grows longer.

Consequently, the time that it takes humans to perceive any harmful outcome of their activities is increased. Ecological disruption is ignored because its full impact is not immediately felt: technology temporarily blunts the impact of its own negative effects. This impedes early resolution of problems, when little harm has been done. Instead problems must become serious before they are even perceived, and action to resolve them will be correspondingly delayed. Sensitivity is reduced.

An added problem is that technology is interwoven with the

structure of society, helping to define who benefits from production as well as conferring power. The social effects of technical change can be very disruptive, as the early years of the Industrial Revolution attest. Change is smooth only when it does not upset the existing social balance or when it favours groups which are already dominant: technological choice is a political process. Amelioration of undesirable side-effects of dominant technologies is improbable.

Existing legal, political, economic and social **institutions** often fail to respond appropriately to human-created environmental disturbances. Indeed, they often ignore them, adding to the time it takes for ecological disruption to be defined as a problem. Thus recognition and acceptance of the fact of environmental disruption do not automatically result in swift action. The political and social ramifications often lead to clashes between interested groups over their preferred solutions. Sometimes resolution of such disputes takes a very long time, and impedes proper scientific investigation of the issue.

MAKING THE LINKS: THE RELEVANCE OF ECOLOGY

One of the most important features of biological systems is the capacity for self-regulation, or homeostasis. Ecological systems are normally in dynamic equilibrium, meaning that while they are constantly changing, there is a balance between the various elements of the system. Thus, if the population of rabbits increases, the number of foxes will rise, limiting the numbers of rabbits by predation. Once hard times come, and the rabbits decline, the foxes will follow them.

Human social systems depend heavily on this very characteristic. Ecological systems — the 'natural environment' — are treated as sources for resources of all kinds, and as sinks for the disposal of waste products. Thus 'wild' resources (such as timber, game, minerals, fish from oceans and rivers, and wild grains and fruits) as well as 'cultivated' resources (such as grain crops, fruits from orchards, and even fish from aquaculture) exploit natural ecological processes and depend on the underlying ecological system. Wastes are discarded, buried, dumped in the ocean, and so on. These additions often result in eutrophication and contamination of food chains: in the Midwest of the USA excessive fertiliser use has contaminated groundwater and consequently town water supplies. Slowly, wastes are 'biodegraded' and recycled through the system. This process can be so slow that archæologists find the rubbish heaps of long-forgotten civilisations one of their most fruitful sources of information.

But humans do not simply treat the environment as a source; instead they deliberately modify it to divert to their own use a greater supply of particular resources, such as food, fibres for clothing, or building materials. Domestication, whether of plants or animals, involves preferring and protecting species useful to humans, which might not be able to survive in such numbers in the natural environment. Often such species are specially bred, lacking resistance to disease or predators.

Where cultivation is not extensive, and there remain vast reserves of 'wilderness' or at least unspoiled tracts, then the resulting impact may not be severe. The residual uncultivated land acts as a reservoir of wild species, both floral and faunal, maintaining species diversity and hence the ecological stability of the whole system. If cultivation is abandoned, in fact, the land will probably revert over time to its pristine state. This ability to restore the equilibrium is known as resilience.

But when the human system overloads the ecological, breakdowns occur. Even in modern times, over-cultivation of sparse, poorly watered soils has created some notorious dust-bowls in the USA and Australia. Many parts of South Asia and the Middle East have been scarred by thousands of years of exploitation, which have depleted soil nutrients, drastically limited species diversity, and sharply reduced the resilience of the natural ecosystems. Even recreation can have an impact: four-wheel-drive and trail bike enthusiasts have devastating impacts on local ecosystems, leading to erosion and changes affecting plant and animal species.

Such problems are aggravated by the interdependence of ecological systems: perturbations at one point may have far-reaching effects. Things cannot simply be 'thrown away': they must go somewhere, sometimes with bad effects.

Thus, even though insulated by technology, humans cannot afford to ignore the resulting environmental impacts. In fact, it may be fatal: ecosystems can restore equilibrium if sufficient resilience remains, but there is no guarantee that the new equilibrium will include any particular species, humans included.

CHARACTERISTICS OF ENVIRONMENTAL PROBLEMS

The long feedback loop between human societies and their natural environments masks the ill effects (or, indeed, the beneficial ones) of a particular action; they are not immediately apparent and may only appear much later. Toxic chemicals or ionising radiation, for example, especially in trace quantities, may result in illnesses such as cancer, but only after 'incubation periods' of 15 to 20 years or more. One result is that societies may ignore environmental problems or postpone their responses to them; this may make them impossible or very expensive to remedy.

Only if an early-warning system feeds in information, and a system of political attention responds in a timely fashion, can such risks be averted. The first system is science — most importantly disciplines such as biology and ecology — and the second is the political system itself. Their poor and inefficient intercommunication create many problems.

These difficulties are exacerbated by certain inherent characteristics of environmental issues, most notably invisibility, irreversibility, indivisibility, and urgency.

Ecological Principles

Ecology is a relatively young science, and is still not widely understood. Most people are ignorant of ecological principles or have only the faintest inkling of their importance. Ecology stresses the importance of interactions among biological processes and **energy and material cycles** within the biosphere. Ecosystem degradation can often be traced directly to human activities; for example, deforestation is thought to reduce the availability of atmospheric oxygen. Agriculture withdraws nutrients from the soil faster than they can be replaced by natural processes or the application of artificial fertiliser; as a result, soil fertility steadily falls. Some parts of the world which are poor today were once rich and fertile: the coastal deserts of North Africa were the granary of Rome 2,000 years ago. Biological resources are *capital* in need of careful management: they are neither cheap nor inexhaustible.

Ill effects frequently arise from ignorance of the most basic features of ecosystems. Environmental problems result from conflict between human action and desires — such as the pursuit of profit or power — and the limitations of the biophysical world. As they affect humans, Barry Commoner summed these up as the 'four laws' of ecology:

> Everything is Connected to Everything Else
> Everything Must Go Somewhere
> Nature Knows Best
> There Is No Such Thing as a Free Lunch.[2]

Chapter 2

Everything is Connected to Everything Else
Ecosystems are vast interconnected networks. An effect at one point may have unpredictable repercussions elsewhere. The several major cycles — such as oxygen, nitrogen, and carbon dioxide — which circulate important elements through the whole global system are imperfectly understood. But interference with the carbon dioxide cycle is blamed for the Greenhouse effect. Animals and plants are linked together in **networks** and **chains** (such as the food chain) which transmit and disperse nutrients or toxins. Furthermore, species depend on each other in complex ways. Birds and insects transmit seeds and pollen; predators depend on prey; and a breakdown at one point can affect individuals and species removed both in time and space. As Commoner says:

> ... the system is stabilised by its dynamic self-compensating properties; these same properties, if overstressed, can lead to a dramatic collapse; the complexity of the ecological network and its intrinsic rate of turnover determine how much it can be stressed, and for how long, without collapsing; the ecological network is an amplifier, so that a small perturbation in one place may have large, distant, long-delayed effects.[3]

Everything Must Go Somewhere
Matter and energy are indestructible; they can change their form, but cannot simply disappear. All matter and energy comes from somewhere and must go somewhere; wastes, in particular, cannot be 'thrown away', because they will tend to come back and haunt the thrower.

During the 1960s, residents of Minamata and the nearby town of Niigata in Japan began to come down with a strange disease, resulting in paralysis and some deaths. Before long, 'Minamata disease' was traced to heavy metals, primarily mercury, discharged into Minamata Bay by a nearby factory. Filter-feeding shellfish were absorbing the metals, greatly concentrating them. Local residents consuming the shellfish — much relished in Japan — rapidly exceeded the safe limits and became very sick.

Biomagnification of wastes also occurred in the Great Lakes in the USA; mercury, dumped by chloralkali plants producing feedstocks for the plastics industry, was thought to be insoluble, and hence to pose no danger to health or to Lake fauna. After cases of poisoning were traced to high levels of mercury in fish it was discovered that anaerobic bacteria on the lake floor were converting metallic mercury into soluble salts, whence it was getting into the food chain. Other heavy metals, such as cadmium, are also pollutants. Such pollution is a feature of the 'plastic age,' because of the greater importance of the petrochemical industries, which use more toxic substances in greater quantities than older technologies.

Nature Knows Best
Natural ecosystems have a logic and structure that cannot (in the present state of human knowledge) be improved by human action. They can be disrupted by overloading and by interference with their processes.

Overloading often occurs with pollution; the capacity of the system to absorb wastes is swamped, and unpleasant side-effects occur. Eutrophication, for example, occurs when pollutants rich in nitrogen — such as sewage — are super-abundant. Bacteria and algae, which feed on nitrogen, increase, raising the biological oxygen demand. This depletes the water of oxygen and kills the fish, sometimes in spectacular quantities.

A well-documented case of human interference is the use of chlorinated hydrocarbon insecticides such as DDT. Invented during the Second World War, and extensively used against malarial mosquitoes in the Asian campaigns, DDT was enthusiastically adopted by farmers and public health authorities after the war. Unfortunately, DDT and its breakdown products are highly persistent over a long period of time; they are indiscriminate in their effects; and they are ineffective as insecticides in the long term.

DDT is concentrated by biomagnification, ending up in the systems of the larger animals particularly. In humans, it accumulates in the liver; in many birds, especially birds of prey, it additionally affects the reproductive system, causing sterility, the laying of thin-shelled eggs, and other effects. It has been responsible for some catastrophic declines in bird populations.

DDT is toxic to all insects and, in sufficiently high concentrations, to vertebrates. It not only kills pests, but also beneficial insects. There have been cases of farmers spraying pests at the exact time that their predators are breeding, thus wiping out both pest and predator. Later, the pests regenerate, but the predators, fewer in number and with their genetic chain broken, do not reappear.

And insects develop resistance to DDT; in each generation, a few individuals survive because they are more resistant than the rest. These are the only ones to breed, passing on their resistance to their offspring. Quite rapidly, under this powerful selective

45

Chapter 2

pressure, the species as a whole becomes resistant. But further up the food chain, larger, slower-breeding species do not develop so much resistance, or as quickly.

Thousands of new substances, many non-biodegradable, are dumped into the biosphere every year. Not only is safety testing inadequate, but even inert chemicals such as synthetic textiles turn up in the oceans, thousands of kilometres from their points of origin, with totally unknown effects.[3]

The impact of these chemicals is uneven, not only between species, but also between humans. Some people are so highly susceptible to pesticides, that their release within a radius of kilometres will bring on potentially fatal reactions. Should the comfort of the majority prevail over the lives of a tiny minority? Or should this be treated as an early warning for the rest of the species? Should any substances that cannot be recycled be produced in the first place? These are questions of **equity**, both **intergenerational** (i.e., between past, present and future generations) and contemporary.

There Is No Such Thing as a Free Lunch

All movement, whether animate or mechanised, and all exchanges of matter and energy occur at some cost of matter or energy. Animals and even plants burn fuels in order to release energy for growth and motion; machines require fuels to function at all. The more energy and materials are required for one purpose, the less will be available for others. Thus a decision to favour the expansion of human populations, or to offer them a higher level of consumption, will, all other factors being equal, withdraw that energy and those materials from use by other species. Habitat destruction is an example. Furthermore, there are always costs associated with any action. Increasing the amount of energy used, for example, has generally required environmental modification (such as deforestation) or the drawing down of stocks of fossil resources. Either action denies the benefits of the resource to others, including the users themselves at a later date.

Significance of Commoner's Laws

Commoner's laws are not scientific, nor, even though most environmental problems involve all simultaneously, do they by themselves explain or help resolve the sorts of political, economic, and social problems which are often posed by environmental issues. It is not possible simply to transpose ecological laws to economics and politics. Their impact has to be understood, via their socially relevant characteristics.

But they are a useful link between science as early-warning network and politics as the response. They dramatise the *implications* of ecological processes in a way which is likely to be grasped even by politicians and others who are ignorant of ecology.

They also help in responding to Commoner's important insight that it is not simply technology or population that cause environmental crises. Rather, technology acts as a multiplier on population, greatly increasing its impact. This is a social synergism, not unlike those that occur in ecosystems. The more powerful the technology, the greater the environmental impact is likely to be and, consequently, the greater the disturbance to supporting ecosystems.

This means that the *choice* of technology becomes very important. The core of Commoner's *The Closing Circle* was his demonstration that, since 1945, technology choices in the USA had consistently favoured processes and products that resulted in increased environmental impacts. The new 'miracle' materials, such as plastics and synthetic fibres, were non-biodegradable, and turned up in quantity all over the world. Pollution from carbon dioxide, as a result of increased use of motor vehicles, had risen steeply. Other forms of pollution were also growing apace. The most spectacular increases were in such things as non-returnable bottles and cans, products of the throw-away economy. What was controversial was Commoner's demonstration that the choices to introduce the technologies concerned were almost always undemocratic, in the sense that they were made by major corporations or other centres of *economic* power which were not responsible to the population at large, and that their primary motivation was to save labour and increase profits. Non-return containers, for example, saved organising a return network for glass bottles. But they massively increased pollution, eliminated the jobs of those involved in returning the bottles ('recycling' in present-day jargon) and wiped out the opportunity for a vast army of small boys to pick up some pocket money by collecting them. The latter task was now transferred, lock, stock, and barrel, to the public purse, in the shape of local government and its litter collectors. Increased profit was bought at the price of higher social and environmental costs. Similarly, soap, a universal, reasonably biodegradable cleaning material, made from widely available materials (such as dead sheep), was replaced by detergents made from petrochemicals, and much less biodegradable. But detergents, though no better at cleaning, are vastly more profitable to the producer than soap.[4]

Commoner was bitterly attacked for allegedly neglecting the

importance of population in creating environmental problems. But, far from downplaying it, Commoner's concern was to show how the multiplier effect produced by the combination of population with technology dramatically increased environmental degradation. And his demonstration went further: he showed that the USA, with five percent of the world's population (approximately 200 million people in 1971 when he was writing; now roughly 250 million) had by far the greatest impact (in fact, the USA uses about half of the world's annual consumption of raw materials). By showing that profitability was the main criterion for choice of technology, contributing substantially to the problem, he made his conclusions politically unpalatable, especially to supporters of the existing system.

Thus Commoner demonstrated not merely that modern industrial society can pollute very heavily, but that the *form* of political and economic organisation could have an impact. Furthermore, he showed how, decisions about issues which impact heavily on ecosystems are very often taken without any consideration of those impacts.

Overview: the Significance of Adaption and Technology

Human reliance on technology is central to the characteristic human mode of adaptation. Technology makes possible the support of elevated populations, which in consequence become highly dependent on it. It also lengthens the feedback loop between human society and the ecosystems on which it depends, diminishing attention to ecosystem changes, even when significant.

The problem is complicated because the reasons for choosing technologies often take little account of environmental criteria, and are frequently highly irrational in ecological terms. Furthermore, those with the greatest power over the choice of technology are seldom those on whom the consequences will fall; they are insulated from the consequences of their decisions. Frequently, too, they will resist attempts to wrest power from them or to change their decision criteria. Thus adaptive change may require social change of a kind which will be resisted, sometimes violently.

Further Exploration

Human adaptation is a well-rehearsed topic, and an outline of the basic principles can be found in almost any good introductory anthropology textbook. Two of the most stimulating are Charles F.

Hockett's *Man's Place in Nature*, and Marvin Harris' *Culture, People, Nature*. Hockett's book has an exceptionally entertaining account of the emergence of typical human institutions such as the state, as well as a highly personal view of the nature and potentialities of human evolution. Harris' work, with its strong emphasis on ecology as a causative factor, has been very controversial; however, he frequently voices points of view which are worthy of careful consideration. For example, he argues that because humans exploit natural resources to survive, no human civilisation can last; inevitably, 'limits to growth' of one kind or another will eventually arrest progress and impose unavoidable decline. Boyden's *Western Civilisation in Biological Perspective* has a more strictly biological perspective; it is convincingly applied to Australia in the more recent book, *Our Biosphere Under Threat*, by Boyden, Dovers, and Shirlow. This is the first of a series emerging from the Fundamental Questions Program (sic) in the Centre for Resource and Environmental Studies (CRES) at the Australian National University.

Perhaps the best book on the 'basics' of ecology is Ehrlich's *The Machinery of Nature*, a remarkably readable outline of modern ecology, with discussions of many major issues, such as population and habitat size. For Australia, Smith's *The Unique Continent* is very useful.

There is a sober and very illuminating discussion of human population growth in *Ecoscience*, by Paul Ehrlich, Anne Ehrlich and John Holdren, at Ch. 5. It is particularly useful for its discussion of exponential growth and its illustration of the effects of major technological revolutions on population, an issue to which Catton also gives prominence.

Boyden, Dovers, and Shirlow also deal with technology and its environmental impact, especially in 'high-metabolism' societies such as modern Australia, where rates of use of energy and materials are high. This is also an important focus of Commoner's *The Closing Circle*, first published in 1971. Commoner's controversial argument that the combination of population with increasingly devastating technology, creating a synergistic 'multiplier effect' greater by far than the impact of either factor alone, makes the choice of technology, and the socio-political forces which govern it, critical. He was violently attacked by Ehrlich and his collaborator Holdren for slighting the importance of population growth, sparking off a vigorous debate which has never quite died down.

Wilkinson's *Poverty and Progress* also adopts an ecological approach, this time to economic development and technological change. The book is essential reading, not only to understand the background to modern times, but also for its insights into technology choice.

Ancient civilisations and ecological failure come under scrutiny in Bilsky's collection *Historical Ecology*, and, most pointfully, in Ch. 5 of Ponting's recent *A Green History of the World*.

The thesis about the impact of the Industrial Revolution, and in particular its impact on population growth and the environment, is most articulately expounded by Catton, whose lively paper of 1976 'Why the Future Isn't What It Used to Be (And How it Could be Made Worse Than it Has to Be)' was followed in 1980 by his book *Overshoot: the Ecological Basis of Revolutionary Change*. Catton places the work of the 'Limits to Growth' school in a historical and ecological context, in the process significantly modifying its emphasis, though not the major conclusions.

The paragraphs on the significance of ecology depend heavily on Dryzek's extremely important book *Rational Ecology*.

For more about 'Minamata disease,' see Tsubaki and Irukayami, *Minamata Disease: Methylmercury Poisoning in Minamata and Niigata, Japan*.

CHAPTER 3

ECOLOGY AND POLITICS

Even in democratic countries, politics is often seen as an activity limited to politicians in capital cities. Citizen participation is confined to elections; only a small minority participate in political organisations and actively engage in attempts to influence government.

Mass media such as television and radio distance people from political activity, reducing it to spectacle and even entertainment. At the same time, they trivialise many issues, reducing them to 30 second 'grabs' and never explaining them effectively. Newspapers display considerable bias, tending generally to favour the conservative side of politics, but, even more importantly, failing to consider issues analytically.

This can lead to bizarre misconceptions, as well as dramatic lack of communication. The major Australian political parties — Labor and Liberal — are almost indistinguishable on most economic issues, having adopted the suicidal 'economic rationalist' policies urged on them by their Canberra economists, out of touch with the legitimate concerns of citizens deprived of livelihood and prospects. In the Australian federal election of 1993 they firmly ignored the problems of the environment and sustainable development, in order to promote rival, marginally differentiated economic policies. Months later, it was apparent that neither the politicians nor the journalists had much idea what had produced the 'upset' result; most simply assumed that their perceptions of the issues were shared by the electorate.

The paradox is that politics seems remote from ordinary people; yet nearly everything is political, and there is little individuals can do about many issues. It is the failure of politics to *tackle* the urgent problems that produces its air of remoteness and irrelevance; politics is not *engaged*.

The Pervasiveness of Politics
Yet one clear implication of human adaptation through technology is that, because of the dominance of cooperative activity in human behaviour, a very large part of human experience is political. Cooperation and coordination involve interaction, the allocation of resources, and the making of authoritative decisions. That is, somebody must decide that everyone is to do *this*, not *that*; and it may actually be more important that everyone *do the same thing* than that they do a particular thing. It is possible, for example, for people to drive on the left or the right side of the road; both rules are adopted, in different countries. What is important about this sort of rule is that everyone does the same thing. If people drove all over the road, there would be too many accidents. The process of deciding that all will drive on the right (or left) and enforcing the rule, is political. Decisions affecting groups of people are by definition political, and there are many, even in primitive human societies.

But cooperation is not the only political activity; conflict is political too. Both involve interaction, gains and losses, stakes (in the sense of some material or symbolic interest in the outcome) and rules and expectations. Even such actions as working for a particular firm may be 'political', not in the formal sense of involvement with the institutions of politics, but in the much more pervasive sense that it supports or undermines a particular set of power relationships.

If politics is so pervasive, then there are at least two questions which have to be asked in relation to environmental policy problems. One is, how do environmental questions impinge on the daily lives of people, as well as on the political system? A start was made on answering that question in Chapter 2; some more aspects of it will be considered shortly. The second question is, how can the political dimensions of environmental questions be understood? In one sense, the whole of this book is about just this issue; but it is possible to make a start at clearing away some of the undergrowth.

ECOLOGICAL KNOWLEDGE AND ITS IMPACT

Ecological systems affect human society mainly through the *feedback* from social actions which disrupt ecological processes. As Commoner's laws predict, such feedback is unpredictable, emerging at points remote from the original perturbation, and depending on scientific principles some of which are as yet unknown. The impact is complicated by **scientific uncertainty, displacement**, and the **problem of entropy**.

Scientific Uncertainty

Ecological knowledge is often incomplete; scientific data are often inadequate or unavailable, and there are many scientific questions to which reliable, predictive answers are not available. For example, the effects of PCBs, dioxin, and other chemicals in the ground and in water supplies are not accurately known. Finding out involves expensive research at the very threshold of the sensitivity of existing tools, both scientific and statistical. When research is pushed to its limits, legitimate disagreements emerge, about the data and how they are interpreted; scientific uncertainty arises.

Nor, even with the best tools, is the evidence always conclusive. Much of the evidence for environmental damage is suggestive and probabilistic; in such cases, the only connections which can ever be demonstrated between the various phenomena are *statistical*: that is, certain things are found in association, and one is therefore assumed to cause the other. A very well-known example is lung cancer, which has a strong positive association with tobacco smoking. While most people accept that the evidence is certainly strong enough to take moves to reduce smoking and protect non-smokers from its impact, some diehards argue that the link has never been *proved*. Technically, in strictest logic, they are quite right, though by the time the link *was* proved, it would be too late for most sufferers. The need to rely on probabilistic evidence has implications for **methodology** — the way the problems are studied — as well as leading to the commonsense conclusion that such evidence must be treated with care.

When urgent answers are needed to practical questions, it may be impossible to find reliable scientific answers quickly. This is why answers to 'what if?' questions assume considerable importance: what would we do in the face of various kinds of crises?

If there is no reliable scientific answer, it may be better just to treat the risks as serious and attempt to minimise them. This approach is sometimes known as the 'cautionary principle'. It involves the adoption of strategies that minimise the damage associated with the worst possible outcome, even at the cost of some sacrifice of present goods.

Problem Displacement
Another problem is that the empirical evidence does not tell the whole story. Because 'everything is connected to everything else', an indicator in one place may show a decline, but can be paralleled by a rise somewhere else. This is called **displacement**.

. In the winter of 1952–3 a 'killer smog' caused over 4,000 deaths from respiratory problems in London. The cause was the near-universal burning of coal for industrial purposes, domestic heating and cooking; fogs, common in winter, mixed with coal smoke to form the thick choking 'smog' (the word is a contraction of 'smoke and fog') which had been common for nearly 100 years. (Conan Doyle's Sherlock Holmes stories of the 1890s, for example, mention 'pea-souper' fogs.) From 1953, the burning of coal was banned in parts of London, the building of higher smokestacks to disperse factory wastes encouraged, and a number of other drastic 'smoke abatement' measures were taken. The effects were dramatic: rapid improvement in apparent air quality, and a marked increase in the number of sunny days. However, private car and diesel truck use grew rapidly, and during the 1950s all surface public passenger transport was turned over to diesel buses. London now has serious petrochemical smog problems, made worse by the increased sunlight! Furthermore, a significant amount of this pollution reaches Europe and is precipitated as acid rain.

London's case is in part an example of *spatial displacement*: building tall chimneys does not lead to elimination of the problem. Rather, it disperses it more widely, so that it comes down where it once didn't. Fish in Norwegian lakes die instead of asthmatic citizens in London.

It is also a *displacement in time,* which typically involves turning a present problem into a future cost: London solved its coal-fire pollution problem but worsened the growing petrochemical smog. Similarly, used fuel rods from nuclear power stations are currently stored in 'ponds' of heavy water, pending a solution to the disposal problem: displacement to the future.

Displacement across media can also occur. London's coal smog problem was transformed into a petrochemical one. Even the deliberate 'processing' of pollutants can transform them into further pollutants. Thus a municipal council running short of space for 'sanitary landfill' — tips — might build an incinerator, use the heat for power or heating, and pass the problem on to the next higher level of government, as air pollution. Transforming pollution does not really solve it.

Political systems often tend to displace rather than solve environmental problems, postponing solution of the underlying problem and leaving no-one better off in the long run. When this is true, the chickens will one day come home to roost, maybe all at once, and very probably as vultures.

Entropy as a Limit

A large part of the environmental debate has revolved about the question of material **limits to growth**: that is, the question whether the materials needed to support human technology will become so scarce that civilisation itself will collapse. In one sense, the answer to this question is obvious: the earth is a small and finite planet, and if human populations grow sufficiently large, it will certainly be impossible to feed, clothe, or house them. If the rate of growth is very rapid, the crisis could come frighteningly soon. Most of the arguments against this position are designed, not to deny the existence of ultimate limits, but to show that they do not pose a serious threat in the foreseeable future.

One of the major arguments for not worrying is that resources will never become scarce because they can be *substituted*, so that as one thing grows scarce, another will replace it. This line of thought assumes that, so long as *some* sort of substitute is available, things never become absolutely short. There are technical objections, in the sense that substitutes for some materials are very hard to find; but the point is to substitute for the *function*. However, this may still not take sufficient account of exponential growth in *demand*. It is quite true that things generally 'run out' by becoming scarce and expensive rather than simply unavailable, but in the limiting case where demand is high and substitutes poor, this may mean very little. If you hit a brick wall at 100 kph, interpolating a 200mm rubber mattress will not greatly cushion the impact.

But underlying at least one version of the substitutability argument is the theme that the only absolute scarcity is of low entropy.

Chapter 3

Entropy is a term from physics, referring to conditions in which there are high concentrations of order and energy, rather than low. Entropy is high when energy is dispersed, low when it is concentrated. Every living thing is an area of low entropy, concentrating and ordering things from nature, essentially for its own use. Humans exploit this ready-made low entropy, converting it to their own use. Obviously, if there is not enough to go round, some may miss out; if the self-regenerating capabilities of the areas of low entropy are damaged, then their capacity to support human social organisation will decline. This is why preserving the productive capacity of the biosphere — particularly plants — is so critical.

Many 'solutions' to ecological disruptions fail to escape from the entropy problem. New energy technologies such as nuclear fusion create problems through their release of heat, with climatic implications. The same goes for beaming power from space. And finding more room for bloated human populations — through migration into space, for example — is likely to require too much energy and resources to be any answer for the bulk of humanity.

Implications

If crises are difficult to detect, then they cannot simply be planned for, because they may occur without warning or even be present already. So they must be treated as limiting cases — things that are not wanted, and which public policy will attempt to avoid. This is called 'negative policy space': it defines what is not desirable. It is also the basis of the 'cautionary principle' or 'anticipatory' policy-making.

Furthermore, if empirical data is always inconclusive, then it cannot be relied upon. For example, if the exact effect of a particular air pollutant cannot be known accurately, then the effect of pollution abatement measures may have to be assessed using (say) a notion of desirable maximum dose, rather than starting from nonexistent 'baseline' data. In effect, other criteria must be used to evaluate policies, perhaps derived from the characteristics of ecology itself; and if the problem is to maintain low entropy, policy choices have to be geared to that imperative: the priority of ecological considerations.

Ecological rules cannot simply be transposed to human societies without modification: the ways in which ecology affects humanity are specific, and they are mediated through the social process of production.

56

SUSTAINABILITY AND RATIONALITY

The idea of behaving in such a way as to maintain the critical ecological processes is often known as 'sustainability'. However vaguely, principles such as 'ecologically sustainable development' (ESD) depend implicitly on the notion of **ecological rationality**.

Rational behaviour is behaviour which achieves a given end most effectively or economically, or is consistent with a set of important principles. For example, those wishing to live a long life and die at a ripe old age would be well advised to avoid sports with high mortality rates, such as motor-racing. Similarly, social organisations claiming to have ecological rationality would tend to promote ecologically sustainable behaviour.

Ecological Rationality

There are many arguments both for and against ecological rationality. Most of them are about self-preservation, though there are also arguments which depend on religion, while others use moral ideas such as the notion that all species have a right to existence. There are also aesthetic pleasures to be derived from nature. But the utilitarian values of ecosystems — specifically for production, protection, and waste-assimilation — are the most basic to ecological rationality, because all the other values rest on them. We cannot appreciate the æsthetic value of forest or wilderness if we are not alive in the first place. The other values may nonetheless be important or even profoundly moving.

A *natural* system which is ecologically rational can resist shocks and stabilise itself. It can provide life support though self-regulation, for the system as a whole. It is stable because it is resistant and resilient. **Resistance** is the ability to maintain system functioning under stress; **resilience** the capacity to restore the system once stress has destabilised it. Both require negative feedback in the system.

Human life support requires more: consistent and effective support specifically for humans. Consistency — long-term reliability — specifically requires maintenance of low entropy. The ecological stability problem for human societies is located at the interface of human and natural systems. Humans have a material interest in the stability both of human and natural ecosystems; the latter is shared with other species.

In most social and political systems, social choices are justified by appeal to some kind of rationality, most commonly economic, legal, or political. Ecological rationality is a relatively new idea,

Chapter 3

and competes with the claims of the others. It is therefore important to compare them.

Economic Rationality

Economic rationality requires the maximisation of **utility** across the whole system. Utility itself is generally taken to be the total of *material* gratifications to the individual, though in fact the original definition counts any gratification as an increment in utility. Rules for the evaluation of social choice under economic rationality require some system for the aggregation of individual preferences: that is, some sort of value has to be put on preferences, so that they can be summed up or compared. Economists make primary use of monetary values, and economic 'rationalists' often define social choice purely in these terms, which excludes many things which are valuable, but cannot be bought or sold.

Economics tends to treat ecological problems as 'externalities', which must be adjusted, for example by price mechanisms, so as to optimise production. Natural resources are seen simply as feedstocks, factors of production, and the primary concern is with the optimal rate of depletion. Both the 'externality' approach and the concern with optimal depletion rates conceal some sweeping assumptions about substitutability of resources. In addition, some economists explicitly argue that the 'best' approach is that which is most responsive to the wishes of the people most directly affected. This last is the sort of conflict that underlies the Tasmanian wilderness issues of recent years: local as against national interests. But what is democratic or responds best to 'consumer demand' may be far from ecologically rational.

Economic actions can be 'rational' even if they are destroying future options and the very chance of survival. This bias is built into economics through the practice of 'discounting' future benefits. Their 'present value' is determined by applying a percentage discount rate for every year the benefit lies in the future; if positive, the discount will result in a reduced 'present value', which is then compared with others. This practice tends to overvalue present consumption by comparison with future. If future benefits will actually be greater than present (as, for example, with an increasingly scarce resource, in a context of exponential growth or depletion) only negative discount rates should be considered, since these will capture the enhanced 'value' of the resources at a later date. Without discounting, economics cannot choose between present and future, and is unreliable. But since its primary goal is to maximise production, whereas ecological rationality might choose protection or some other value, it is clearly in conflict.

Legal Rationality

The essence of legal rationality is conformity with a system of rules. These should be — but often are not — internally consistent. The rules are applied in judging individual cases, either disputes between individuals and governments, or between two or more individuals. Very often, the entitlements of each party are defined as 'rights' and cases are treated as a conflict between them.

Democratic countries often place great emphasis on the 'rule of law': the notion that everybody is equal before the law, and no-one is entitled to special treatment. This is an important guarantee of civil rights and a minimum condition for equality. But it can lead to the rules of the legal system being treated as more important than any other, even when they come into conflict with common sense or ecological necessity. In Australia, the Liberal Party has frequently reiterated support for 'states' rights' in the face of serious threats to dwindling wilderness, over which it was clear that states would not take action. Such use of the law as an excuse for inaction is particularly spurious when indulged in by those who make the laws in the first place.

Law has nothing to say about environmental questions in general, though it does in some cases assign rights in relation to water, land, and other environmental goods. Nor do any of the warring philosophies which underpin legal thought have ecological relevance. Legal rationality is not necessarily in conflict with ecological; more often it is simply irrelevant.

Political Rationality

It is very difficult to disentangle the notion of political rationality from simple expediency. Often, in common speech, a course of action which advantages a political actor will be described as 'politically rational'; in that usage, it is understood to mean that an advantage — short or long term — is gained. Obviously, an action will be politically rational if it can be shown to achieve a given political goal effectively or efficiently. Since almost any objective can become a political goal, political rationality casts a wide net. Hitler's political goal was to exterminate the Jews; Lenin's to industrialise Russia; that of the anarchists to abolish government altogether. The political rationality of each has to be evaluated in terms of its effectiveness in attaining its own goal; but few would wish to make that the sole judgemental criterion.

Political rationality has also been defined generally as effective-

Chapter 3

ness in solving the collective problems confronting the system. This could lead to some argument about which solutions were effective, and what constituted effectiveness, however. For example, displacing a problem might well 'solve' the immediate political problem; but it certainly could not be described as effective in the long run, and perhaps not even in the short. It might even be that everyone in the political system agrees that the problem is 'solved' when it actually isn't.

Like economic and legal rationality, political rationality has no underlying ecological touchstone; it may be neutral, beneficial, or positively destructive.

The Priority of Ecological Rationality

The ecosystemic productive, protective, and waste-assimilative capacity underlies the enjoyment of any value whatsoever. Thus claims for the priority of any other value must be weaker.

The question then is, how much priority should ecological rationality have? Other rationalities can generate tradeoffs; for example, using aerosols for convenience, against a notional deterioration in the ozone layer, may be economically rational. The problem is that to permit a series of individually rational tradeoffs of this kind can quickly lead to a situation in which the sum of them all is environmental destruction. Thus such trade-offs could be permitted only if shown to be ecologically rational: that the sum of them would *not* be destructive. In short, ecological rationality must have priority; other values may be addressed only when it has been substantially satisfied.

Environmental Argument

Most socio-political questions require judgements of various kinds. This happens in part because data are often incomplete — as the Love Canal and Kingston cases demonstrated — and subject to legitimate differences about interpretation. These are largely scientific issues. But socio-political issues often require the assessment of possible courses of action, and some thought about their probable consequences. (The latter is particularly important with policy.) Science, of course, also requires judgements, but most have to do with the interpretation and assessment of data rather than with policy questions. Policy injects a greater element of subjectivity. This in turn means that claims, both about matters of fact and about the consequences of future actions, have to be examined carefully and *critically*. Environmental argument frequently mixes the scientific with the political; they have to be disentangled before they can be analysed dispassionately.

This is a serious potential pitfall for any kind of environmental evaluation, and it is particularly treacherous where policy questions are involved. Thinking and arguing about the environment must be done with care.

Environmental questions are frequently controversial, both because of scientific uncertainty and because their resolution may challenge either received wisdom or actual pecuniary interests. Few disputes end in bloodshed; instead, various arguments, more or less shoddy, are used to justify the positions taken up. Frequently, an argument is not *ipso facto* the reason for a particular position; in fact it may be a justification, used to rationalise a position reached on other grounds.

Scientific Method

Though there is much argument about scientific method, a key feature of all scientific knowledge is that, though reliable, it is not final. Scientific theory is open to continuous revision in the light of new knowledge, and experiments and observations must always be capable of replication by others. One theory of scientific progress is even based explicitly on the idea that science proceeds unevenly, by a series of 'revolutions', each triggered by the inability of the preceding scientific paradigm to explain known anomalies.

As many questions of ecology are at or near the frontiers of knowledge, both uncertainty and the 'open' nature of science itself are relevant. Many of the examples used in this book are of cases where nobody knows for certain what the extent of a problem is, or what the outcome of a specific action is likely to be.

Two points follow. Firstly, uncertainty, though it can make decisions difficult, is not necessarily a bad thing. It means that differing views are legitimate, so long as interpretations of the data they draw on are reasonable. Secondly, many decisions about environmental policy questions are based on *judgements*, in the light of the available data, about what the best course of action might be. Those judgements may well turn out to be wrong, even if they were the best that could be made at the time. Thus care has to be taken to avoid irreversible policies and hasty conclusions.

That in turn means that considerable weight is thrown on the arguments for particular courses of action or interpretations of the data.

CHAPTER 3

Argument as Justification

Arguments are not always *reasons* for action; in fact argument is often used to justify positions reached on other grounds. This phenomenon was remarked in the seventeenth century by the philosopher Thomas Hobbes, who argued that humans are passionate and rationalising rather than rational. What he meant was that humans often choose courses of action because their emotions dictate, and then seek arguments ('reasons') to justify their actions. This is called *ex post facto* (after the fact) justification.

Thus people may have strong personal loyalties to particular groups or organisations, and they may support their ideas or policies without much thought for their actual effects. This can often account for the long persistence of particular political philosophies or policies, such as Marxism or conservatism, even when obviously outdated. Often they are at the height of their *political* power when already fossilised and discredited *intellectually*. (A good place to find fossilised ideas is a daily newspaper.)

Under such conditions, people will obviously be tempted to support their views with incomplete, improperly interpreted, and selective data. They may also adduce arguments which have central fallacies: for example, conservatives often argue that nothing can be done about inequalities and injustices in society, and that therefore it is not worthwhile to try. The fallacy in this case lies in the first assumption: that society cannot be changed. Sometimes, however, the false assumptions are more subtle. Goodin has exposed three such fallacious approaches in his book *Political Theory and Public Policy*: 'Impossibility as an Excuse for Inaction,' 'Risk as an Excuse for Maldistribution,' and 'Uncertainty as an Excuse for Myopia'. He attacks the assumption of the first approach by showing that actions initially thought unfeasible may prove possible under careful and analytical inspection; in the second case he points out that individuals often cannot assess for themselves the risks to which they are exposed, especially in a complex society, and that hence social action is needed to protect them; in the third that it is often important to look well into the future when weighing courses of action.

It is important to be on the alert for bad arguments or self-serving justifications. This can be done if something of the structure of argument is understood.

Thinking About Environmental Problems

Environmental problems, while complex, are not different in *kind* from other problems. The same analytical techniques may be used to understand them. Arguments about environmental questions can be subjected to the same analysis as arguments about any other matter.

Arguments and How They Work

To think analytically about a problem involves reducing it to a series of component parts, and attempting to detect the interrelationships between them. It is rather like taking a watch apart and trying to work out what the various bits do. This is one of the things science does, but it is also fundamental to plain old-fashioned common sense.

In deciding what to do *about* something — for example in making policy — it is also necessary to think synthetically. That is, having pulled the watch apart to see what is wrong, it must then be put back together and made to work again.

Both processes rely on **arguments**: statements about the logically necessary relationships between things. The simplest form of argument is a **syllogism**:

> Premise 1: *All cats are grey in the dark.*
> Premise 2: *It is dark at present.*
> Conclusion: *Therefore any cat I see will appear grey.*

The conclusion to this syllogism depends on the truth of the two premises. If all cats indeed appear grey in the dark, and it is dark at present, then any cat seen must appear grey. If the second premise is untrue — i.e., it is not dark — then cats may well appear as tabby, black, white, ginger, etc. And if the first premise is untrue — that is, cats exist which do not appear grey under dark conditions — the conclusion will also be untrue. This particular first premise is a **generalisation**; i.e., a statement which takes the form 'all Xs are Y', where Y is an attribute of some kind. Such statements can always be disproven by producing a **counter-instance**: an example of an X which does not follow the rule stated in the generalisation. If, for example, cats existed which glowed orange under dark conditions, they would not then appear grey, and the first premise would be invalid. *One* authentic counter-instance is sufficient to disprove a generalisation.

In political discourse, the premises of arguments are often

Chapter 3

unproven assumptions; it is important to isolate them and see if they are really true. These assumptions frequently take the form of generalisations, the truth of which can also be critical. If it were indeed true that society cannot be changed, then (and then *only*) it would follow that attempts to change it would be futile; but that initial assumption is critical to the whole argument.

The second thing is that there must always be two or more premises. If there is only one, then either the conclusion is true by definition, or there is a premise which has been suppressed or omitted. A statement which is true by definition is called a **tautology**: for example 'all bachelors are single men'. Most tautologies are trivial, but they can have instructional value; 'all insects have six legs' is tautological because possession of six legs is a defining characteristic of insects, but it would be a useful way of teaching someone to distinguish between insects and (say) spiders.

Suppressed premises are more serious. For example:

> *Labor has made an awful mess of governing Australia.*
> *Therefore, to avoid the mess, I will vote Liberal next time.*

The suppressed second premise to this argument is that the Liberals present an alternative. Stating the second premise clearly exposes the flaw: what if the Liberals are offering essentially the same policies as Labor, which, for example, they did in 1986-93? Then it would make no sense to vote for them; if Labor's policies result in a mess, and the Liberals offer the same policies, then the mess ought to be about as bad or even worse. Avoiding it will require identifying genuine alternatives. Omission of the second premise in this way is often called the fallacy of the **excluded middle**; it is very easy to do, and should be carefully avoided. Avoidance of excluded middles is particularly important in goal-directed policy evaluation: changing policies is always costly, but if no genuine change results, it is clearly wasteful too.

Simple syllogistic arguments can also oversimplify; a common trick of bar-room debaters is to batter their opponents into submission with arguments which seem perfectly logical but which in fact ignore important issues. Consider:

> *Salination is the worst environmental degradation problem in the Murray-Darling Basin.*
>
> *Therefore if salinity is cured, the environmental degradation will be cured.*

This looks like a tautology ('Don't be bloody silly, of course things will get better if the worst problem is cleaned up.') but in fact the conclusion is not true by definition, nor is it necessarily true at all. This time, the trouble is that salinity is not the only problem, just the most central. So if salinity is cured, it is reasonable to expect that the suite of problems directly related to it will also be cured. But other problems may change little, and could be worsened. If, for example, curing salinity meant that more water could be diverted for irrigation, then problems due to reduced flow and the loss of the seasonal flooding might well be exacerbated. This difficulty has to be borne in mind when evaluating **remedial** and **meliorative** approaches to policy-making, such as incrementalism. Having killed the dragon, how does Ivan, seventh son of Ivan, cope with the marriage-hungry maiden and her father?

Syllogisms Are Too Simple
Few real-life problems are simple enough to reduce to a syllogism. But the example teaches two useful precepts. Firstly, *always look for the assumptions on which a conclusion is based, and check their validity.* Secondly, *check the logic*. Does the conclusion follow from the stated premises, or is there some hidden assumption which has not been stated? Do the premises require the conclusion? (If only some cats are grey in the dark, the conclusion that the next cat we see under poor lighting conditions will appear grey is not warranted.) Surprisingly often, policies are offered which allegedly attack specific problems, but which actually fail to do so. Sometimes this is the result of sloppy or careless thinking; sometimes the motives are more sinister. In either case, pointing out the flaw can be the first step to a more effective resolution.

Sloppiness or intellectual sloth often appears in the evaluation of policy measures. Often governments will point (with pride or indignation, depending on the circumstances) to measures they have taken to alleviate some problem, be it salinity or unemployment. But they frequently fail to establish the *effectiveness* of the policy by reference to appropriate indicators. If 1,000,000 people are unemployed, creating 100 jobs, while laudable, cannot honestly be said to 'tackle' it; the cure is quite inadequate to the scale of the problem. Similarly, if reducing salinity in the Murray failed to resolve problems of species loss or soil degradation, then it could not be accounted a success in relation to environmental degradation as a whole, however effective it was as a solution to the salinity problem taken alone.

A more sinister example was the Bjelke-Petersen Government's policy on national parks in Queensland. This permitted mining in national parks, and imposed no controls other than the usual procedure of applying for mining leases through the Mining Warden. In other words, a national park, putatively a place in which conservation values are uniquely protected, had no such protection from mining. This exploited common expectations; creation of national parks, by giving the appearance of protection, left the State Government free to approve mining if it chose, while deflecting criticism in the meantime.

This type of behaviour is often called **tokenism**, or **'symbolic politics'**: that is, people concerned about an issue are rewarded, not with effective action on the question, but the *appearance* of action. Politics is often about symbols rather than reality; ideas like 'national pride', 'race', or even 'a mandate for our policies' frequently have no referent in the real world.

Avoiding Argument

The use of symbols is a powerful way of avoiding action, and often of avoiding argument as well. Humans often respond violently and irrationally to symbols. Things such as national flags or particular ideals ('democracy', 'socialism' or 'freedom' for example) often stimulate strong responses which have little to do with the actual problems at hand. They are thus very useful to politicians and others who wish to avoid argument, or deflect attention from dodgy premises and excluded middles. Sometimes these efforts can be quite ludicrous, as when the then US President, George Bush, presented himself for re-election in 1992 on a platform of trustworthiness and probity — just as his complicity in a secret, and very shady, arms-for-hostages deal with Iran was becoming increasingly difficult to deny. Similarly, in 1992–3 few observers took seriously the Serbian claim that resistance to an 'Islamic fundamentalist state' was the reason for their wholesale butchery and torture of their Muslim neighbours, beside whom they had been living peaceably for 40 years.

Evaluating Actions and Arguments

It is not necessary to be a highly trained philosopher to sift out dishonesty, prevarication, and even fundamental errors in logic. Very often, simple tests, asking how effectively government is responding to a particular issue, can quickly puncture official rhetoric and evasiveness. For example, how do governments respond to important environmental issues? Do they commission serious research to find out the best answer — or do they pay a public relations firm to manipulate public attitudes?

A furore of this kind occurred in Britain in 1990. Producers of beef cattle had adopted the practice of supplementing their diet with the minced-up remains of sheep killed for mutton or lamb, to economise on expensive grain, and to yield a higher level of protein in the resulting beef. But the beef cattle started to go mad, and it emerged that they were succumbing to a brain disease, previously unheard of in cattle but one that frequently affected sheep and related species; its source was the ground-up sheep remains. The British Government attempted a cover-up, first denying that a problem existed, and then attempting to minimise its significance. But sales of domestically-produced beef fell catastrophically, as consumers avoided it for fear of contracting Mad Cow disease themselves.

In this case, the outcome was very striking, since the action was not only ecologically irrational — it makes more sense to solve a problem than deny it — but also economically irrational, since it hurt the beef producers, and politically irrational, as it exposed the Government as a pack of buffoons.

Sometimes the outcomes are less dramatic. The USA, a major importer of beef from Australia, adopted strict limits on pesticide residues in imported beef at about the same time — 1972 — that it banned the domestic use of DDT. State governments in Australia did not legislate accordingly; instead, they continued — through advice tendered by their Agriculture Departments — to encourage the use of chlorinated hydrocarbon pesticides. Several times since, Australian beef has been denied entry into the USA because it contained excessive pesticide residues. Here, government failure to legislate conditions suitable for a major export market was politically irrational, and the farmers' ignoring of overseas regulations was economically irrational; both attitudes were short-sighted.

In cases such as these, as well as in many others, some simple, straightforward questions can be asked; the answers will quite often be rather revealing. Some examples are: Are there programmes for effective environmental management, which actually tackle the serious problems, with measures which are on a sufficient scale to check or reverse the damage? If not, is the government or the proponent of that particular line of argument indulging either in tokenism or displacement? Is the government (or other proponent) taking all *relevant* factors into account? If not, is the omission damaging to the proposed course of action? What are the reasons for this neglect?

Chapter 3

In short, it is often possible to evaluate the effectiveness of government policy, as well as the honesty of political party promises, by applying simple, 'commonsense' tests. Very often such tests will discover 'window-dressing' rather than real policies, or serious ignorance of the nature and scale of the problems in question.

One reason for this is that most institutions and practices are built about imperatives other than sound ecological management. This is most evident in the political economy of state societies, the dominant form of social organisation in modern times.

Overview: Thinking Rationally

Politics seems very remote to most ordinary Australians. Though constantly bombarded with images of it, they do not participate extensively, and their perceptions are filtered through mass media which trivialise and reduce their understanding of it. Furthermore, as political parties focus increasingly on technical issues — such as their management of the nation in a narrow economic sense — politics becomes more and more divorced from the concerns of ordinary people and from the burning issues of the day.

Yet the very nature of human adaptation, with its highly cooperative, collaborative organisation of productive activity, means that daily life is permeated with politics. Far more activities are political than most realise; rather like the 'dull ache' of making a living by selling labour, many people neither notice nor resent them.

Poor knowledge of ecological principles means that many people are unaware of the ecological problem, which is in any case slippery and difficult to pin down. Scientific uncertainty, problem displacement, and maintenance of low entropy all create problems. It becomes necessary to adopt behavioural strategies which make allowance for uncertainty and change.

Ecological rationality is fundamental to sustainability. It requires a concern for the maintenance of the productive, protective, and waste-assimilative capabilities of ecological systems. These are the aspects that impinge most directly on human societies, and which are susceptible to disruption by human activity. Ecological rationality, because it is fundamental to survival, has to take precedence over other rationalities.

Trying to make choices about environmental problems also involves arguments about what the issues and problems are, and how best to respond to them. It is important to examine arguments carefully, and in particular to attempt to detect flaws in logic, unexamined assumptions, and self-seeking or self-justificatory arguments. The use of symbols as substitutes for argument is also a problem.

FURTHER EXPLORATION

Ecological rationality is (very similarly) defined and its implications explored in papers by Bartlett and Dryzek; the latter's book *Rational Ecology* includes an excellent discussion of the reasons why ecological rationality must be afforded near-lexical priority over other forms. Goodin's *Political Theory and Public Policy* is indispensable for those who wish to follow up ethics, morality, and the place of prescriptive political theory in public policy.

There are remarkably few intelligible books about argument and how it can be used; while the philosophers have a large literature, most of it is inaccessible to the lay reader. Nosich's text *Reasons and Arguments* is a healthy antidote; it is couched in elementary language and is easily accessible to even a beginner. Its analysis is more complex than that offered in this chapter; but, especially in Chs 5 and 6, it offers a similar approach. Altogether more sophisticated, Barry's *Political Argument* is an attempt both to classify the kinds of arguments used in politics and to suggest some conclusions about the way they are used.

Part 2
The Murray–Darling System

Many, perhaps most, Australians are substantially ignorant of the unique characteristics of the country's ecological systems. They are often unaware how substantially they have been modified, especially in the last 200 years. Many such modifications have taken place without adequate knowledge, and sometimes despite expert advice. The adoption of irrigation in the Murray–Darling river system preceded the necessary scientific research and ignored warnings about the likely problems. As a result, salination, land degradation, species loss, and ecosystem disruption have plagued irrigation schemes since their inception.

Many of these problems have persisted for 100 years and more; some of them are the direct consequence of the political organisation, rather than the physical and biological constraints, of the catchment area.

CHAPTER 4

SALINITY IN THE MURRAY–DARLING RIVER SYSTEM

The Murray–Darling is Australia's largest river system; the total length of its rivers is 3,780 kilometres. The catchment covers 1,000,000 square kilometres, one-seventh of the total area of the Australian subcontinent. While it drains only five per cent of Australia's runoff, it supplies 'almost three quarters of all water used for domestic, industrial and agricultural purposes in the nation'.[1] The basin has one quarter of Australia's cattle and dairy farms, roughly half the sheep, lambs, and cropland, and three-quarters of the irrigated land. Its annual production is worth about $10,000,000,000. Two million people live within the basin or depend on it for their water supply.[2]

Significant environmental damage and consequent losses of production are in part a consequence of the basin's distinctive physical, biological and climatic conditions.

Hydrology and Geography

In terms of area the Murray–Darling basin is the fourth largest river system in the world. There are over 20 major rivers and hundreds of smaller creeks and streams. Its average annual flow is disproportionately low by world standards, at 12,000,000,000

Chapter 4

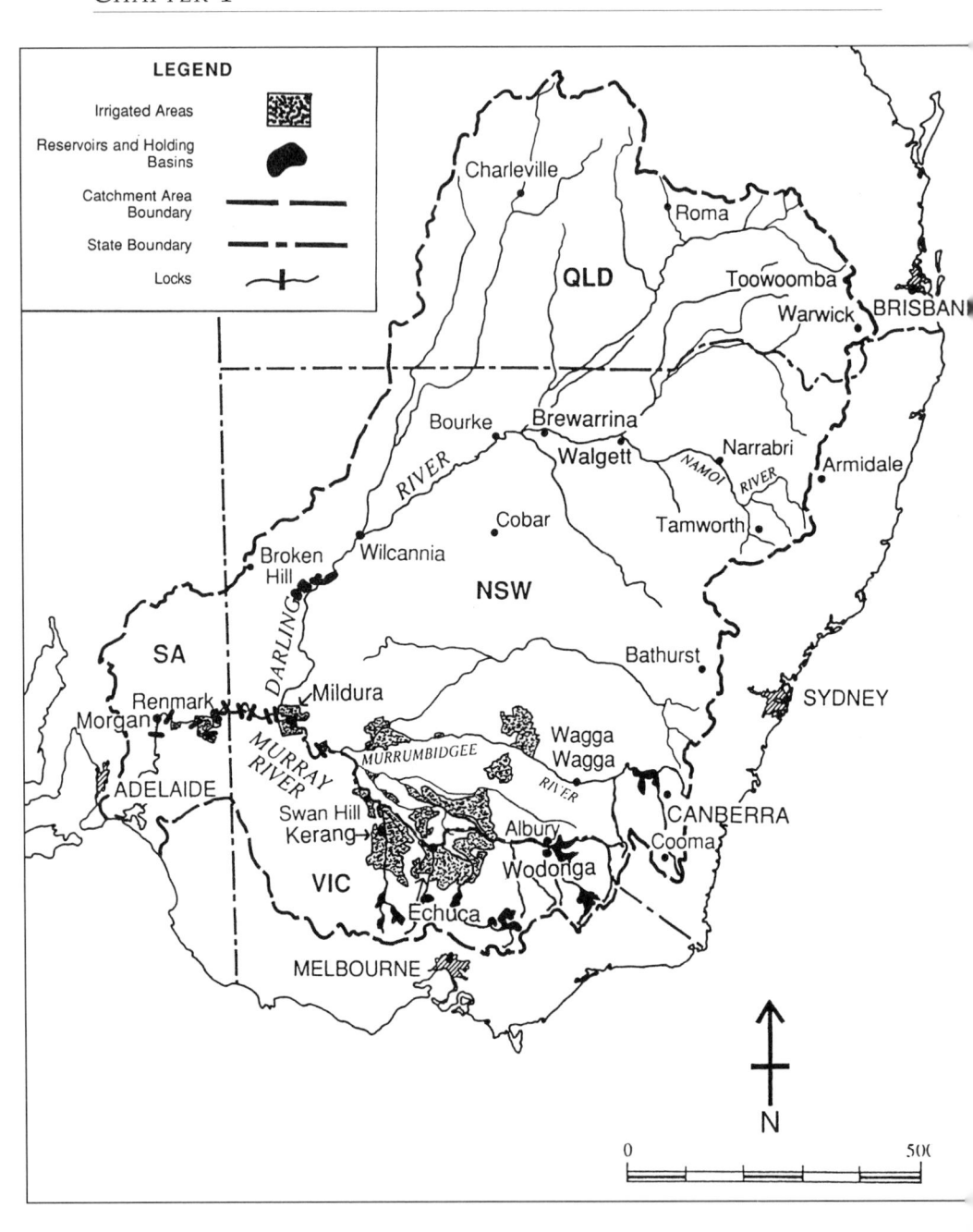

Fig 4.1
The Murray–Darling Basin.

Salinity in the Murray–Darling

cubic metres, reflecting the aridity of the catchment. Table 4.1 shows the world's nine biggest river systems in terms of flow; fourth in terms of area, the Murray rates ninth in flow.

Table 4.1
Nine world river systems

Rank	River	Country or region	Catchment area (million km^2)	Average annual flow (million m^3)	Flow per unit area (m^3/km^2)
1	Yang-tze	China	1,950	895,000	459
2	Danube	Europe	833	282,000	339
3	Ganges	India	1,530	180,000	116
4	Columbia	USA	617	176,000	285
5	Indus	Pakistan	937	109,000	116
6	Nile	Africa	2,860	89,000	33
7	Sacramento-San Joaquin	USA	117	40,000	342
8	Colorado	USA	638	20,000	31
9	Murray–Darling	Australia	1,060	12,000	11

(Source: Frith & Sawer, *The Murray Waters*, p. 33.)

The low flow, in relation to catchment area, of rivers traversing tropical and subtropical regions is very noticeable. This underlines the importance of the Murray–Darling basin's location, between latitudes 24°S and 38°S: most of the world's hot deserts occur between latitudes 15° and 35°. The basin is very flat, most of it being below 200 metres above sea level. The drainage pattern of 98% of the region is typical of arid and semi-arid environments, restricting surface runoff. On its eastern and southern borders, the Great Dividing Range is an important source of water, but is too low to develop permanent snowfields for a year-round supply. However, the two percent of the Basin's area which lies in the highlands contributes 37% of the flow.

Rainfall in the basin is extremely variable, from 1400 millimetres per annum in the highlands to 300 millimetres in the northwest plains. Significant annual and long term fluctuations also occur. Drought year flows, often only 3,700 million cubic metres, are 13 times less than flood peaks of 48,000. High summer temperatures mean that in most parts of the basin evaporation rates regularly exceed rainfall.

Geophysical Development
The Murray basin, in the south, is divided from the Darling basin

Chapter 4

by a low range of hills. A good deal is known about the geological history of the Murray basin, but the Darling is less well studied.[3] As the bulk of the flow, and the major salination problems, occur in the Murray basin, the omission of the Darling from what follows is not of great significance.

Sedimentary and volcanic rocks formed in the early Palaeozoic era, about 570–450 million years before the present (BP), underlie the greater part of the Murray basin. The sedimentary rock was laid down under an ancient sea which covered the area at the time. About 250 million years BP the region underwent major geological upheavals. It rose above sea level, creating a network of rivers and creeks which later became an important component of the underground drainage system.

The basin's two major watertables were formed between 60 and 40 million years BP when the central area began to subside below sea level. The northern area forms part of the Great Artesian Basin and is now overlain by the Darling River and its tributaries. To the south, the Murray Groundwater Basin lies under a large part of the modern Murray river catchment, but is no longer coextensive with it. The south-western corner of the Murray Groundwater Basin was submerged again during the Miocene period (25–12 million years BP). Two million years BP the sea once again retreated, leaving behind deposits of silt, sand and limestones.

The Murray basin has two geologically distinct zones. The *riverine plains* lie mainly in the eastern zone, and largely consist of alluvial soils derived from sediments from the surrounding highlands. Major sedimentation finished some 40,000 years ago at the end of the 'prior stream era'. In contrast the *mallee* zone, further downstream, is underlain by the sediments deposited during the Miocene inundation. These sediments are highly permeable and saline, making most mallee zone groundwaters highly saline, and accounting for most of the salt which enters the lower Murray. For this reason, the salination problem in the Murray Valley below Albury–Wodonga is markedly more serious.

In the last 30,000 years glaciations formed a thick fertile deposit of alluvial soil over large areas of the Darling Downs and the riverine plain. During this period many of the early stream channels were buried in sand and gravel, producing an interlinking network of underground groundwater channels which are very important to the hydrology of the region.

The Ecology of the Murray–Darling Basin

Because of its vast size and long geological history the basin is ecologically diverse; its 75 biophysical regions range from alpine to arid. Before European settlement vegetation distribution was largely determined by rainfall and soil type. Tall, dense *Eucalyptus* forests grow in the mountains along the south-eastern and eastern parts of the basin; *Eucalyptus* and *Acacia* woodlands develop further inland. To the north brigalow and gidgee woodlands predominate, mixed with saltbush and Mitchell grasslands. Central New South Wales and Victoria are covered by a variety of bow and *Callitris* woodlands. *Stipa* (desert speargrass) grasslands occupy an area from central New South Wales to western Victoria. Along the major water courses *Eucalyptus* forests, including the famous red gum and river box, thrive.

Native vegetation and wildlife have adapted to an environment punctuated by floods and drought. Before the regulation of the river for irrigation and navigation purposes, extensive flooding occurred nearly every two years and widespread inundation could last up to six months. This created many billabongs and wetlands throughout the basin. The River Murray alone has over 222,717 hectares of wetlands, which act as sanctuaries and breeding grounds for much of its animal life and provide a year-round source of water. Flood events are important in the life cycles of many organisms, providing conditions both for breeding and for seed dispersal. Wildfowl breed on wetlands in flood years, and riverine vegetation is especially adapted for flooding.

In turn, the vegetation has many important effects on hydrology. The saline groundwater formed earlier in the geological history of the basin slowly leaches out into the river system, regulated by an unusual floral adaptation. Deeply rooted native trees — such as mallee scrub — draw excess water out of the watertables, leaving the salt behind. They also have high rates of evapotranspiration, meaning that fresh water is returned to the atmosphere from the aquifers. The combination of deeply rooted plants and high evapotranspiration rates has kept groundwater levels well below the surface, and saline discharges to moderate rates in most years. Prior to agricultural development saline groundwater rarely reached levels high enough to cause salt scalds. The salinity of the river system could become quite high in periods of drought, but the riverine vegetation and aquatic life is adapted to survive this form of stress.

European settlement, however, has had a severe impact on the ecology of the basin. Fifty-four per cent of all vegetation has been cleared. Eighty per cent of the remainder lies in the arid and semi-arid areas of Queensland and New South Wales, but heavy grazing pressure from stock and feral animals is a major obstacle to its

Chapter 4

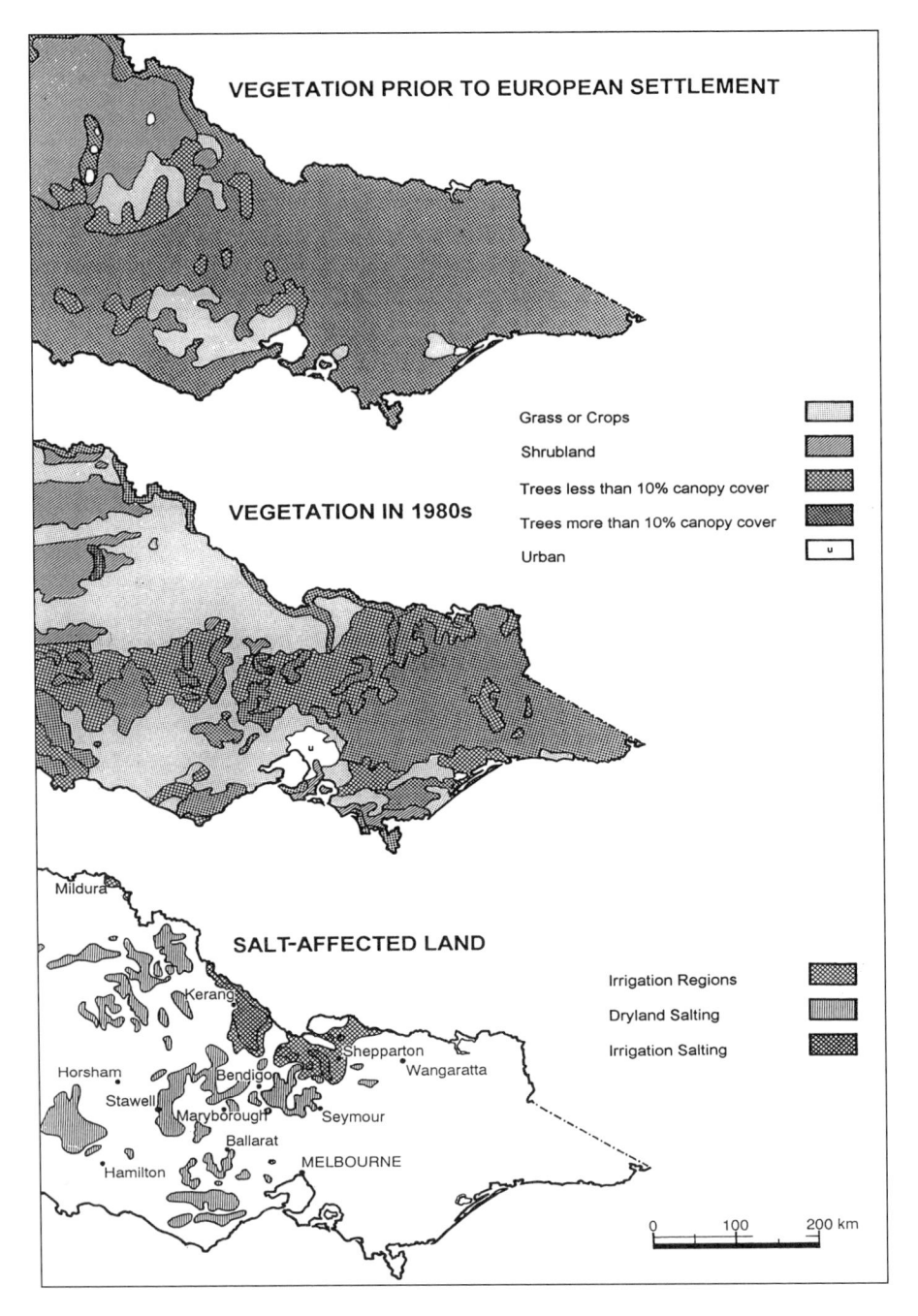

Fig. 4.2
Clearing and salination in Victoria. Note how the Barr Creek–Kerang area is the most affected by irrigation salting, and has also suffered the most extensive clearing.
(Source: *Salt Action*, p. 7.)

regeneration. Victoria has cleared the most native vegetation, with only 16% remaining. Salinity in the basin is closely linked with clearing. As Figure 4.3 shows, this is strikingly evident in the Barr Creek–Kerang area of Victoria, one of the most heavily saline regions in the basin.

The wheat-sheep belt is the most extensively cleared area in the basin, with less than nine percent of its original vegetation. Clearing still continues. Of 2000 species of vascular plants, 10 are now considered extinct and 37 are endangered. At present only 2% of the basin lies within wildlife reserves; of the basin's 75 biophysical regions, 48 (64%) have less than 5% of their area included in such reserves.

The fauna of the basin have also been adversely affected. At European settlement, the region had approximately 85 species of mammals, 367 species of birds, 151 species of reptiles, 24 species of amphibians and 20 species of fish. Since then, 20 species of mammals have become extinct and 16 others are now endangered. Thirty-five species of birds are also considered to be endangered. Between 1860 and 1910 the basin held the world record for the rate of extinctions among native fauna. The animals most affected were small, ranging in size from 0.5 to 5.0 kilograms. Most extinctions have occurred in shrub and grassland habitats. The major causes for these extinctions were extensive clearing, increased predation from exotic carnivores such as cats and foxes, and increased competition from rabbits, cattle and sheep. Exotic organisms released into the basin include 12 species of mammals, 12 species of birds and 9 of fish.

Problems

The Murray–Darling basin's problems, though generally more serious, are typical. For example, about 52% of Australia's soil is estimated to be degraded.[4] Intensive land use aggravates the problem. Dryland salination, common in the Murray–Darling, also occurs elsewhere; many of the lessons now being learnt about the retention and encouragement of native vegetation are applicable in the basin.

Agricultural development has entailed large-scale modification of the region's natural ecosystems. Woodlands were extensively logged and cleared, and grasslands grazed. The river system has been greatly altered to provide water for irrigation and to assist navigation. Dams, weirs and irrigation channels have been built to store flood water and regulate and distribute river water for irrigation purposes. But these modifications have created many serious problems

1. *Reduction of flow*. Present flow is about one third of that in the 'natural' state. The deficit is caused by diversion of water for irrigation, town water supply, etc. This affected South Australia most seriously, because of its concern with navigation and with water supply for Adelaide. Native flora and fauna, adapted to the natural flow, have also suffered.

2. *Salination* first appeared near Cohuna in 1900; according to official figures (which are probably underestimates) 80 hectares were affected by 1911, 400 in 1913, 300,000 by early 1930s. The problem is now widespread throughout the region: 620,000 hectares (0.6%) of non-irrigated soils and 122,000 hectares (12%) of irrigated soils are now affected. Even if effective remedial action is taken immediately salination will worsen for many years to come.

3. *Land degradation* stems firstly from soil exhaustion as a result of nutrient decline due to heavy cropping, overstocking, clearing, and wind and water erosion; and secondly from salination.

4. *Pollution* results mainly from nitrogenous fertilisers, urban sewage and pesticides, which are washed into the river or infiltrate groundwater.

5. This in turn affects *water quality*. Both salinity and nitrogen runoffs affect the suitability of the water for town water supplies, as well as for irrigation. This affects Adelaide, which depends very heavily on the Murray for water. Riverside communities, such as Mildura, Renmark, and Berri, have an additional burden because extensive sewage treatment is essential (and now compulsory) to avoid further degradation of water quality. Turbidity of water from as far away as the Condamine and Balonne rivers is also damaging, especially to aquatic life. Serious concerns arose during 1991–92 over extensive algal blooms in the Darling.

6. Impacts on *native species* have also been deleterious. River red gums have been affected badly by flow regulation. They are adapted to occasional floods mixed with dry periods; but nowadays those near dams often stand permanently in water and are killed, while others never experience flooding. Waterbirds also rely on flooding, especially for breeding opportunities. Native fish such as the Murray trout and cod, crayfish, etc., are adapted to specific salinity levels (lower than the current ones) and are also in decline. As these are prime commercial species, direct economic losses follow. The whole riverine ecology is adversely affected by blue-green algae.

Table 4.2
Important resource degradation issues in the Murray–Darling Basin

Land	Water	Other environmental	Cultural
Wind & water erosion	Over-commitment of water resource	Clearance & decline of native vegetation	Maintenance of Aboriginal heritage sites
Dryland salinity	Deteriorating quality	Degradation of wetlands	
Irrigation salinity & waterlogging	• salinity • turbidity • nutrients • microbes • pesticides	Degradation of tourist & recreation sites	Maintenance of historic heritage sites
Soil acidification			
Soil structure decline	Deterioration of groundwater resources	Habitat destruction/ modification	
Pests, plants & animals		Species decline & extinction	
		Destruction of natural heritage sites	

(Source: Murray–Darling Basin Commission, *Draft MDB Natural Resources Management Strategy*, p. 4.)

The economic effects alone are serious. The Murray–Darling Basin Ministerial Council estimates the cost of lost production due to salinity and soil acidity at $220,000,000 per annum. The 45,000 ha of agricultural land lost to dryland salinity in Victoria could yield wheat worth over $8,000,000 per annum were it still in production. Estimates of costs to eliminate the problem run as high as $785,000,000. Some other costs, such as loss of potential tourism, are less 'visible' but still serious; they have not been quantified.

There are also social effects. Farm bankruptcies and loss of production affect local communities, which experience declines in population and economic activity. Decreased production leads to unemployment, which is directly linked to petty crime such as

CHAPTER 4

theft; suicide rates in country areas tend to rise markedly in times of economic hardship. These and other anti-social effects can be costly to the community as a whole.

All of these things are consequences — direct or indirect — of irrigation; and salination is the central problem, as well as the most intractable.

SALINATION IN THE MURRAY–DARLING SYSTEM

Salination is associated with all large dam schemes, and is most acute in arid zones. One of the world's earliest irrigation schemes, in Mesopotamia, diverted water from the Tigris and Euphrates rivers. Beginning on a very small scale about 5500 BC, irrigation spread until by 3000 BC the Sumerian empire thrived on a massive agricultural surplus. This 'land of milk and honey' was famous for the 'hanging gardens' of Babylon. In less than 1000 years, however, soil productivity fell drastically, rendering the empire vulnerable to conquest. By 1700 BC, wheat, previously the staple crop, was not grown at all in Mesopotamia. Subsequent irrigation schemes encountered identical difficulties, including soil salination, high groundwater tables, and surface scalds. Populations fell catastrophically. Salination had transformed the Tigris and Euphrates valleys into the predominantly desert landscapes of what is now Iraq. Biologically, Mespotamia is no longer self-supporting. Only imported food and manufactures, paid for by oil exports, preserve Iraqis from catastrophically low living standards.

Many salination problems appear far more rapidly. In the 'five rivers' region of the Punjab, now split between north-western India and Pakistan, the British constructed massive irrigation works a century ago and settled millions of farmers on land made newly productive. Today the whole area suffers problems identical to those found in the Murray basin.[5] Similar problems have arisen in the USA, especially in the more arid western states.

In Australia, salination affects some 32,400,000 hectares, 4.3% of the continent. Nearly 87% of this land is naturally salty: coastal marshes, inland salt pans and flats, and large inland areas. The rest is the product of European settlement and accounts for 4,200,000 hectares, 0.53%. Most of this area, 3,200,000 hectares, is found in the arid and semi-arid regions of Australia, the predominant form of damage being scalding. The salting of the remaining 1,000,000 hectares is largely the result of saline seepage and occurs in the southern parts of the continent. Much of this land is in the Murray–Darling basin and is highly fertile and potentially very productive.

From a political ecology perspective, salination is the key problem besetting the Murray–Darling basin. It is closely related to all the other problems present, affecting the inhabitants of three of the four States contiguous to the basin. Knowledge of the causes of salination, and of measures for treatment and prevention, is essential for determining how it arose in the first place, as well as why governments and farmers have failed to control it. An assessment of present control measures requires an appreciation of its causes and the range of its effects.

Salination Processes

The Murray–Darling basin's aridity reflects its low rates of flow. High temperatures and consequent high evaporation rates mean that even in their natural condition the rivers of the basin are mildly saline. Salt stored in the soil and in groundwater is mobilised by the hydrological cycle. It appears in the river as a result of:

1. inflows of salt in groundwater;
2. inflows of salt in drainage returns;
3. the concentration of salts by evaporation from the Murray and the reservoirs along it;
4. the flushing of billabongs and swamps where salts have been concentrated during periods of low flow;
5. inflow of salt from tributaries;
6. addition of wind-borne salts in rainfall; and
7. addition of salt by weathering of mineral soil particles in river sediments.

As part of the natural processes of evaporation, precipitation, drainage, and groundwater flow, salt enters the river from the atmosphere, the soil, groundwater, and billabongs and swamps.

European settlement and subsequent irrigation disturbed the equilibrium of these natural salination processes, which had stopped salt from reaching the soil surface and kept inflows of salt to the rivers to a minimum, by keeping the watertable of the underlying saltwater aquifer some metres below the surface. Three processes greatly increased the rate of salination: land clearing, erection of river structures, and irrigation itself.

Clearing of native vegetation for agriculture and irrigation

CHAPTER 4

removed deep-rooted trees such as mallee scrub. Transpiration was reduced, while groundwater recharge continued at the normal rate (and even increased in some cases), forcing watertables to rise. Numerous undesirable effects followed: river salinity increased due to discharge of saline groundwater; where watertables remained high, pastures became waterlogged and groundwater was discharged via the surface; and soil salination occurred, with increased salinity via high rates of evaporation and erosion of salt-affected soil into the river.

River structures such as dams, weirs and locks are often located above saline aquifers. This creates **overpressures,** forcing watertables higher. This groundwater may flow directly into rivers and streams, or contaminate surrounding land by leaching.

Finally, **irrigation** itself can result in overpressures, often creating a 'groundwater mound' and forcing more water into the river system. Most irrigation water is not utilised by crops, but is wasted by evaporation or percolation through the soil and entry into an aquifer. The *rate of accession* to the groundwater system is determined by the lining of the channels used for distribution and drainage, among other factors.

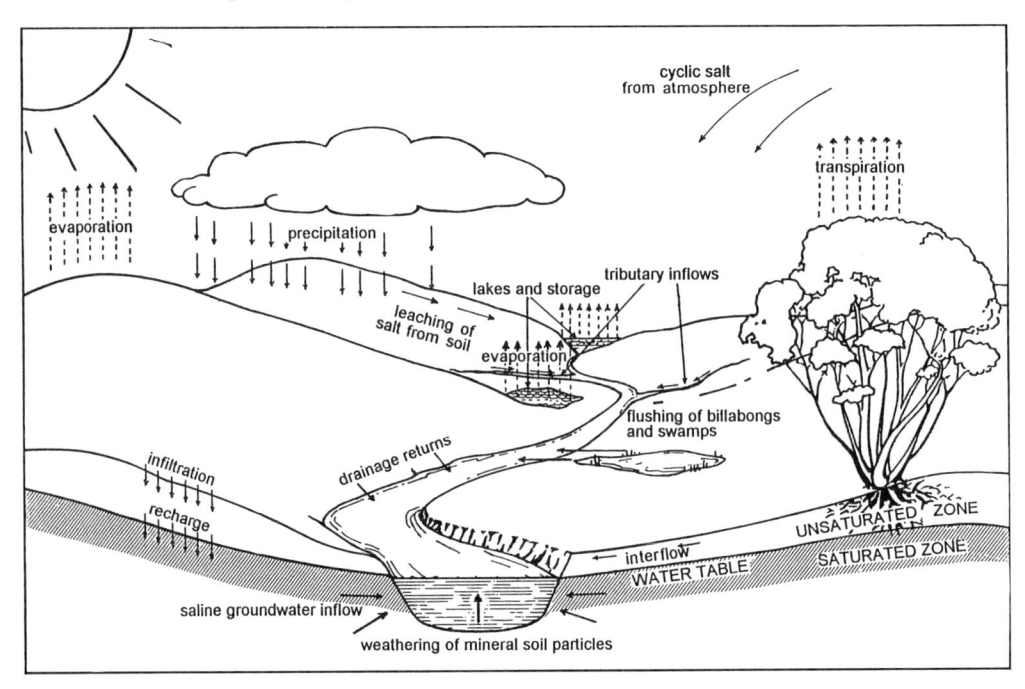

Fig. 4.3
The hydrological cycle and natural sources of salinity.
(Sources: ASTEC, *Environmental Research in Australia: Case Studies,* fig. 4, p.9; Jakeman fig. 2.4, p.47.)

Salinity in the Murray–Darling

Leaching frequently follows clearing, especially in soils where the new water table is close to the surface. Salt springs appear in gullies below that level, often draining directly into the river system. When the watertable reaches a critical distance from the surface (typically two metres or less), capillary action causes the salty groundwater to rise to the surface, where evaporation slowly increases its concentration. The soil is then contaminated with salt residues. Plants are poisoned by salt and die, leaving the land bare. Erosion quickly follows, producing scalds. Runoff washes surface salt into river systems, increasing water salinity and reducing water quality .

Irrigation increases accessions of groundwater due to clearing, infiltration, and overpressures, creating 'groundwater mounds' with accompanying saline seeps, leaching of salt to the surface, and saline scalds.

Technical and 'Natural' Solutions

Since salination occurs when saline water rises to the surface, the available techniques for managing salination rely on control or prevention. They involve two basic approaches:

1. subsurface drainage of saline water; and
2. minimising the accession of water into aquifers.

Subsurface drainage structures intercept saline groundwater, so as to dispose of salt before it can reach the root zone and do perma-

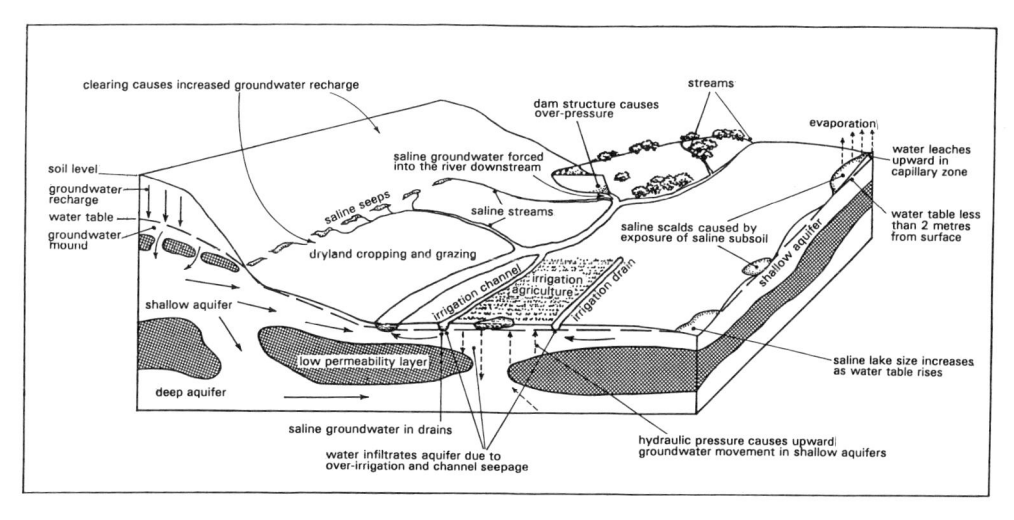

Fig. 4.4
Sources of salinity after clearing and cultivation.
(Source: modified from a diagram in *Salt Action*, p. 6.)

nent harm. One method is to pump groundwater into evaporation pans on the surface, where it is evaporated to leave a salt deposit. This lowers the level of aquifers by removing water from them. Pumping mimics the action of native trees, but fails to leave the salt below ground. The consequent risk of seepage of highly saline water from evaporation pans to rejoin the aquifers increases costs, and in the worst case can completely negate the effects of pumping.

A second method is to install tiled drains one or two metres below the soil surface, leading saline water off into streams or evaporation pans, instead of contaminating soil. Subsurface drainage is expensive and therefore economical only in areas which produce high value crops. Its usefulness is also limited because it produces a saline effluent. As with evaporation pans, careful disposal is required to prevent salt contamination of land or water.

There are several ways to minimise groundwater accessions:

by improving irrigation systems;

by improving irrigation scheduling;

by improving surface drainage;

by sealing drainage channels to prevent leakage; and

by improving land use and crop management.

Accessions can be minimised if less water is applied to produce a given crop. Improvements such as 'drip' irrigation, or more precise control of irrigation by rationing or scheduling water use, increase water efficiency. Reducing the amount of water lost from the irrigation system to aquifers can be achieved both by sealing drainage channels and by ensuring that water is used when it will be most effective, through scheduling.

'Natural' techniques include such measures as 'retiring' land, maintaining a cover of natural vegetation, such as mallee scrub, or other measures that mimic the natural processes which restrain salination.

Most drastic is 'retiring' land, which means withdrawing it permanently from production and allowing the natural vegetation to re-establish itself. Land retirement is unpopular with farmers, who dislike seeing 'unproductive' land, and therefore is often unacceptable politically. But there has been considerable involuntary retirement of land due to the farm bankruptcies of the 1980s, resulting in some revegetation. Because it is quite unplanned it has often not been in the appropriate places.

Maintaining natural vegetation, and even deliberate revegetation of selected areas, is becoming popular. Although the effects depend critically on the local hydrology and soil conditions, quite small amounts of native vegetation can in some cases be highly effective in controlling the watertable, by improving evapotranspiration rates. Additionally, native vegetation is now known to have a positive impact on soil fertility.

Among other measures to limit salination, the use of crops which behave similarly to the native vegetation is often suggested, and is now being explored in Victoria. Dates, for example, are deep-rooted, and transpire water while leaving salt in the aquifer in much the same way as mallee. Since they are also drought-tolerant and set fruit best in a dry atmosphere, they have been seen as a suitable alternative crop for times when water is short and conditions harsh. The principle of using crops that have similar ecological effects to native vegetation, or which complement existing cultivation patterns, has considerable potential. 'Saline agriculture' of this kind is now seen as the only option for some areas where salinity is especially intractable.

OVERVIEW: AN UNRESOLVED PROBLEM

Salination is continuing to worsen and will do so for many years even if all harmful practices were to stop overnight. It has already persisted for the better part of a century. There are both technical and political reasons why the problem is not under control.

The efficacy of specific techniques depends greatly on local conditions. Although salination is extremely widespread, the precise processes by which it occurs vary considerably. Their impact depends on local hydrology, soil permeability, extent of clearing, and numerous other variables. Consequently, estimates of the extent of the problem are imprecise, and the data from which they are derived are generally incomplete. In each area the problems are different and require different solutions, sensitive to the local conditions. Differing *technical* requirements mean that solutions effective in one area will not necessarily work in another. For example, planting deep-rooted native vegetation is more effective in preventing the watertable from rising where the *lateral* permeability of the soil is high; low lateral permeability can limit the beneficial effects to a very small area.

Such variations entail localised, highly problem-specific research. They require targeted local management techniques, closely adapted to local conditions.

Because salination can be greatly reduced if land owners use irrigation water carefully and economically, and maintain adequate tree cover on their properties, improved land management is important. Because poor land management may cause or exacerbate salination problems on neighbouring properties without similar effects occurring on the land of those causing the problem, land users in a region must cooperate if soil salination is to be avoided. The political and administrative problem is to engender and maintain such cooperation.

Lack of cooperation, both among landholders and among governments, explains the long persistence of salination and other ecological problems, despite their harmful effects. The reasons for poor cooperation lie in the political history of the Murray–Darling basin.

Further Exploration

There is a plethora of reports and investigations of Murray basin salination or various aspects of it. Williams' paper 'Salinity and Waterlogging in the Murray–Darling Basin', in the ASTEC collection, *Environmental Research in Australia*, is one of the most readable. The paper by Jakeman, Thomas, and Dietrich, (*Water Resource Management in the River Murray: Models of Salinity Travel Time and Accession and their Application*) has a very clear outline of the processes of salination (pp. 13–37), drawn largely from the 1987 Maunsell Report. The Victorian Government's draft strategy of 1987, *Salt Action*, also includes some excellent explanations of the problem, though primarily focused on Victoria.

For the broader context, a short but comprehensive outline of the various problems facing the Murray Valley is McConnell's chapter 'Problems Surface: the Murray Valley', in the Birrell, Hill, and Stanley book *Quarry Australia?* A very complete overview is to be found in the Murray–Darling Basin Ministerial Council's *Murray–Darling Basin Environmental Resources Study*, published in 1987. It has copious tables, diagrams, and statistical data, and is an essential reference source. There is also a very useful comprehensive bibliography.

Innovative approaches to land management are canvassed in a number of works. Barr and Cary, in *Greening a Brown Land*, offer case studies of both disaster and success with irrigation and salination occupying considerable attention. More homely are Breckwoldt's two booklets, *Wildlife in the Home Paddock*, and *The Last Stand*.

CHAPTER 5

IRRIGATION: A HISTORY OF CONFLICT

The use of water in the Murray–Darling Basin has been controversial since the 1840s. Divided jurisdiction and constant conflict involving four States has deeply influenced the major institutions and laws of the region. Failure to appreciate the ecological limitations on 'development' in the area, especially the problems and disadvantages of irrigation under arid conditions, has been a central problem.

COLONIAL CONFLICTS

Until Federation in 1901, Australia was a collection of semi-independent Colonies. When Victoria was separated from New South Wales and granted self-government in 1850, difficulties quickly arose in defining the exact border, especially where New South Wales met Victoria and South Australia. Encountering difficulty collecting customs revenue, New South Wales successfully lobbied the imperial government in Britain to pass legislation giving it complete jurisdiction over the Murray from its source to the eastern boundary of South Australia. This departed from the common law principle of drawing boundaries in the *middle* of the stream, and implicitly extinguished any rights Victoria might have had to draw water from it. But the British Parliament's Act of 1855, implementing this decision, did not remove all the ambiguity. Victoria's right to water remained a serious issue for many years to come, although the first dispute to arise was territorial.

> ## BOX 5.1 THE AUSTRALIAN COLONIES
>
> The Australian colonies were at first ruled direct from London and treated as part of Britain. Because of the vast distances and poor communication — in the early nineteenth century, voyages to and from Britain often took six months or more — a Governor was appointed. His functions included maintaining order, and in the early days of the Sydney colony, even securing supplies of food and other necessities.
>
> The Governors initially enjoyed absolute power, which was used wisely by some, though not all. But the need for a codified system of law led to the introduction of a **Legislative Council** — in 1824 in NSW — which had the function of *enacting laws*. That is, the Council would be presented with a proposed law, and after debate, would agree to accept or reject it.
>
> These early Legislative Councils were appointed by the Governor, and were often hand-picked to support his policies. Before long, however, election was introduced, initially with a very limited **franchise**: that is, only certain designated classes of people — mostly wealthy property-holders — were allowed to vote.
>
> Self-rule and partial independence from Britain began with Victoria and South Australia in 1855, followed by New South Wales in 1856, and Queensland in 1862. Western Australia was the last, in 1890. Britain retained control over foreign affairs and the Privy Council was the final court of appeal in legal matters, but each State levied customs duties (or **tariffs**) on the produce of the others. Each now raised its own taxes, had its own Parliament, and administered its own laws. With self-government, a **Legislative Assembly** was added to each Colony's political system, initially under a limited franchise. Proposed laws now had to be considered by both Houses of the Parliament, **Upper** (Legislative Council) and **Lower** (Legislative Assembly), which collectively made up the **Parliament** of each Colony.
>
> Most States granted full adult franchise and abolished plural voting — the granting of extra votes to some favoured categories of voter — at some time between the granting of self-rule and the end of the nineteenth century. But their Legislative Councils, by

> and large, remained bastions of conservatism, with electoral systems which over-represented country areas, and through their restriction of the franchise to property-owners, over-represented the wealthy and powerful in society. Some states — notably Victoria, Tasmania, South Australia, and Western Australia — still over-represent country areas in their Legislative Councils. Historically, this has often delivered a veto power to conservatives, since control of the Upper House has enabled them to block legislation perceived by them as inimical to their interests.
>
> This dominance of landed interests — among its other effects — made 'closer settlement' of squatter lands difficult to achieve, and thus made irrigation attractive as an alternative.

Squatters had leased land on Pental Island since 1845. After separation they continued to pay their annual rents to the Victorian Government because the island had been part of the old Division of Port Phillip. In 1859 New South Wales claimed the island — and the rents — because it lay within the River Murray. Following claim and counter-claim, both colonies finally requested British adjudication as to the intent of the 1855 Act; Victoria was awarded the island.

This conflict dramatised the lack of a mechanism for settling disputes between the colonies. Without some sort of mechanism for resolving conflict, intercolonial cooperation and coordination within the Murray–Darling basin was so difficult as to be nearly impossible. The limited common law basis for riparian rights made the task of the law courts difficult. And worse still, by granting one colony ownership of the river and its water, the British Parliament, though reluctant to involve itself deeply in colonial affairs, made riparian disputes inevitable.

NAVIGATION

Beginning with a home-made steamboat launched in 1853, South Australian navigators had by 1860 established a thriving river trade throughout the Murray–Murrumbidgee–Darling river system, reaching Albury on the Murray, Wagga Wagga on the Murrumbidgee, and Bourke, Walgett, and Brewarrina, 1600 miles (2500 km) from the sea, on the Darling.

River trade was profitable, and vastly quicker than overland travel with bullock-carts or pack animals. But it was risky, due to

Box 5.2 Riparian Rights and Common Law

Common Law, in Anglo-Saxon legal systems, is derived from long-established practice. It relies on **precedent**: that is, the pattern established by earlier legal decisions and by established custom. Where no specific **legislation** — that is, a law or laws passed by a competent **legislature** — exists, common law is the basis of all legal decisions.

Individual landholders enjoy **riparian rights** under common law, to draw water from a stream on or contiguous to the property, and to use that water for domestic purposes or for watering stock (i.e., cattle, sheep or other livestock). Other, downstream landholders enjoy the same rights, so implicit in the notion of riparian rights is the understanding that upstream landholders will not divert the stream or adversely affect the quality of the water. Under common law, downstream landholders have the right to sue an upstream landholder for **damages**, and to seek **injunctions** requiring them to desist from diverting or fouling the water.

Between independent states (or nations) there is no common law. As international law is rudimentary and almost always unenforceable, riparian issues are normally subject to explicit agreement in the form of **treaties**, which are agreements concluded between nations. Frequently, treaties can be enforced only by resort to war, but are often respected voluntarily by the nations concerned.

The Australian Colonies, before Federation in 1901, were in the position of sovereign nations: there was no law and no rules existed as to riparian rights. This was the main reason for the conflict among them: essentially, any agreement had to be negotiated. Failure of any one party to participate could (and did) scuttle any attempt at negotiation.

Within each colony, there was initially no legislation in relation to riparian rights. Only the common law applied. The Victorian Irrigation Act of 1886 sought to extinguish riparian rights at common law, creating instead property in a **'water right'** attached to the land itself. However, this water right was a right to a supply of water, and the legislation did not make effective provision to deal with disputes between landholders or Water Trusts created under this or earlier legislation. It was also unclear whether the 1886 Act applied to holdings created before that date. This situation was not resolved in Victoria until the 1905 Act, which eliminated all vestiges of riparian rights at common law, and created the State Rivers and Water Supply Commission.

For further discussion, see Sandford Clark's 'The Murray River Question,' and J.M. Powell's *Watering the Garden State*.

the variable flow of the rivers, and unforeseen delays made transit times uncertain and at worst very lengthy. The longest recorded was 37 months from Morgan to Bourke, beginning in 1883; a cargo of building materials arrived long after the pub it was intended for had been completed, using materials brought in on the new railway. In a countryside where the fall of the rivers was a few inches in as many tens of miles, heavy rainfall could create instant floods, which dispersed as rapidly. Steamers sailing downriver with the flood often cut across flooded paddocks to save tortuous navigation round many bends; but some were left high and dry, trapped behind the trees that line the rivers. Ian Mudie celebrates this practice in his line about taking a steamer 'seventy miles out of the Darling on a heavy dew'.

New South Wales and Victoria saw the river navigation as diverting 'their' trade and profits to South Australia. New South Wales was legally responsible for upkeep of the river within its borders, yet gained little from river trade: in fact, it was an avenue for smuggling. In practice South Australia carried out most river maintenance; this involved 'snagging' — the removal of dead trees and other debris clogging the channel — the provision of locks (or 'locking') and preventing diversions of water that might reduce the flow. In 1863 and 1865, negotiations with New South Wales to share these costs failed.

To intercept the river trade, Victoria and New South Wales both developed railways to ports such as Echuca (1864), Wodonga (1873), Albury (1881) and even Bourke (1885) and Walgett. Rail transport, though more costly, eliminated the long delays and uncertainties of river navigation. Quoting low rates to and from river ports — so low that goods must have been carried at a loss — Victoria and New South Wales quickly captured most of the trade.

But navigation was also challenged by irrigation, which competed directly for the Murray's waters.

IRRIGATION BEGINS

During the 1850s, some landowners in western Victoria had begun to irrigate. Calls for government involvement were soon made. In 1857 it had been suggested that the Murray and tributaries be dammed and irrigation canals built to supply water throughout the basin. The 1864 drought inspired widespread calls for such developments. By the mid 1860s drought was thought to be the biggest single threat to continued agricultural development, especially in Victoria. Irrigation was seen as a solution. Responding to

CHAPTER 5

growing farmer pressure, Victoria passed legislation in 1865 permitting the establishment of local bodies to supply water and irrigation in their districts. These bodies were to pay their way by levying rates on the properties they supplied.

Increasing public interest and attention saw New South Wales prepare to follow Victoria's lead. The pressure was reinforced by a major drought in 1877–81, which caused considerable hardship to farmers in the north of Victoria.

'Closer Settlement'
Both Victoria and New South Wales faced a major political problem. The gold rushes of the 1850s and 1860s triggered massive rises in population. As the easily-won gold was worked out, gold mining declined. Pressure to find alternative employment for large numbers both of miners and others dependent on mining increased. An attractive option was to settle them on the land, but this was prevented by two major factors.

Firstly, most good land had already been taken up by squatters, who were politically powerful. Attempts to resume large estates and subdivide them for 'closer settlement' had been defeated by subterfuge and by watering down the relevant legislation. Secondly, at that time none of the land (even that held by squatters) was particularly good, and the range of possible produce was very limited. Dry land agricultural productivity was declining due to ignorance of soil conditions and the ill effects of clearing and introduction of exotic species. The scope for growing food crops was limited because perishable foods could only be sold in Australia. Exports had to be able to survive a long journey to Britain, often of up to six months. Apart from gold, Australia's only viable export was wool. But the shortage of available land meant that the size of farm which could be made available to settlers was too small to be economic. Irrigation offered a way out: smaller holdings, with irrigated pastures to permit more intensive wool production, seemed an attractive, politically feasible way to settle surplus population.

Irrigation Wins Out
By the 1880s, South Australia, New South Wales and Victoria had each developed a distinctive position concerning the Murray's waters:

> New South Wales claimed exclusive use of all waters above the South Australian border, since the border with Victoria

was on the *south bank* of the river, the channel proper lying in New South Wales. It claimed that any concessions it made to the other Colonies were acts of grace, revocable at will.

Victoria claimed exclusive rights to all waters of tributaries in Victoria, plus riparian rights on the upper Murray.

To maintain the navigability of the rivers, South Australia was opposed to any diversions of water; for example, for irrigation.

Interstate Conferences in 1857, 1863, and 1865 had been unable to resolve these differences. In 1881 South Australia again attempted, unsuccessfully, by correspondence with Victoria and New South Wales, to gain recognition of its position.

By 1885, Royal Commissions in both Victoria and New South Wales were enquiring into irrigation development. After a Victorian suggestion for a joint Royal Commission to include all three Colonies fell through, the Victorian and New South Wales Royal Commissioners, at a joint meeting, agreed to two resolutions:

(a) waters of the tributaries of the Lower Murray may be diverted and used by the respective colonies through which they flow;

(b) the whole of the waters of the Lower Murray 'shall be deemed to be the common property of the Colonies of New South Wales and Victoria' and each of them was entitled to divert one half of the available water.[1]

Though both Colonies indicated a willingness to make some water available to South Australia in compensation, Victoria moved within days to legislate for irrigation, and to create water rights for the irrigators. South Australia's response, especially to creation of water rights for private irrigators in 1886, was increasingly bitter and strident, reflecting concern about its repeated lack of success in negotiating with New South Wales and Victoria. As the downstream colony, it could not retaliate against harmful actions by the upstream colonies, but there was no legal framework within which to pursue its case.

By making common cause, Victoria and New South Wales had found attractive solutions to their problems, but at the cost of cooperation with South Australia.

Box 5.3 Royal Commissions

A Royal Commission is an investigative body, nominally appointed by Parliament, though the decision to create it is really in the hands of the government of the day. **Terms of Reference** are normally specified in the enabling Act of Parliament. The writing of Terms of Reference in such a way as to produce a desired result from the enquiry is a well-developed political skill in Australia.

The one or more **Commissioners** are typically judges, senior lawyers (very often **Queens Counsel**, or QCs) or senior public servants. They are seconded (temporarily transferred) from their regular work. Choice of commissioners may significantly affect the outcome; in one well-known case, a prominent but notoriously boozy Commissioner recommended a sweeping liberalisation of Victoria's liquor laws!

The enabling Act normally specifies the **powers** enjoyed by the Commission, which may include the power to call for submissions from the public and interested parties; to hear evidence; and to **subpoena** documents and other evidence. (That is, to require the production of the materials, under threat of legal penalty: a person failing, without reasonable cause, to produce evidence in their care could be gaoled or fined.)

On completion of its investigations, a Royal Commission reports its **findings** to Parliament. It may find as to matters of fact — for example, that certain actions happened or did not happen — or it may make recommendations on matters of policy. The 'WA Inc.' Royal Commission reported in 1992 that there had been mismanagement of public funds by senior politicians; the Commission of Inquiry into the Conservation Management and Use of Fraser Island and the Great Sandy Region (the second Fitzgerald inquiry) reported in May 1991 with a range of policy options for the Queensland Government to consider.

A semi-judicial enquiry such as a Royal Commission can examine an issue with much more thorough attention to the issues than is possible in the rough-and-tumble of Parliamentary debate. This can also appear to 'take the issue out of politics' by handing it to an apparently impartial

> tribunal. However, it should be remembered that both the findings and the recommendations of Royal Commissions can be political dynamite, ruining careers and sparking off fierce controversy. In addition, many enquiries have been criticised for alleged bias or favouritism. Despite these shortcomings, Royal Commissions have remained important tools of government in Australia.

IRRIGATION DEVELOPMENT IN VICTORIA

By 1885 irrigation was politically dominant. It was the most important prospective use of the Murray Valley. But the many technical objections to it were known *before* the major irrigation developments began. As so often happens, they were ignored by the enthusiasts and politicians, especially in Victoria.

Technical Doubts

In 1880 the Victorian government appointed a Water Conservancy Board, consisting of the former Chief Engineer of Water Supply, George Gordon, and the Assistant Surveyor-General, Alexander Black, to make proposals for the management of existing water resources in the north of the State, and to investigate the potential of irrigation development in general: the first serious investigation of water resources in Australia. The Board's second Report, in 1882, highlighted the main environmental differences between Australia and other countries and identified the major obstacles facing irrigation.

In countries where irrigation was successful the domestic populations were large. Water was plentiful in the *dry* season because their rivers were fed by large snowfields, which melted all year round. Relatively small storages could then smooth fluctuations and divert water at low cost. Irrigation in these countries was therefore cheap to provide and water was available when it was in greatest demand. But in Australia snowfields were scarce; there was no reliable supply of water from melting, especially when most needed. To smooth fluctuations, large storages were essential, but would have been unavoidably exposed to evaporation rates far outstripping precipitation. In consequence, irrigation infrastructure would have been so much more expensive that the general economic benefits of irrigation were uncertain, making it too risky for the state, private investors, or cooperatives.

Despite these pessimistic conclusions, the Report suggested that trusts consisting of local farmers would be best able to implement irrigation schemes. But the Water Conservancy Board's warnings were largely ignored. In 1881, the Victorian Parliament had already enacted the first *Water Conservation Act*, which empowered the formation of urban or rural Waterworks Trusts to supply water for domestic use or livestock. *The Water Conservation Act 1883*, which followed, authorised the formation of Irrigation Trusts where sufficient local support existed. Some of these Trusts were for supplementary irrigation, rather than full-fledged irrigation farming, and the emphasis on locally managed Trusts was in part a response to political pressure from interested parties, especially wheat farmers.

A number of serious difficulties emerged almost immediately. Many early irrigation schemes were poorly engineered. Little account had been taken of the suitability of soils for irrigation, nor had dams and channels been well located. Lack of a detailed hydrographic survey — not the first nor the last time that political enthusiasm was to outrun available scientific knowledge in Australia — impeded both choice of land to irrigate and location of channels. The result was a waste of water, coupled with inadequate supplies when most needed. The many localised Trusts and private schemes failed to cooperate, leading to many disputes over riparian rights, drainage, and water supply.

Political Enthusiasm

Under the premiership of Alfred Deakin, later one of the Founding Fathers of the Australian Federation, the Victorian Government became very enthusiastic about further irrigation. A Royal Commission, headed by Deakin himself, was appointed in 1884; its report, setting aside the objections offered by Gordon and Black, endorsed further irrigation. *The Irrigation Act 1886* 'effectively nationalized Victoria's surface waters,' preventing the establishment of further riparian rights (though without clarifying the existing ones).[2] It also provided for the State Government to build and maintain major irrigation works, and made State Government loan moneys available to Trusts for construction of lesser works. The loans were to be repaid by water charges imposed on irrigators. A further Act, the *Waterworks Construction Encouragement Act*, provided for private schemes; it was intended specifically to make provision for the Chaffey brothers, Canadian irrigators who established the towns of Mildura and Renmark.

This occurred at the height of Victoria's land boom, which peaked in 1888, only to collapse two years later. Huge paper fortunes were backed by the 'security' of land valued at prices so inflated that some Melbourne suburban blocks did not reach their 1888 prices again until 1962! When the 'Land Banks' which were the major agency of the boom collapsed, many Victorians lost fortunes; but the well-connected were enabled to make 'secret compositions' with their creditors, some paying as little as a farthing in the pound. Political jobbery — the doing of favours for friends and colleagues, often at public expense — was rife. Unremunerative railways were built so that speculators could make money from land sales, and by various other devices, public funds were transferred into private pockets. In this atmosphere of major scandals and blatant political corruption, there are well-founded suspicions that the *Waterworks Construction Encouragement* Act was yet another favour to friends.

The 1886 Act ensured the continuing extension of irrigation, but did little to address its problems. The 26 Trusts established under its provisions were soon in debt, and the Victorian Government was unable to recover capital or even interest on monies advanced, nor could it recover the cost of providing facilities through water charges. Many trust members, uninterested in irrigation, had been joining to increase the value of their land, by capitalising the water right created under the provisions of the 1886 Act, and then cynically selling out for a quick profit; they refused in some cases even to pay for water, insisting that a right to water supply in perpetuity was included in the price of their land. Land speculation reduced the ability of most irrigation projects to generate returns sufficient to cover the initial investment. Of about ninety Trusts in existence by the late 1890s, most were in deep financial trouble. A further Royal Commission, appointed in 1894, reported in 1896. The *Water Supply Advances Relief Act 1899* attempted to remedy the situation by writing off 75% of the Trusts' liabilities.

The Chaffey brothers, too, ran into financial difficulties. Disputes between Trusts continued. Public resentment of irrigation, for soaking up public funds which might have been used for other more worthwhile purposes, grew. Salination destroyed some settlements, and serious seepage affected many channels. The latter was made worse by an unexpected ecological effect: the running water was attractive to yabbies, which burrowed into the banks, worsening seepage.

CHAPTER 5

The State Rivers and Water Supply Commission

Finally, in 1905, the Victorian Government legislated to abolish the Trusts, centralising the management of irrigation in the State Rivers and Water Supply Commission, and extending its powers under further legislation in 1909. Its second chairman, the American engineer Elwood Mead, pursued a vigorous programme of 'closer settlement', in some cases compulsorily repurchasing and subdividing land. Well-engineered schemes supplied large numbers of small blocks supporting intensive 'horticultural' production of high-value crops at high densities. This programme was successful in engineering and agronomic terms, transforming the riverine environment by the application of the latest technical knowledge. The economic, ecological, and political consequences, however, had not been calculated or foreseen.

The most important administrative change embodied in the 1909 Act was to the water right assigned to each parcel of land. The existence of a water right inseparable from the land had improved its value, in part because it was an entitlement to a subsidy: a supply of water at less than the real cost. But after 1909, to ensure that the water was utilised, and that farmers not irrigating would move aside and make room for those who would, landowners had to pay for the water whether they used it or not. This successfully eliminated speculation in irrigated land, but led to waste: farmers, forced to pay for water, would use it whether it was needed or not. Waterlogging of soils and denial of water to those in greater need resulted.

The other effect was a consequence of closer settlement. As substantial local populations grew up, they developed considerable political strength, mainly through the ballot box. This made it very difficult for the Government to raise water charges, even when economically justified by increased costs. In consequence, water charges became increasingly unrealistic, encouraging further waste.

Such setbacks did not shake faith in irrigation; instead, despite their fundamental weaknesses, the irrigation schemes became the great showpieces of economic development in the Murray Valley.

IRRIGATION IN PRACTICE

But farm sizes were too small to be viable and prices for produce too low for profitability. Early schemes in South Australia were so unsuccessful that, of eleven cooperatives established in 1893, only one was still in existence in 1905. New South Wales, learning from

Victoria's mistakes, maintained centralised control from the inception of its Murrumbidgee scheme of 1912; by buying up land, subdividing and selling at an enhanced price reflecting the value added by irrigation, it sought to avoid financial failure. But by 1914 the same pattern as in Victoria was emerging: capital investment could not be recovered, debt could not be serviced, and water charges were unpaid. Much of the capital was lost irretrievably, as the works concerned were not included in later schemes, and in many cases were simply abandoned.[3]

This did not prevent the massive expansion of irrigation in Victoria and the New South Wales Riverina, in two major bursts of investment during the mid-1920s and mid-1950s. Smaller but significant flurries of activity occurred in the first years of the century and between 1934 and 1943.

Irrigation Economics

As B.R. Davidson points out, it is the investment after the 1920s that is most puzzling. In the early days, there was considerable — though misplaced — confidence that, once initial difficulties had been overcome, irrigation farming would be profitable. Improving markets, underpinned by a series of technological breakthroughs, bolstered that confidence. In the last decade of the nineteenth century, refrigerated ships opened the British market to Australian meat and dairy produce. Dried fruits and other high-value produce rose in importance, and in the first years of the twentieth century, new hybrid wheats, the discovery of the need for trace elements, and better agricultural machinery made wheat generally profitable. But hard experience showed that none of these things made irrigation successful.

> All of the earlier schemes had been founded in the belief that farmers' returns would be large enough for them to receive an adequate return on their capital after paying the operating cost of the irrigation schemes and interest on the capital invested by the State. With the failure of the Murrumbidgee scheme in 1915 this belief was abandoned. From then until 1966 (when the Commonwealth Treasury insisted on benefit/cost type analysis before schemes were commenced) irrigation was considered a success if gross farm revenue gave a satisfactory return to State capital.[4]

The State, in other words, could not expect a return on its investment: farmers could only profit if they were relieved of all but water charges, themselves too low to cover maintenance costs. No repayment of the initial capital invested or interest on it could ever be made. The Victorian Royal Commission of 1936 explicitly accepted that a continuing loss to the State was unavoidable, justifying it as the price to be paid for rural development. Though the engineering problems had been overcome, irrigation was an economic failure and had degenerated into a subsidy to farmers.

Because of the drought-prone climate, water was assumed to be the factor of production in shortest supply, to be used as efficiently as possible by maximising the volume of production per unit of water. But Davidson contends that the factors in shortest supply were capital and labour, and the existence of ample land suitable for dry land agriculture and pastoralism, at much lower capital cost per unit of output, meant that irrigation could not compete. Nor could irrigated land act as a reserve pasture for agistment of drought-affected stock; because of the high capital cost, pastures had to be fully stocked at all times, leaving no surplus capacity. Thus the appeal of irrigation, especially after 1920, was as a solution to a *political* problem, not an economic one. It made possible 'closer settlement', a perennially attractive goal, and one which was difficult to implement with dry land farming due to the political power of the squatters, which had given rise to the paradox that suitable land for 'closer settlement' was hard to find, despite the fact that, in economic terms, land was abundant.[5]

Irrigation expansion had virtually finished by the 1970s. Only the Namoi region has been the subject of significant investment in recent times. By 1969, the total irrigated area was 1,031,000 hectares. Of this, 70% was under pasture, 16% was cropped for cereals, and 9% was devoted to horticulture (including orchards and vineyards). Pastures are relatively unproductive, and are only economic if water charges are low. Horticulture, by contrast, is highly productive: in New South Wales, the 6% of the irrigated land used for horticulture returns 24% of the income. Davidson concludes that, even allowing for the profitability of horticulture, the irrigation works as a whole have never afforded a return to government capital invested. Even if horticulture were to be encouraged, it is clear that other, less worthwhile uses might advantageously be scaled back.

These questions are of particular importance in the 1990s, as much of the old capital investment in irrigation wears out, and the question of replacement becomes urgent. Past errors can reasonably be treated as 'sunk costs' and written off; but new investment should be subjected to proper economic evaluation as well as to careful ecological consideration.

Irrigation Ecology
Irrigation intensified many of the ecological effects of European settlement. It introduced alien species, of plants as well as animals, even more rapidly than dry land farming. It transformed the land more radically, because of its more intensive nature, and the need for engineering works such as dams, channels, and drains.

In particular, irrigation demanded the regulation of the river's flow. Floods are now roughly half as frequent as in the 'natural' state, when flooding occurred in two of every three years. But the indigenous species, adapted to the variable flow, used it in various critical ways. Some, including important commercial fish such as the Murray Cod, bred at flood time, using the floodwaters to reach areas which might normally be out of reach. Waterfowl also used the vast temporary lagoons created by flooding to mate and raise offspring. Flora such as the River Red Gum also relied on periodic floods to breed; but cannot tolerate the permanent inundation or prolonged dry periods of the present régime.

Irrigation structures also adversely affected the mobility of fish: only two weirs, and none of the dams, have fishways. The salt-tolerance of riverine species, especially fish and crayfish, reflects the historical highs and lows of the basin's rivers. Irrigation, by making the river more saline, exceeded these tolerance levels at times.

Many faunal and floral species have been in decline since as long ago as 1912, and substantial simplification of the native ecosystems has taken place. Introduced and feral species have had significant and often highly detrimental effects on the ecology.

The prosperity of the basin depends on the health of its ecosystems. When these break down, production and even survival may become difficult. Ecological sustainability is a necessary precondition for continued exploitation of the basin's resources.

Effects of Overcapitalisation
Effective environmental management and the profitable operation of farms can all too easily conflict. Ecological rationality does not necessarily coincide with economic. A very important reason for this is overcapitalisation of farms, which occurs through capitalisation of the water right: when an irrigated farm is sold, the seller charges the buyer a price which reflects the expected value of the water right's contribution to increased production. The buyer then has to recover the extra capital cost by increasing the production

of saleable goods. The more heavily capitalised a farm, the more intensive its exploitation must be if it is to turn a profit. This has led to unecological land use practices, making it harder to find an ecologically sustainable pattern of use for the Murray–Darling basin. If irrigation had been less ambitious and smaller in scale, a lower-intensity, more 'natural' pattern of use, with more sensitivity to native flora and fauna, may have been possible.

But that would have required more scientific investigation, in order to minimise the impacts of introduced species and farming practices on the basin. This was never done, and even now, research is less than adequate. Taxpayers as well as farmers are still paying for this ignorance, as well as the costs of harmful property speculation from the 1880s. Levels of capital investment needed to be lower and more selective, given the unprofitability of many forms of irrigation agriculture.

OVERVIEW: LESSONS FROM EXPERIENCE

The water resources of the Murray–Darling basin have caused conflict and controversy from the outset. The governments responsible were unwilling to cooperate in managing the basin; instead, each defended powerful local interests. Poor planning and misguided adherence to the principle of decentralisation led, in Victoria particularly, to conflict at the local level as well. Public funds were wasted due to lack of elementary coordination among irrigators, poor design and siting of works, and lack of knowledge and understanding of soils and vegetation, as well as unrestrained speculation in land and water rights. Adequate knowledge was available for evaluating the likely success of irrigation, but this was ignored as successive governments succumbed to political pressure and enthusiasm for 'development'. Even when it became obvious to all that irrigation was uneconomic, governments still continued to promote it and eventually acquired all the financial and managerial responsibilities for the schemes.

The economics of irrigation in Australia have always been poor; it was undertaken more as a way of intensifying production — through 'closer settlement' — than as a profitable activity. Produce from irrigated areas, like other Australian exports, has been vulnerable to world prices, which have not offered good returns. Public money invested has never been repaid; following poor results from the early schemes, governments ceased to evaluate returns to capital, seeking returns sufficient only to cover running expenses. There was no serious attempt to consider the cost-effectiveness of capital investment in irrigation from 1915 until 1966. Effectively, irrigation agriculture and the communities it created have been subsidised by the Australian taxpayer. Thus the supposed benefits must be offset against this subsidy, and against a whole series of problems. Among the most important of these are the ecological disruptions, now coming to a head as a series of threats both to ecosystem integrity and to production.

Irrigation has many problems in common with environmental issues elsewhere, such as:

1. inadequate scientific data;
2. ignoring of what data did exist;
3. untested, dogmatic assumptions, in this case about the benefits of irrigation;
4. lack of effective planning;
5. poor cooperation among interested parties, in this case at local level as well as among States;
6. inadequate economic evaluation;
7. political expediency: politically attractive policies were often pursued regardless of their technical, ecological, or economic flaws.

Irrigation began from the recognition that drought was common in Australia. By seeking to 'smooth out' water supply fluctuations by storing water and applying it as required, it attempted to adapt the climate to European agriculture. This ignored the difficulties of water supply in dry seasons, and the implications of high rates of evaporation. The assumption was that climatic restrictions could be circumvented; irrigation was adopted despite evidence that this assumption was faulty. Its subsequent history has vindicated the critics: not only has it failed to solve the drought problem, it has created novel environmental stresses, some of them familiar from overseas experience, and others specific to the Australian environment.

From the 1850s to the 1980s, the management of the Murray–Darling basin remained one of Australia's major public policy failures. Lack of a legal framework for assignment of water rights and protection of riparian rights catapulted the problem into politics. Over 50 years, conference after conference failed to resolve the conflict.

CHAPTER 5

FURTHER EXPLORATION

The early history and legal problems of Murray basin management are dealt with in Clark's papers on 'The River Murray Question' in the *Melbourne University Law Review*. Mudie's poem 'They'll Tell You About Me' is in many anthologies; see, e.g., *The Penguin Book of Australian Verse*, pp. 156–8. Mudie has also written a history of the Murray navigation, *Riverboats*. A useful, but quite conventional account of irrigation development in Victoria is to be found in Chapter 10 of Powell's *Environmental Management in Australia, 1788–1914*, a source of much of the factual data for this chapter. The heretical view, that irrigation was an economic disaster, is most succinctly expressed in Davidson's chapter 'Irrigation Economics,' in Frith and Sawer, *The Murray Waters*. It is Davidson who calls attention to the Water Conservancy Board's warnings about climatic constraints on irrigation. The contrast between Davidson's handling of this report (p. 193) and Powell's (p. 128) is very revealing. Davidson enlarged on his views in *Australia Wet or Dry?* Mining development in Australia, including the gold rushes, is the subject of Blainey's *The Rush That Never Ended*; restrictions on the profitability of crops due to transport problems and remoteness of markets are detailed in the same author's *The Tyranny of Distance*. For an account of the Victorian land boom of the 1880s, its subsequent collapse, and the highly ingenious and varied forms of corruption it spawned, see Cannon's *The Land Boomers*. Pope's paper on development policies, though concentrating on the Commonwealth government and on immigration and soldier settlement, is illuminating for its revelation that government policies were being formed in ignorance of physical and environmental constraints.

Those unfamiliar with the *institutional* processes associated with irrigation, such as legislation, Parliamentary procedure, the functions of government, and so on, should consult a standard text on Australian politics. At an elementary level, Stewart and Ward, *Politics One*, is good; more advanced treatments can be found in Emy and Hughes, *Australian Politics: Realities in Conflict*, or Parkin, Summers, and Woodward, *Government, Politics and Power in Australia*.

CHAPTER 6

GRIDLOCK AND LANDSLIDE

The early experiments in irrigation, especially in Victoria, embodied a conviction that independence and self-help were virtues which any broad development scheme should encourage. These ideas derived from the Liberal values of the day, but they were not very successful in practice.

The small scale of the early irrigation trusts was one of the problems. Their inability to carry out scientific investigations or to finance major engineering works led to poorly coordinated, piecemeal works. These created conflict over stream flow, drainage, and so on, reproducing in miniature the squabbles between the Colonies. This lack of cooperation or even coordination was the reason for State takeover of the Trusts and the writing off of their debts, despite the considerable cost to the taxpayer. It was also the cause of the abandonment or expensive reconstruction of so many early irrigation works.

Thus experimentation with a highly decentralised institutional structure for irrigation had led each State independently to conclude that a centralised structure, in which the state government played a direct role, was unavoidable.

While the resulting centralised water authorities were signally

Chapter 6

BOX 6.1 LIBERALISM

During the nineteenth century politics in the English-speaking world (and in many other countries as well) was largely a contest between **Liberals** (or **Whigs**) and **Conservatives** (or **Tories**). Organised labour, and with it Trades Unions and a Labour Party, were still developing: they did not gain power through a popular vote until, in Queensland in 1899, the world's first Labor government was elected. One reason for the slow growth of Labour, despite the horrendous working conditions of the Industrial Revolution, was that for most of the century voting was restricted to the better-off classes.

Conservatives tended to emphasise respect for authority, the importance of tradition, and the idea that some people were fit to rule, while others were not. They were the party of the rich, the landed gentry (in Australia, the squatters) and they stood for social stratification.

Liberals inherited a political philosophy which had been forged in the struggle between Parliament and the King in Britain in the seventeenth century, and since enriched by such great thinkers as John Locke, Jean-Jacques Rousseau, and John Stuart Mill. It asserted the importance of the individual, and based its political theory on the idea that there existed a **social contract** between the rulers and the ruled. Rulers had the duty to rule according to law, and the ruled had the duty to obey those laws, provided that they were properly enacted, and enforced equally on all.

Liberalism believed in the formal equality of all individuals before the law, and supported this belief with a doctrine of **human rights**, which were thought to attach to all individuals equally. Liberalism also embraced the *laissez-faire* economics of Adam Smith, which emphasised the element of rational self-interest in economic relations, and tended to glorify the individual entrepreneur. It was for this reason that small, self-governed Irrigation Trusts, with responsibility for their own affairs, were attractive to the Victorian government of the 1880s.

successful in solving the engineering problems of irrigation, they never solved the economic problem, and their ecological impact was devastating. Nor did the imposition of state control resolve the intercolonial conflicts.

Federation

Intercolonial conflicts were a central issue in the constitutional debate which preceded Federation in 1901. The outcome was a victory for no-one, and left many issues unresolved. Section 100 of the new Constitution of the Commonwealth of Australia forbade the Commonwealth to 'abridge the right of a State or of the residents therein to the reasonable use of the waters of rivers for conservation or irrigation', which was a victory for New South Wales.[1] South Australia gained less: Section 98 granted the new Commonwealth power over navigation. Though some 87 vessels still plied the Murray in 1901, navigation was rapidly succumbing to fierce, often highly subsidised rail competition.

The creation of the **High Court** and provision for an **Interstate Commission** were expressly designed to help resolve Murray waters issues. However, one of these institutions was destined to destroy the other: in the *Wheat Case* of 1915 the High Court effectively stripped the Interstate Commission of any power.

Meanwhile, popular discontent had finally compelled reluctant politicians to a conference, convened at Corowa in 1902 by the Murray River Main Canal League. The resulting Interstate Royal Commission (*not* the same body as the Interstate Commission mentioned in the new Constitution) failed to find a solution to the water distribution problem. Demands for Commonwealth control over the Murray also came to naught. Bills to establish the Interstate Commission were introduced in the Commonwealth Parliament's first (1901) session, and in 1909, but the governments of the day were not greatly interested, for which they were much criticised. Motions for a constitutional amendment to place the Murray and its tributaries under Commonwealth control were put forward in 1910 and 1912, but were withdrawn once the Interstate Commission Bill passed in the latter year.

Meantime, amid continuing bitter controversy and threats of litigation, the States themselves were attempting to reach agreement. The Premiers' Conferences of 1905 and 1906 both agreed on provisional formulæ for water allocation, but each time, agreement collapsed. The 1906 Conference, however, recommended the establishment of a standing commission to administer water distribution. In 1911 a committee of engineers from each State was estab-

Chapter 6

lished. It reported in November 1913 that irrigation development was having a negative impact on river flows and navigation, but both navigation and irrigation could coexist if appropriate measures were taken. This made possible a major advance in interstate relations: the River Murray Waters Agreement of 1914. It was intended that the Interstate Commission should implement the Agreement, but this became impossible upon its collapse. Instead, a new intergovernmental body, the River Murray Commission, (RMC) was created.

The River Murray Commission

Despite limited powers and a highly restricted jurisdiction, the River Murray Commission was to become the dominant body in the administration of the Murray system for 70 years. However, the initial success of the Commission in making possible exploitation of the basin for irrigation degenerated over the years into stalemate. The RMC's task was distributive: it was created to coordinate engineering works and to oversee the allocation of irrigation waters according to formula. It could only refer policy questions to the constituent governments; it had few powers of independent action. Its lack of regulatory power led to a steady deterioration of the riverine environment which was not, until late in the Commission's life, seen as a concern.

The Work of the Engineers

The River Murray Waters Agreement and the creation of the River Murray Commission would never have been possible without the groundwork of the engineers. Though not themselves immune to parochial bickering, their work on rates, variability and volumes of flow, and on evaporation, created an important basis of factual data from which agreement could be negotiated. Beginning in 1902, a series of engineering investigations not only established the essential information, but also determined the pattern of dams and weirs required if effective regulation of the river's flow for irrigation was to be achieved.

The work of the engineering team appointed by the 1911 Premiers' Conference, which reported in 1913, was especially important. It eliminated some of the uncertainty which had complicated the reaching of agreement among the States. Once the volume of flow was known, debate could focus on possible formulæ for the allocation of water, rather than on postures. A 'contract zone', within which bargaining was now possible, had been created.

Rapid progress followed. It was quickly agreed that Victoria and New South Wales would share equally in the flow at Albury, and that South Australia would be guaranteed approximately 1.25 million acre-feet of water annually, equivalent to half the flow at Albury plus inflows from tributaries below Albury. This would provide a navigation channel five feet deep during the eight to nine month navigation season, as well as allowing some irrigation and meeting needs for stock watering and domestic supply. In principle agreement was reached on a programme of construction for dams, locks, and weirs.

The key was the engineers' proposal to guarantee South Australia a minimum flow of water at all times. From the South Australian point of view, the compromise was acceptable, though not ideal. New South Wales and Victoria came off well, because they gained additional water in good years, and no longer had to contend with continual political and legal challenges from South Australia. It was now possible to embark on a construction programme which promised both to enhance navigability by building locks, and make more water available for irrigation by constructing storage dams.

Agreement on allocation of water, in other words, made possible cooperative measures to *enhance* all the various uses over which the States had previously been in dispute. The engineers had found something for everybody — except, perhaps, the long-suffering taxpayer, who had to foot the bill.

The River Murray Waters Agreement and Commission

The River Murray Waters Agreement was passed by the Commonwealth, Victorian, New South Wales, and South Australian Parliaments in 1915. It was an example of legislation under the 'concurrent powers' of the new Constitution: no one level of government was solely responsible for the Murray, so in order to act, it became necessary for the Parliaments of each of the States involved, plus the Commonwealth, to pass complementary enabling legislation; each Parliament's Act would authorise those actions over which it had jurisdiction, so that all the Acts taken together would authorise the agreed action. When passed by several governments in this way, such legislation is also known as **concurrent legislation**. Its purpose is to establish a framework for coordinated action.

The Agreement ratified the formula for water allocation, empowered the construction of the locks and storages, which was to be carried out by the States themselves, imposed navigation charges, and — very importantly — provided for the new River Murray Commission to establish a system of uniform gauging of the Murray and its tributaries. More recently, procedures were established for the release of water for saline dilution.

During the currency of the Agreement, from 1917 to 1987, it was regularly altered because of changing conditions. In 1923, in recognition of the declining importance of river trade, it was changed to give priority to irrigation. Altogether, the Agreement was amended eight times, the last time in 1981. What never changed was the fundamental allocative principle: the agreement was founded on the *distribution* of water among the competing States. This could not be disturbed for fear of upsetting the arrangements which had been built onto it.

1917 saw the establishment of the River Murray Commission to administer the Agreement. All four contracting governments — New South Wales, Victoria, South Australia, and the Commonwealth — were represented on it. The RMC commenced construction of a grand system of 15 weirs, 13 locks, and two major dams, the Hume and Burrinjuck. During the Depression of the 1930s, the Agreement was amended to reduce the number of weirs and locks. The Hume reservoir above Albury, then one of the world's largest dams, was completed in 1936, and the barrages at the mouth of the river in 1940. In 1948, the Snowy Mountains scheme, which had an irrigation component in addition to the electricity supply element, was set up.

The major focus of the RMC was to manage water flow for navigation and irrigation; salination was not then considered a major problem.

Personnel and Operations
The RMC consisted of four Commissioners, one from each State and one from the Commonwealth. Each State representative was a senior public servant with experience in water engineering. Until the 1970s, the president of the RMC was always a federal Minister with an appropriate portfolio. The unanimous consent of all the Commissioners was required for any major decisions. The cost of all new works was shared equally among the four partners; maintenance costs were shared equally among the three States.

For the greater part of its existence, the staffing of the RMC was minimal. It had only two full-time engineers, who were responsible for a wide range of tasks. The staff grew from three in the late 1960s to sixteen by 1984, in response to expanding responsibilities.

However, the RMC successfully contained conflict through the formula guaranteeing South Australia a fixed amount of water, and dividing the remainder between New South Wales and Victoria.

Powers

The RMC's powers were limited to:

1. gauging of river flows (a function it never exercised);

2. approval and supervision of works (actually, supervision has always been done by States);

3. fixing of navigation charges (negligible even by 1914, as railways steadily took over the river trade);

4. oversight and verification of water distribution according to formula;

5. undertaking of some minor flow-conserving works; and

6. oversight of measures by Victoria and New South Wales to protect the Hume catchments.

This meant that functions outside its jurisdiction included the planning of irrigation schemes; land use planning in the Basin; coordination of water resources development; the authorisation of water diversion for individual use; river pollution; forestry; fisheries and wildlife management; water-craft (i.e., boats and other vessels); river structures, apart from those under the Commission's control; and flood control or flood warnings. It also had no jurisdiction over the major tributaries of the Murray, in particular the Murrumbidgee (which fell wholly within New South Wales) and the Darling. This meant that it had little opportunity to control sources of pollution or salinity lying on these streams, nor could it control their flow.

Worsening problems of salination and land degradation during the 1960s and 1970s made the inadequacies of the existing system increasingly evident. As the need for regulation, rather than distribution, became ever more apparent, the RMC attempted to cope with the need.

Expanding the RMC's Role

The Commission responded by expanding its role to include identification and reporting of problems. But policy matters remained under the control of the States, which also undertook the major works. The administration of the basin remained fragmented, with no one authority having sole power to manage it.

In response to these constraints the RMC was able to *reinterpret*

Chapter 6

some of its powers to tackle the new problems, though they had not been intended for the purpose. The powers to regulate stream flows were used for periodic flushing, to control water salinity. Water levels were also adjusted to protect wildlife and for other environmental reasons. And the RMC commissioned a study of salinity in 1967, engaging engineering consultants for the purpose, though this action was not strictly within its powers.

In addition, the RMC used its unique position to lobby each State to implement coordinated policies. Cooperation between the States and their respective departments was essential, though not always forthcoming.

The Dragon Stirs

This lobbying was not futile. In 1973, at the instance of the Premier of South Australia, the then Labor Prime Minister, Gough Whitlam, convened a meeting of heads of government in order to discuss the future of the Murray, with special reference to salinity. A working party of government officials was established, reporting to a steering committee of Ministers in October 1975 — only a month before the Whitlam Government was dismissed. A year later, this working party report was adopted by the four governments, the working party itself disbanded, and the RMC given the task of drawing up revised legislation. The RMC's proposals went forward in July 1978, but Ministerial meetings in October 1979 and February 1981 failed to agree on them. The main opposition came from New South Wales, still concerned that agreement would impose excessive costs on it. A new agreement was finally concluded in 1981, after some hard bargaining and tough legal wrangling. It was generally regarded as a failure.

Salinity problems continued to grow worse, leading to important changes in political perceptions. In particular, Victoria began increasingly to see itself as a victim of salination, rather than feeling that it would not be greatly affected. This led to a realignment of forces: now South Australia and Victoria confronted New South Wales.

Changes in Victoria

Despite perceptions of the new Murray Waters Agreement of 1981 as a lost opportunity, one seemingly minor provision sparked some important changes in Victoria. The new Agreement, for the first time, limited the permissible levels of saline discharge into the Murray.

This posed a particular problem for Victoria, because Barr

Creek, near Kerang, is the most saline of all tributaries to the Murray. As far back as 1975, proposals for massive evaporation pans at Mineral Reserves basin and Lake Tyrell had been formulated, but quickly fell foul of local farmers, fearful of additional salt infiltration into groundwater. Not content with political opposition, they launched legal moves to head off the schemes.

The new Cain Labor Government, on its election in 1982, saw a need for political action. It is significant both that it was a new government which saw the need for change, and that local opposition was a trigger. The entrenched definition of irrigation and water management as an engineering problem, dating to before the turn of the century, and reinforced since the 1909 Act by the ascendancy of Elwood Mead, led to conflict. The solution, seen as the application of the best available technology, displayed little concern either for the goals sought or for the side effects.

It was the latter that motivated the farmers: afraid that the proposed evaporation basins would worsen the problem of salt infiltration into groundwater, rather than relieve it, they actively opposed the plans. This may have been the trigger for the Victorian Government to ask the other question, about the overall goals of the process. Certainly, immediate remedial action was required; the question was where and how. In addition, there loomed the whole question of irrigation economics and efficiency, especially as much of the infrastructure was nearing a time at which replacement would have to be considered.

The Parliamentary Committee
The outcome was an all-party Parliamentary Select Committee, which, after extensive investigations, reported finally in October 1984. It received more than 90 submissions, from local councils, other State governments, and many interested parties, ranging from the Australian Conservation Foundation to farmer action groups. It interviewed 170 witnesses. Salination emerged as the primary concern of the majority. The Committee also visited the key areas on tours of investigation.

In addition, the Committee launched a substantial research effort. It appointed a Director of Research in February 1983, following this with secondments from the Rural Water Commission and the Department of Agriculture, to a total of four staff. These were supported by an advisory committee drawn from Universities, the Commonwealth Scientific and Industrial Research Organisation (CSIRO), farmers, and private consultants. A number of reports on particular facets of the problem were commissioned from consultants.

Chapter 6

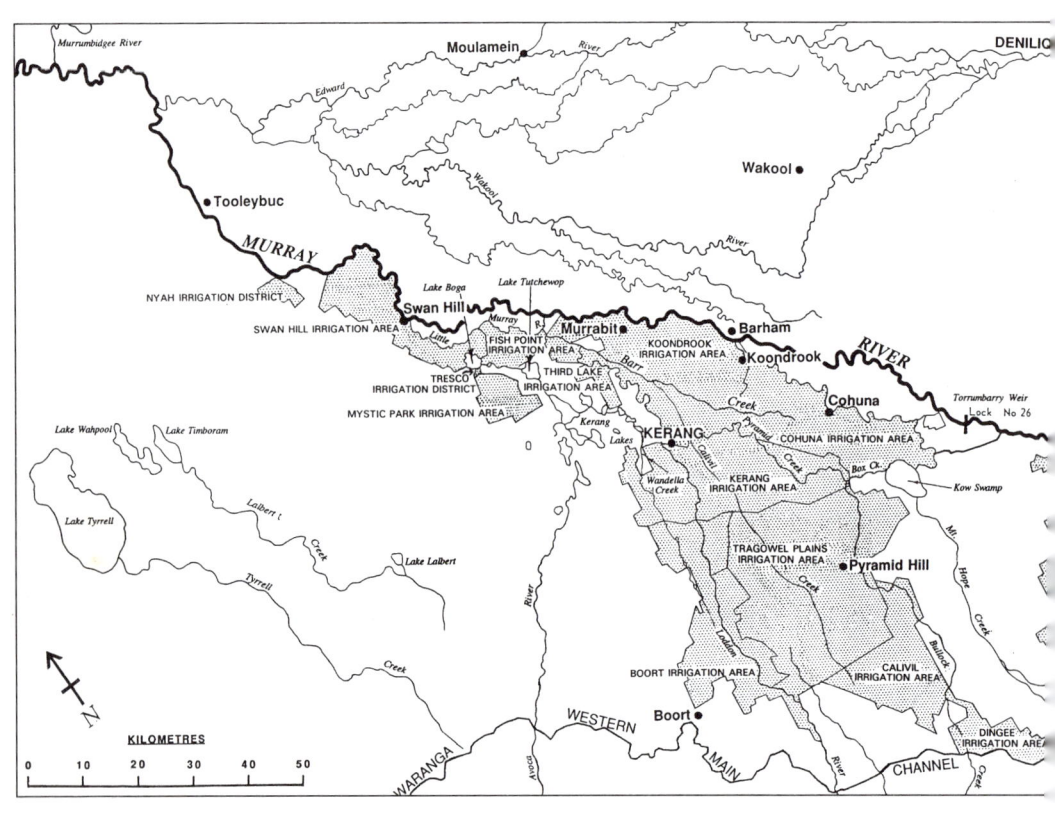

Fig.6.1
The Barr Creek–Kerang Area. Barr Creek itself lies between Kerang and the Murray.
(Source: Victoria State Rivers and Water Supply Commission. *Salinity Control and Drainage*.)

The Committee's final report, *Salt of the Earth*, concluded that not only was salination extensive, but it was worsening rapidly. Production losses were considerable, and water available for irrigation was often itself saline.

Implementing Salinity Controls

Cabinet's first response to the all-party Committee's report was to set up a task force from its own ranks to consider courses of action. Rather than adopt the Committee's recommendation to pass the 'lead' role on salinity control to a reorganised Department, Cabinet decided to set up a Salinity Control Unit (later Bureau) within the Department of Premier and Cabinet. This was a response both to interdepartmental rivalry and to the strongly expressed preferences of one Minister and his senior advisers. The Unit, which was to function for a fixed period before review, was supported by a Departmental Salinity Liaison Committee, linking ten departments and authorities. This in turn contained a Strategy Development Working Group. This tactic of centralisation of policy-making largely eliminated interdepartmental rivalry, helped by a strong task orientation and the refusal of key figures to fight departmental battles.

A pilot programme was then launched, covering the Goulburn and Broken River catchments. A Regional Salinity Team consisting of four persons was established, reporting directly to the Ministerial Task Force via the Salinity Unit, and supported by a Salinity Program (sic) Advisory Council representing local interests. This programme successfully drew on local expertise for the elaboration of a detailed plan, tailored to local conditions and needs.

Administrative Reform
Salt of the Earth recommended significant administrative reform. Labor, on assuming government in 1982, had already embarked upon a major reform of State Government administration in general. It wasted no time in tackling water administration.

The problems it faced were substantial. There were serious interdepartmental conflicts over control of various activities and responsibility for them. At the same time, there were considerable overlaps in the responsibilities involved. In consequence, interdepartmental rivalry was undermining cooperation on important questions, such as the design and implementation of public works or management programmes. Much time was wasted attempting to resolve these 'artificial' conflicts, which were a consequence not of the salination problem, but of the poor organisation of the Victorian governmental structure.

The Victorian Government reorganised both water and land administration. An important influence was John Mant, who had advised on similar reforms in Tasmania, served as Whitlam's private secretary in 1975, been a department head in South Australia, and later organised the 1984 Australian Institute of Political Science (AIPS) Conference on the Murray. His proposals for Victoria were designed to eliminate interdepartmental conflict by internalising issues within a single administrative unit, and emphasising results rather than internal bureaucratic goals. By reducing divisions of responsibility and opportunities for interdepartmental conflict, Mant sought to get greater 'productivity' from the system in terms of addressing and solving problems.

The Ministry of Water Resources and Water Supply was replaced by a new Department of Water Resources; at the same time, the State Rivers and Water Supply Commission was transformed into the Rural Water Commission. Local and regional water authorities were restructured, more than halving the total number. Uniform requirements for reporting and annual accounting were introduced, with the aim of decentralising administration and local decision-making, but securing better financial accountability.

A new Department of Conservation, Forests, and Lands took over the functions of the Departments of Planning and Crown Lands, the Ministry of Conservation, and the Forests Commission. The latter had formerly been autonomous, its primary goal being to maximise production from forests. Eighteen new regional offices were established, each being a point of contact with the new Department's full range of activities, where previously services had been centralised and fragmented. Within the new Department, a Land Protection Service was formed by amalgamating of the Vermin and Noxious Weeds Authority (from the Department of Crown Lands and Survey) with the Soil Conservation Authority, formerly part of the Conservation Ministry. Further centralisation occurred in 1990, with the creation of the new Department of Conservation and Environment.

Changing Perceptions
Victoria's self-perception as an 'upstream' user inflicting costs on South Australia but reaping only benefits for itself crumbled rapidly once the severity of the salinity problem was appreciated, and its response was very interesting, especially in terms of the criteria established in Chapter 1. The Victorian Government made a serious attempt to understand the salination processes which were going on. It commissioned scientific research designed to clarify the problem. It sought to tackle the institutional inadequacies in the system, initially by domestic administrative reform, and subsequently by supporting change at the national level. And, as Kellow shows, the foresight and political will existed to identify a long-term problem and seek solutions. True, as Kellow also shows, some issues were deemed too politically 'hot' to handle. But the reforms were sufficiently substantial to make a considerable difference to the problem in Victoria, and they met the criterion of ecological sensitivity — by building in regionally-specific data and plans — quite effectively. This latter capability, Kellow has suggested in another context, is one of the advantages of state government. The Commonwealth, more remote from the problems, has to content itself with a more 'broad-brush'

approach, which may not be as effective in the finer grain.

Furthermore, the Cain Government committed itself to a major revamp of the State's environmental programmes, of which the salinity control measures were only a part. In 1986, the then Minister for Conservation, Forests, and Lands announced the LandCare programme. This was a self-help programme for farmers, which provided subsidies and technical assistance for better all-round land management. It was a considerable success in Victoria and was later adopted by the Federal Government for a national programme. Its political utility lay in the fact that it was a mechanism to link farmers to conservation measures, and salinity to broader conservation issues.

> The scope and seriousness of the salinity problem (and other problems) must be seen as having facilitated these changes in administrative structures, since it underscored the interrelatedness of environmental problems of this nature and reinforced the view emerging in the 1980s that the old 'engineering' and 'development' orientations of commissions insulated from ministerial control were no longer appropriate. In Lowi's terms, these changes reflected changes in the function of the state from distribution to regulation, with a concomitant change in bureaucratic structure. This process was completed in 1990, with the amalgamation of the Department of Water Resources, Department of Conservation, Forests and Lands, and the 'environment' part of the Ministry of Planning and Environment to form a Department of Conservation and Environment.[2]

Nationally, changes were occurring too.

THE NATIONAL SCENE

Ex-Prime Ministers do not simply disappear, however much their opponents might wish they would! Especially if long-lived, they frequently play an important behind-the-scenes role both in their own parties and in politics at large. Gough Whitlam, pitchforked prematurely from office in 1975, was no exception. He maintained a strong interest in public affairs; and not surprisingly, was especially active in those matters he considered 'unfinished business' from his own incumbency.

This factor became particularly important at the beginning of the 1980s, as Labor won power in State after State, as well as gain-

Chapter 6

ing office in Canberra in 1983. Especially since 1982 had seen the Cain Labor Government's accession to office in Victoria — considered by many the hardest State for Labor to win — the time was ripe for action. Labor was now in power in all the basin States but Queensland; in principle, the climate for cooperation could scarcely have been better. Working behind the scenes, Whitlam persistently drew attention to this fortunate situation, suggesting that the time was ripe for mutual cooperation and agreement on a number of unresolved issues, the Murray included.

A conference of the AIPS in 1984 — organised by John Mant — served further to focus interest, bringing together experts and knowledgeable parties from a number of disciplines. In the same year, the RMC published its own review, *The River Murray Salinity Problem*.

Political alignment was reinforced by a series of changes at the administrative level. Positive and constructive Federal Government involvement, and some important behind-the-scenes politicking, facilitated substantial changes in the States' positions. With surprising rapidity, a major breakthrough occurred on the management of the Murray-Darling basin as well as the salinity issue.

Interstate Links

Labor's presence in office in the major basin States as well as federally brought into play some further personal factors. Mant's links in the federal government sphere, as well as across three states, greatly eased the attempts by Hopgood, a senior Minister in the South Australian Cabinet, to gain some agreement. Similarly, Dr John Paterson, who became the first Director-General of the Victorian Department of Water Resources, had had previous experience in New South Wales, as well as having been active in Labor politics for many years.

A network of politicians and administrators, many of whom knew each other, and some of whom had already worked successfully together, was now in place. As often happens, this made possible the creation of new policies.

Much of the groundwork had already been laid elsewhere. Mant had advised the Tasmanian Government on the reform of land management; he applied essentially the same principles in Victoria.

CHANGES IN NEW SOUTH WALES

Before his appointment to the new Victorian Department of Water Resources, Paterson had been instrumental in the reorganisation of the Hunter Valley Water Board in New South Wales. There, while making substantial economies and reorganising administration, his primary emphasis had been on a move away from 'engineering' criteria to the use of economics techniques for evaluating water use and costs. In particular, he had abolished pricing policies which encouraged consumption, replacing them with one which stressed the effective use of scarce capital and water.

Paterson applied this experience when chairing a Working Group which submitted a comprehensive report to the first meeting of the Murray-Darling Basin Ministerial Council (MDBMC).[3] He sought to replace the existing system of fixed quotas for water and salinity with a flexible but quality-oriented system. Upstream States would be allowed increased saline discharges, but had to compensate those downstream (in practice, mainly South Australia) by 'dilution flows' of fresh water. Since 1981, with quantitative controls on saline discharges, it had been difficult to drain areas such as Barr Creek. The new rules meant that highly saline, waterlogged areas could immediately be drained. At the same time, a standard for salinity in the river itself was to be adopted, and continuous water accounting was to be introduced in New South Wales and Victoria. Additionally, a number of improvements in water management, cost sharing, investment, and the institutional arrangements were proposed.

Paterson had found incentives for New South Wales in particular to join the system, by demonstrating that cooperation could lead to gains for it as well as the other States. This change in payoffs was crucial for New South Wales, since the policy could then be represented as being to the State's advantage: critical when dealing with farmers and voters. Novel information technology — linear programming — was used to optimise the various costs and benefits.

> Of crucial significance in this process was the way in which the salinity problem was redefined as a positive-sum game, with gains for all states, whereas the prevailing perception up to this point had been that the problem was zero-sum, with any action (and expenditure) by New South Wales and Victoria being seen as acts of pure altruism, for the sole benefit of South Australia.[4]

Not only did Paterson's work show that benefits could accrue to New South Wales and Victoria through cooperation; it reduced the essential asymmetry of the situation. This served both to diminish the suspicion with which South Australia viewed the motives of the upstream states, and to make New South Wales considerably more willing to explore compromise and trade-offs.

But the further implication was that the new MDBMC would not be able to start with a clean slate. An agreement along the lines proposed by Paterson was an implicit condition of New South Wales' participation. The whole process of political convergence was already beginning to define a plan of action, the context determining the policy.

Institutional Innovation

Such a plan, in turn, implied a series of highly significant institutional changes, with far-reaching consequences. As serious intergovernmental talks began, the outlines of policies to replace the flawed 1981 compromise were already becoming clear.

The Murray–Darling Basin Ministerial Council

In 1985 the Murray–Darling Basin Ministerial Council was established. Whereas the RMC had been an administrative body, the MDBMC was the first permanent *political* institution concerned specifically with the Murray or its tributaries. It was a Cabinet-level body, drawing together the relevant Ministers of all the governments involved. It was intended from the first for decision-making, rather than 'taking the issue out of politics' by administrative routine.

The MDBMC's charter states that it is to:

> promote and coordinate planning and management for the equitable, efficient and sustainable use of the land and environmental resources of the Murray–Darling Basin.

A draft action plan for the basin was drawn up, and legislation put in train for the formation of a new Murray–Darling Basin Commission, to replace the River Murray Commission. This was to have more extensive powers, especially for research and enforcement, but policy was to be decided at the political level, by the Ministerial Council. However, Queensland refused to participate until 1990, and the Australian Capital Territory (ACT) was left out. The former, serious omission, was rectified after the Goss Government came to power in 1989, and amended legislation to incorporate Queensland was prepared during 1991.

The Commission

The Murray–Darling Basin Commission (MDBC) is responsible for advising the Ministerial Council and implementing its decisions. Established in 1988, it consists of two members from each of the participating States, and has a staff of 39. Decisions are by majority vote, except for water allocations, which must still be unanimously agreed by the participants, a very demanding rule which makes it easy for one member to 'hold up' the others. There was no change in the formula for apportionment of water use between the signatories.

The Commission administers three major programmes:

1. The Salinity and Drainage Strategy, implemented in 1988, seeks to reduce infiltration and runoff of saline groundwater, at the same time improving drainage in irrigated areas to reduce waterlogging.

2. The Natural Resources Management Strategy was inaugurated in 1989. Some 40% of its budget is devoted to research into aspects of natural resource management in the Basin; the remaining 60% supports various works and initiatives designed to improve resource management. Many of these are intended to emerge from the activities of local groups.

3. In August 1990 the MDBMC approved the adoption of a Nutrient Management Strategy; after extensive preliminary research, this was implemented during 1993.

Local Participation

As in the Victorian salt management strategy, there was a large role assigned to local participation. Through a Community Advisory Committee linking State and special interests, local initiatives were to be fostered, and local participation sought in developing regional plans for implementation within the broader, basin-wide framework. The model was to be the arrangements already adopted in Victoria for the Salt Action programme and LandCare. Despite some early conflict, the Committee, reorganised in 1989, was successful in encouraging 'Communities of Common Concern' in affected areas.

Participation was encouraged by a carrot-and-stick policy, in which help and finance were available to participants in approved drainage and land management schemes.

Chapter 6

Overview: from Distribution to Regulation

The old colonial disputes had created a climate in which any cooperation was difficult. Governments were unwilling to relinquish even a part of their power and autonomy to bodies outside their own State boundaries. Moreover, past disputes had demonstrated that there was little legal obligation on States or their institutions to cooperate with each other. New South Wales, least affected by flow and water quality problems, was especially reluctant to cooperate; it had little incentive to implement potentially unpopular policies if the benefits were to flow to South Australia or Victoria. South Australia suffered soonest, bringing it into conflict with Victoria and New South Wales at an early stage.

The resulting standoff pitted the one downstream State, South Australia, against two upstream ones, Victoria and New South Wales. While South Australia's desire to maintain navigation dwindled steadily, its concerns over water grew. Adelaide became progressively more dependent on the Murray for its water supply, and irrigation also developed in the Renmark area. Thus it required a sufficient *quantity* of water, at an acceptable level of *quality*. This became more and more difficult to secure, as upstream diversions increased. The 1914 compromise did not address the water quality issue.

The River Murray Waters Agreement and the River Murray Commission had embodied a once-off compromise. By treating the problem as a technical one, and passing it to the engineers, the politicians had created themselves an escape hatch. It did not resolve the conflicts among the States, but offered a formula for distributing water which was acceptable to all. Once that had been agreed, it was possible to formulate a programme of engineering works, and to create an administrative body to implement them, as well as to oversee the routine of water allocation. This contained, though it did not resolve, the nineteenth century tussles, by striking an agreement about the *shares* of Murray waters to which each State was entitled. It was an unfair stalemate, but it endured for 70 years because the alternatives were worse.

The structure, personnel, and functions of the RMC reflected the central core of the Murray Waters Agreement, about quantities of water; within the established developmentalist framework of Australian politics, it sought to create a stable framework for resource exploitation. It could be seen as the instrument of an immature resource régime, still faced with abundant scope for exploitation. In terms of the scheme developed by the US political scientist Theodore Lowi, the political and administrative framework was distributionist: a 'frontier' régime concerned primarily with the location and exploitation of resources. The RMC was essentially an agent of distributive politics, applying engineering solutions to a problem which was itself defined in distributive terms.

But the Agreement did not provide a forum for political resolution of the interstate conflicts; effectively, it put them on hold. New South Wales and Victoria continued to emphasise irrigation, South Australia navigation and — later — water quality. The Commonwealth, initially supportive of irrigation for its claims of extensive agricultural development, became increasingly disillusioned with it. Little political or administrative change was possible, not least for fear of upsetting the fragile consensus. South Australia did succeed in establishing a claim to additional waters diverted by the Snowy Mountains scheme, and later played an important role by insisting on the Chowilla dam scheme. The supersession of Chowilla by the far superior Dartmouth storage in Victoria involved a guarantee to South Australia of a share in the impounded water as compensation for abandoning Chowilla; both this move and the establishing of a claim to Snowy waters were helped by the RMC's mediatory role.

But, in general, reliance on technical means to resolve the water problem simply ossified it, by setting the whole issue in concrete. The result was to be 70 years of 'institutional gridlock', during which quality issues — especially salination — became increasingly urgent.

Nonetheless, Kellow concludes that 'the Murray Waters Agreement proved to be a reasonably successful and extremely resilient instrument for governing the Murray'.[5] When pollution and salination problems became serious, the RMC drew attention to them and to its limited powers to deal with them.

Consequently, by the early 1970s there was widespread awareness of the problems besetting the basin. But it was still necessary to create an atmosphere of cooperation, and to find political, institutional, and administrative arrangements which made it possible to address these problems. The first attempt failed: the 1981 Agreement made few changes and failed to tackle the most pressing problems, especially salinity.

It was not until the mid-1980s that effective reform, at a startlingly swift pace, was made possible by a series of circumstances which were quite fortuitous. They included the presence of Labor governments in the four jurisdictions critical to agreement; but they also included the existence of policy communities capable of

Chapter 6

pulling together, and policy networks in which particular individuals — and their interactions — were of critical importance. The effect was not simply a redefinition of the problem, but the laying of foundations for future action.

The important changes included a switch from distribution to regulation as the major focus of administration, and with it a recognition of the importance of environmental factors in river management.

By contrast with the RMC, the intent of the MBDMC is regulatory. As Kellow points out, little further irrigation development is possible or likely in the basin; most existing dam sites are utilised, and irrigable areas are in short supply. Under a more realistic economic régime, some contraction of irrigated area may occur. Thus the resource régime is now 'mature': there are no new frontiers. The primary need now is not to meet an ever-growing demand for water, but to promote its more efficient use, prevent wastage, and ensure water quality. These functions are primarily regulatory, despite an enduring distributive component, and thus require different institutions.

Further Exploration

The ins and outs of the legal and political wranglings over the basin are chronicled by Clark in vastly greater detail than can be followed here; interested readers should dip into his articles, if only to gain an appreciation of byzantine intricacies of law and politics in what was still a very small nation. The constant shifts in attitude of the States, sometimes small and sometimes large, are particularly fascinating. Attempting to reach agreement was akin to building a house on sand. Powell's *Watering the Garden State* deals usefully with the early years of the RMC and with the Snowy and Chowilla schemes.

From the 1970s on, the above account necessarily relies heavily on Kellow's *Saline Solutions*, though the emphasis is slightly different. Kellow also has far more detail of the political manoeuvrings, and is essential reading for an understanding of the developments of the 1980s. The Victorian Government's *Salt Action* is also well worth reading, both for an excellent and detailed account of Victoria's salination problems and for its explanation of the measures put in place to deal with the problem. Kellow's chapter 'Environment, Federalism and Development,' expounds the view that state governments can get 'closer' to problems than the Commonwealth. The successive *Annual Reports* of the MDBC are a useful source of up-to-date information, as well as documenting changing emphases in research and implementation. They are full of useful tabular data.

Lowi's distinction between regulatory and distributive policies is to be found in his review article on American business, as well as later works.

CHAPTER 7

TAKING STOCK

The creation of the Murray–Darling Basin Ministerial Council was an important political breakthrough. The Council itself is a novel response to the Murray–Darling Basin management problem, since no permanent political institution had previously been created. Its programmes are the first attempt to tackle problems that have been festering for decades. But these successes, important though they are, still leave a lot of ground to be covered. And some problems continue to worsen.

Failure to deal effectively with technical or ecological problems may result either from the nature of the problems themselves, from the way in which government addresses such problems, or a mixture of both.

Crabb's Criticisms
Dr Peter Crabb, of the Centre for Resource and Environmental Studies at the Australian National University in Canberra, has put forward several criticisms of the 1987 Murray–Darling Basin Agreement, the framework for the new legislation and institutional arrangements. Crabb feels that cooperation and consensus are the correct approach to the management of 'interstate river basins', and that the Agreement was the 'most that could have

been achieved'. But he wonders whether it is adequate to its task, or whether it is just a sop to the environmental lobby. He assesses the Agreement in terms of six basic principles:

1. *The river basin principle:* there is now greater recognition of the need to manage the basin as a whole, but the Agreement is probably still inadequate to cope with the border rivers.

2. *Limited sovereignty:* the States have made some concessions, but there is still doubt that the Council and Commission could act effectively in an emergency, and whether effective quality standards could be set and enforced. States would be very unlikely to further relax their sovereignty.

3. *Equitable participation:* there have been significant improvements, but equity remains highly dependent on better knowledge, trust, goodwill, and cooperation.

4. *Equitable apportionment:* there has been no change to the water-sharing arrangements. Unresolved conflicts exist on the Queensland-New South Wales border rivers. Only Victoria and South Australia have arrangements for sharing groundwater.

5. *Protection and non-abuse:* environmental protection and the non-abuse of the natural environment remain the major unresolved problem, unrecognised by the practical arrangements.

6. *Interrelationships* between water users, between States, and between regions are inadequately recognised both in the new mechanisms and in the non-participation of Queensland.

In short, Crabb's view is that there is still considerable scope for States to squabble over riverine resources, and for some to sabotage sound river management principles either by refusing to cooperate or by overexploiting resources in order to 'beat' other States to them. The 1987 Agreement is not a measured solution to identified *ecological* problems, but — as so often happens — a response to *political pressure* from farmers, scientists, environmentalists, and others. Crabb fears that it attempts to head off the problem by making the minimum amount of adjustment to existing arrangements, and that this will mean that the new Commission has inadequate powers to carry out its tasks, and the new Council — the decision-making body — will be unable to reach agreement on policies.

Some of Crabb's criticisms can be granted immediately. It is true that the new Agreement falls short of ensuring complete pro-

CHAPTER 7

tection to the basin from an ecological point of view, and questions of equity are, at the least, incompletely addressed.

A difficulty is that the pressures arising both from the strategic positions of the 'players' and from the structure of the federal system make some issues exceptionally difficult to address. As noted above, an equitable outcome to the conflict between New South Wales (and, to some extent, Victoria) and South Australia is hard, if not impossible to achieve, due to the asymmetry of their positions. The best that could normally be hoped for is a stable, but unjust solution in which everybody gains something from enhanced cooperation. Similarly, Crabb's view implies a need for greater centralisation of the basin's administration, difficult to achieve in a federal system.

But both Kellow and Paterson view centralisation as less important than effective coordination, from the point of view of successful management. In fact, Paterson argues against administrative overcentralisation. He feels that the extinguishing of common-law riparian rights and state control of water supply led to an economically irrational situation in which only the vagaries of government policy determined such questions as water pricing and entitlements. The resulting water policy was both confused and ineffective: it failed to identify economic costs and benefits clearly and to target subsidies effectively. Instead, all irrigators benefited from unrealistically low prices subsidised by the taxpayer, regardless of need. Paterson accuses Australians of confusing prices and subsidies, and in consequence losing sight of the proper purpose of subsidies. On this view, Crabb's first principle is disastrously wrong: Paterson's argument is that the policy deficiencies of Australian irrigation were a consequence of overcentralisation and the politicisation of water pricing. Correcting them implies greater decentralisation and more clearly defined rights to resources in juridical terms.

Crabb is probably right that the performance of the MDBMC and Commission in an emergency is likely to be inadequate, in particular due to a fairly cumbersome decision-making process, and the continuing limitations on the powers of the Commission. But this may be inevitable: Clark, in discussing the 1981 Agreement, makes the point that intergovernmental compacts are legally dubious and often vague as well. What makes them work is political will, and in particular an ability to see the problem through the same lens. In this regard, the work of the Commission and (earlier) of the Victorian Parliamentary Joint Committee may be of far more significance than any draconian concentration of powers. Emergencies may best be met by a willingness on the part of participant governments to collaborate: even quite strong institutions could collapse if one State became recalcitrant.

Crabb's subsequent points are in some degree open to the same objection: if there is a will to cooperate, these problems will be solved. If not, formal agreement can do little. In fact, on the environmental protection issue, events have overtaken the criticism: action by the Commission on both nutrient management and natural resource management shows that there is a greatly changed policy environment for the basin.

TECHNICAL AND ECOLOGICAL PROBLEMS

One common characteristic of the technical and ecological problems is that frequently — though not always — there are solutions to them, sometimes very well known. Quite often, the difficulties impeding their adoption are political or administrative, rather than technical.

Blue-Green Algae

The Darling River, for example, was afflicted by serious outbreaks of blue-green algae in both 1991 and 1992. Blue-green algae are toxic to riverine species, livestock and humans. They cannot be removed from municipal water supplies, which means that 'clean' water has to be transported in from elsewhere, at great expense. But the causes of such outbreaks are reasonably well known: they are due to excessive levels of phosphates and nitrogen in the water. The major sources of these contaminants are agricultural fertilisers and municipal sewage. The problem can be reduced or eliminated by reducing the volume of agricultural fertilisers used (or at least entering runoff), and by tertiary treatment of municipal sewage to reduce phosphate and nitrogen content, rather than dumping direct into the river. The resulting sludge is suitable for re-use as fertiliser, which can reduce the amounts of artificial fertilisers applied.

At the end of 1992 the Federal Government announced funding for improved municipal sewage treatment in major Darling River towns. This will reduce the flow of contaminants, though whether it will be sufficient is unclear. But the politically more difficult option of attempting to change farmers' patterns of fertiliser use has not been taken up. If such sustained, large-scale efforts need to be made, they will be more difficult than offering grants to municipalities: they will involve regulation, not simply distribution of incentive funds.

The Murray–Darling Basin Commission had anticipated these

problems with its proposals for a Nutrient Management Strategy, which undoubtedly provided a useful framework for channelling Federal Government monies. Nonetheless, there is clearly a long way to go before the problems can be declared 'solved'.

Allocating Resources
A further problem is the balance between the resources allocated to the various programmes. In the past, the bulk of the money and effort has gone for engineering works or measures of some kind. This tendency has not entirely disappeared; the major focus of the Murray–Darling Basin Commission, like its predecessor, is still the provision of engineering works. In addition, it has few staff in comparison to the scale of the problem, and to secure additional resources, must negotiate with other government bodies for their cooperation and support.

For example, the Salinity and Drainage Strategy of August 1988 envisages lowering the level of saline aquifers by pumping water up into evaporation pans. Saline infiltration into groundwater would be reduced by tiling drainage channels. The plan's whole approach is based on engineering and budgetary considerations. The expected cost is $27,000,000; operating costs will be $1,700,000 per annum. This plan fails to tackle several important issues, especially those most concerned with riverine ecology, such as gums, bird breeding, and fish stocks. Issues such as dry land salination are also neglected, as is reafforestation.

The Murray–Darling Basin Natural Resources Management Strategy (NRMS) of 1989 sought to overcome some of these deficiencies, and to address criticisms of the engineering approach. There is provision for restoration of native species, but reafforestation was initially dismissed on the ground that information on its effectiveness is unavailable, and that it would be unlikely to have a significant salinity reduction effect for 'hundreds of years'. This view has now changed, with especial emphasis on the role of recharge areas.

The NRMS relies heavily on education and community action. Land users are encouraged to form special groups called 'Communities of Common Concern'. The Ministerial Council planned to spend $45.2 million on implementing the NRMS between 1989 and 1991. While this appears to be a significantly larger sum than for the Salinity and Drainage Strategy, it is spread much more thinly. The NRMS covers the whole basin, whereas the SDS applies to a relatively small area in the southern irrigation districts.

However, a drift away from purely engineering measures does seem to be developing. There is a renewed interest in mallee scrub, and in the reafforestation of the recharge zones. This shift in emphasis, while slow to emerge, is very significant.

Matching Administration to System Characteristics

One very important lesson from the Murray Valley salination problem is that solutions which work perfectly in one place will often give poor results elsewhere. This is true both for technical measures and for administrative arrangements.

Salt Action points to major variations in the groundwater flow systems, even within Victoria. In some areas, the groundwater system corresponds closely with the surface catchment; flow paths are short. By targeting the recharge zones, salinity control can be achieved in as little as five to ten years. Local action is sufficient in these cases, involving collaboration among relatively few individuals and groups, in a small area.

But 80% of the dryland area in Victoria has regional flow systems, in which the groundwater systems may extend to whole drainage basins. They cross regional boundaries.

> ...salinity that results from rising regional watertables can only be controlled by a regional reduction in watertable levels. Even if widespread action were taken now, results might take many generations to appear. In some cases control may not be feasible.[1]

If it is possible at all, salinity control in these much larger systems will require coordination over a whole region, involving a large number of individuals and groups, as well as a long-term commitment to solving the problem.

Thus technical and administrative requirements will differ greatly. In the one case, a relatively short programme designed to identify the correct technical measures and educate the local population in their implementation and maintenance would be sufficient. Local, largely autonomous organisation would be sufficient, perhaps with some material incentives, such as loans, grants, or tax concessions, to encourage participation. Since results can be expected fairly quickly, a limited life for any organisation can be contemplated.

By contrast, regional organisation may need to link many different areas, and it may be necessary to develop both strategies for the whole groundwater flow system and smaller tactical plans for

CHAPTER 7

each specific area. Since the plans will be useful only if they are dovetailed to the major strategy, large-scale coordination over a long term is necessarily implied.

Working 'with' the Ecosystem
Thus the existing programmes, though a huge step forward, are fragmentary and fail to address some important problems. 'Engineering' approaches in particular may fail to harness the self-regenerative capabilities of the riverine ecosystems themselves.

The underlying processes causing salination make it very difficult to do much about overpressures from engineering structures; but the other processes are amenable to human intervention. *Reduction* of infiltration is cheaper than attempting to dispose of raised watertables by pumping, evaporation basins, or tiled drains. Methods include much more accurate determination of water needs, laser levelling of land to eliminate 'pooling' of irrigation water, and application via pipes or other protected means rather than open channels: the philosophy is to substitute accurate 'drip' irrigation for 'flood' methods. By reducing the need for drainage, and hence infiltration, the problem is solved more cheaply.

To the extent that it springs from an understanding of the processes which cause the problem, restoration of native flora may very effectively get at the causes, rather than trying to treat the symptoms, as engineering measures do. It can increase transpiration and lower watertables, especially in recharge zones, where it can also reduce infiltration. The strong emphasis on restoration in *Salt Action* contrasts strikingly with its dismissal in the MDBMC's *Draft Salinity and Drainage Strategy*. While replanting native species is not without cost, it can be done by individual landholders, is highly flexible, unlike major engineering works, and need not be a huge burden on taxpayers. Native flora are also self-regulating: they do not require expensive, ongoing human intervention.

A further option canvassed by *Salt Action* is 'saline agriculture', which uses salt-tolerant crops, involving 'fine-tuning' the productive system so as to work close to the limits of salinity tolerance. Deep-rooted and perennial crops and pastures are also an option.

And some at least of the problems with native species could be eliminated simply by not regulating the flow of the river excessively. However, it is important that local ecological variations are understood and taken into account; such factors as lateral permeability of soil and local groundwater flows can drastically modify the effectiveness of measures such as tree-planting.

In some areas, therefore, low-cost, ecologically sensitive measures can show rapid improvements; in others, there may be no alternatives to 'land retirement' — that is, abandoning the land for agricultural purposes — or to expensive engineering measures if there are compelling reasons for keeping the land in production. Those reasons are as often political as ecological.

Thus planting of native flora, and especially restoration of deep-rooted species like the mallee, may take a long time, but are essential if even a partial return to 'automatic' regulation of the watertable is to be achieved. In turn, this may require (a) land retirement so as to permit replanting; and (b) on-farm planting programmes. In some areas, the latter are under way; some sources claim that a 10–15% cover of suitable species may lower the watertable sufficiently to substantially restore soil fertility. This may mean that less land is available for irrigation: but since irrigation at present levels is not sustainable in the long term, and since marginal production is currently subsidised, the costs of restoration to the community at large are actually negative. That is, the community as a whole would probably be better off removing some irrigated land from production and reducing the overall irrigated area.

Similarly, imitating 'natural' conditions by 'flushing' the system — periodically releasing enough water to simulate flooding — is beneficial, not only to river salinity levels, but also to native species. This may reduce the amount of water available for irrigation; but might encourage the more efficient use of what remains. State adoption of more realistic water pricing policies would probably contribute.

LIMITATIONS TO REFORM

The effective reform of the administration of the Murray–Darling basin in the 1980s suggests that there is scope for administrative change to deal with such problems, but that it needs to be backed with political determination: 'where there's a will there's a way', in the words of the proverb. But it also demonstrates that there are strict limitations to the effectiveness of administrative reform and even of strong government support in resolving environmental problems.

Budgets and Engineers

The primary tool of government in approaching a problem such as land degradation is to 'allocate resources' to deal with the problem. The primary *mechanism* for this is budgeting: the allocation of money, which is then used either by existing government agencies — departments, bureaux, extension services, etc. — or by organisations specially created for the task, to implement the relevant policies. By its nature, such an approach tends to favour, firstly, techniques which can be costed and for which money can be allocated; and secondly, techniques which are amenable to administration.

Furthermore, there are advantages, both for speed and coordination, in centralising both policy and its administration. Voluntarism — especially where it involves reliance on dozens of uncoordinated groups — can often lead to delays, failures, and incomplete implementation of policy. The early history of irrigation in Victoria was an example of just this sort of problem. Voluntary action almost always involves some sacrifice — if only of time — by those involved. LandCare, a very popular programme with farmers and others in Australia, has gained government financial support, and offers a series of effective strategies for improving productivity and at the same time conserving native flora and fauna. But, although it promises long-term prosperity, it frequently involves short-term sacrifices, in particular the giving up of a percentage of the land. While it has made striking progress in well-off areas such as the Western District of Victoria, LandCare has been markedly less successful in poorer, marginal, and drought-stricken areas, where the immediate sacrifices are difficult or impossible for hard-pressed farmers.

Government's ability to require compliance, and to implement policies even against opposition, is therefore a compelling reason for keeping policy implementation in the hands of government bureaucracy. The difficulties tend to arise when policy is implemented: unpopular measures may be resisted, reinterpreted, and even deliberately subverted. Thus the intentions of the policy-makers may be frustrated by the very mechanisms they choose to implement them.

The monetary approach also tends to favour the breaking up of policy measures into discrete 'projects', which can be costed as a complete 'bundle' and which lend themselves to the processes of annual budgeting and accounting for the funds expended. This trend also favours 'engineering' rather than 'ecological' solutions, since they are less open-ended and much easier to cost: they are more 'visible' administratively. The long record of cost overruns and under-estimation, let alone the ignoring of the real social and environmental costs, seems to be no deterrent.

Comparison between the State Rivers and Water Supply Commission's 1975 plans for salinity control in northern Victoria, and those embodied in *Salt Action*, issued by its successor body in 1987, brings out the point very clearly. The former, single-purpose body's plan is solely concerned with engineering measures, primarily drainage and evaporation. Its only reference to environmental issues is to check which areas are good bird habitats in order to plan saline groundwater disposal at other locations. The latter, a part of the broader post-reform administrative structure, places markedly more emphasis on the prevention of groundwater accessions, 'saline agriculture', replanting native vegetation, and so on.

The bias toward visible, 'budgetable' projects is part and parcel of the traditional developmental policy trend in Australian public policy. This trend is very deeply embedded, and forms a major barrier to rational management of Australia's resources.

Economics

'Projectism', with its lean toward engineering, has some very important economic implications. Irrigation has been largely the province of the engineers; government failure to apply economic criteria during the period 1915–66 led to continuing, substantial subsidy. While subsidies — for example to establish 'infant' industry — can be justified in special cases, there was little attempt to do so for irrigation.

Australian irrigation has never been able to pay its own way from farm gate prices: those paid to the farmer for produce. For example, Victoria charges $10 per megalitre for water; an economic charge would be closer to $25. In consequence, water is very inefficiently used. Economists such as Davidson and Paterson have argued that one very simple way to force more efficient use of water (and hence lower demand) would be to raise charges to an economic level. This would penalise the least efficient farms, but by lowering the demand for water, it would alleviate salination problems, and could help to reduce competition for water and the consequent tensions.

The other question, very much unfinished business, is reinvestment. Much of the irrigation infrastructure in the Murray–Darling Basin will need replacement in the period 1990–2010. If the economics of irrigation continue poor, government will have to consider whether to reinvest in maintaining existing irrigated acreage, or whether to significantly scale down the irrigation effort to include only schemes which are viable in *present-day* terms.

CHAPTER 7

Questions such as these are difficult for politicians to confront, especially as any decision they take is likely to be unpopular. But economic decisions are not the only ones creating serious political strains.

Political Minefields

There are numerous areas of policy which are 'no go' for politicians: in other words, because of political opposition or the fear of electoral backlash, they are unwilling even to consider the issues in question.

Kellow draws attention to the non-controversial, non-conflictual nature of the recommendations of the Victorian Parliamentary Committee. There were no recommendations, for example, for controls on clearing of land, mainly because many farmers were still anxious to clear new land and bring it under cultivation. Yet such controls were already in place in South Australia, and were working fairly well. The Committee contented itself with pious hopes that farmers themselves would eventually come to see the need.

A further example is 'land retirement', a euphemism for the temporary or (more probably) permanent removal from production of areas of land, typically those severely affected by salination. Land retirement is unpopular with farmers, who see it as an attack on their livelihood. They fear that compulsory retirement may fail to compensate them adequately for lost production, and are hence strongly opposed. Despite these attitudes, much land has already been retired simply because it is no longer productive; frequently the farmer has gone bankrupt, and has abandoned it. The resultant natural revegetation is not always where it will do most good. Yet the open discussion of land retirement as a political option is taboo, even though it is regarded as essential by many responsible commentators. Such political paralysis, while rational for the politicians concerned, can often result in a worsening of the problem at issue.

Similarly, political and administrative boundaries often arbitrarily cut across natural ecological regions. Under these conditions, coordinating management policies can be difficult and time-consuming. The boundaries themselves create political and legal problems, some of them very difficult to resolve. For example, Queensland's refusal to join in negotiations or participate in the new Council and Commission meant that numerous problems had to wait until that State's government could be induced to recognise them.

OVERVIEW: ENVIRONMENTAL GAINS

The new Murray–Darling basin agreement has been criticised on the grounds that it still fails to bring the whole basin under a single management régime, and still lacks enforcement powers. These criticisms are themselves open to criticism, both on grounds of impracticability, and on the ground that the central need — coordination — has been confused with the issue of control.

The reorganisation of the Murray–Darling basin's administration which took place during the late 1980s was clearly a success in political terms. For the first time, serious attempts were made to address a suite of problems which were — literally — 100 years old. It paid good dividends for the governments and politicians involved, because it was a solution which was widely acceptable, and which promised improved benefits to farmers and others in the region, at the same time as offering efficiencies which would please voters, taxpayers, and administrators. The newly cooperative atmosphere offered renewed hope for the future.

The recognition of a need for comprehensive management was a gain for the natural environment. In the first place, it led to more regular, quantitative assessment of salinity levels and other problem variables. Secondly, it made more explicit the various factors affecting riverine flora and fauna. It also highlighted the benefits of sensitive exploitation of resources, and lent support to programmes designed to encourage better water management, replanting of native vegetation, and so on. However, it did not completely break away from traditional rigidities in the conceptualisation and implementation of major projects. Furthermore, some problems — such as 'land retirement' — were ignored, mainly because they were politically controversial.

Whether these changes will turn out to be worthwhile in ecological terms remains unclear. In terms of the major problem — salinity — it will probably be necessary to wait some years before the trend in groundwater levels becomes clear, and a picture of the effectiveness of the various measures taken emerges. The same is probably true for other issues, such as flora conservation, faunal species management, and the numerous other goals included in the Murray–Darling Basin Commission's three major strategies.

The question which then remains is whether successful compromise and successful institutional innovation are enough. In particular, do they adequately address the environmental problems? The litmus test of the effectiveness of what was undoubtedly a dramatic set of political and administrative changes must be favourable ecological change.

Further Exploration

Crabb's paper, as indicated, should be read in conjunction with Kellow (*Saline Solutions*) and Paterson (in *Australian Quarterly*). The economics of irrigation are becoming more important as the question of replacing existing capital investment looms. Davidson's 'Irrigation Economics' chapter remains one of the most succinct expositions of his position, by now far less controversial than when first enunciated in the 1960s; for a more comprehensive understanding of his position, *Australia Wet or Dry* remains indispensable. *Salt Action* and recent reports of the MDBMC are essential for an understanding, both of the continuing problems and of current moves to deal with them.

Part 3
A Dash of Theory

Many of the problems reviewed in the preceding chapters arose from a lack of cooperation or coordination. They also reflected self-interested behaviour by the parties.

Scarcity frequently begets competition, which is often wasteful and destructive. In turn, a very large literature, dating back many centuries, has attempted to prescribe forms of organisation, suggest patterns of behaviour, and tackle the problem of conflict resolution. Only in modern times, with an improved understanding of the nature of collective goods and of the interactions underlying them, has real progress been made in resolving some of the traditional puzzles of political theory. In particular, the underlying motivations for human cooperation are slowly being elucidated.

Collective goods, central to environmental issues, are nowadays defined in terms of joint production, joint consumption, or both. Often they are lumpy, displaying threshold effects. They are also fuzzy and difficult to classify.

Three major bodies of theory deal with the resulting problems.

Game theory deals with the logic of interaction. Social choice theory attends to the problems that arise in 'counting heads'. And decision theory attempts to explain what happens when decisions are made or avoided.

CHAPTER 8

SCARCITY, COMPETITION, AND COLLECTIVE GOODS

Though gradual, the change of resource régime in the Murray–Darling basin from a broadly distributive to a much more regulatory basis was nonetheless striking. It was essentially a response to increasing scarcity of a number of key variables. Good land was becoming scarce, as a consequence of the threats to agriculture posed by land degradation and salination. Similarly, water quality was falling, as the number and variety of end-uses increased, and the size of the populations dependent on the water supply grew.

Scarcity-driven changes of this kind are well documented: resources were first squandered, then fully utilised, and finally conserved, in the process of colonising and 'developing' the USA. The 'squandering' stage resulted in dramatic waste and major extinctions. Nearly half the timber cut in Michigan was wasted. The passenger pigeon, once so numerous that flocks were millions strong, was extinct by 1917; the bison was also nearly extirpated. Excessive clearing and unsuitable land use resulted in early soil exhaustion, deforestation, and numerous other problems. Waves of 'development' surged successively westward, opening up 'frontiers' and stimulating massive population growth. A subsequent period of consolidation saw adoption of more intensive but less wasteful forms of exploitation. The final phase, not yet fully developed, is 'conservationist' from sheer need.

In Australia, because the lack of good land in the interior made a steadily expanding frontier an impossibility, from a very early stage government-sponsored and financed 'development' became the norm instead. But abundant resources were nonetheless squandered in the early years of European impact. The seal and whaling industries, the cedar-getters — selectively cutting coastal forests far north into what was to be Queensland long before official settlement — miners defoliating whole regions for scarce fuel, the ringbarking farmers and graziers, all took their toll. Political appreciation of the extent of the damage was long in materialising.

Scarcity and Competition
Scarce resources are by definition critical. Economists expect resource use to be optimised with respect to the scarcest resources: B.R. Davidson's observations on irrigation show how, if judgements about scarcity are mistaken, misallocation of resources will occur. In the case of irrigation, water was assumed to be the scarcest resource, when in fact labour and capital were scarcer. Politically, it was the scarcity of good land that made irrigation attractive: a case of different rationalities pulling in different directions.

Scarcity leads both to competition and to a need for political decision-making. Competition arises as soon as a resource is no longer so abundant that someone exploiting it can leave 'enough and as good' for the next comer. As the resource grows scarcer, competition becomes more intense. For example, in sleepy country towns, shoppers can drive their cars right up to the front of the store, walk in, and do their shopping. In larger towns parking space is rationed, either by others using it, or by rules limiting the duration of each use. In very large towns, parking is often completely prohibited, while some European cities with very narrow streets forbid vehicular access altogether.

Economic 'rationalists' make a lot of fuss about the alleged advantages of competition; unfortunately, most of them can be realised only by price competition in 'well understood' technologies, where inputs (raw material and labour) are infinitely abundant. Where there is only a single firm in the market (monopoly) or a very small number (oligopoly), direct or indirect complicity among sellers is normal. The sellers then become 'price-makers', imposing their own terms of trade. Often, monopoly firms retain control of their technology, and once again can dictate price.

Where critical raw materials are scarce, suppliers may 'corner' the market and drive up prices. Furthermore, most economists assume that competition will be gentlemanly: it is OK to ruin competitors by smart market moves, but to drive them out of business by the use of machine-guns is 'gangster tactics'. Economists implicitly rely on a framework of social restraints which their theories do not take into account.

Most important from the point of view of ecological rationality, competition may encourage irresponsible depletion of resources which will later be scarce. Competition can be wasteful and even destructive. One of the best-known examples is the 'game' of Chicken. In its archetypal form, two hoodlums driving junk cars hurtle towards each other on a narrow road, and the first to swerve is 'chicken'. If neither swerves, the cars collide and lives are lost. Users of environmental resources sometimes attempt to 'beat the competion' to use the resources, destroying them in the process. For example, some owners of tracts of rainforest in north Queensland, fearing expropriation if the government was to declare a national park or attempt to control clearing, indulged in 'shock clearance' of their land, to ensure that it was not fit for National Park or World Heritage status. It is known that some rare and valuable species were thereby destroyed; what is worse is that the landowners gained little or no benefit from destroying their own property. Similarly, the small independent irrigation schemes of the first wave in Victoria competed among each other for resources, especially water, and built unplanned, ill-coordinated, wasteful engineering works.

Victoria suppressed competition by imposing more centralised control, and insisting on the planning and coordination that the small settlements had been unable or unwilling to provide. In the process, all benefited, even those who had previously refused to cooperate, since more resources were made available. Similarly, the engineers' report of 1913 that led to the Murray Waters Agreement had pointed out that irrigation and navigation could co-exist, if only planned use were made of the water resources.

Importantly, creation of a collective good frequently increases the well-being of the participants in ways which could be achieved by no other method. Consequently, it is often better to make use of resources cooperatively or collaboratively rather than to take a 'dog-in-the-manger' attitude. It is no accident that human social and technological development was built on cooperation and specialisation, not some Social Darwinist form of tooth-and-nail competition.

But many situations arise in which competition for resources becomes the dominant fact determining interactions among the participants. Nations, for example, often compete for resources in the shape of land, minerals, or population; frequently competition takes the form of war.

The analysis of competitive situations is therefore important: it may suggest ways of inducing cooperative behaviour and even of converting conflictual situations into more productive ones. This requires the provision of common goods.

Environmental goods are nearly always collective goods. More often than not, they must be jointly produced, jointly consumed, or both. Such goods cannot be simply parcelled out, and they require agreement among those producing or consuming them if they are to be maintained.

COLLECTIVE GOODS

Common goods are also called **collective goods**. They cannot be provided by individuals working alone. They require *cooperation*. The environment — or more accurately, ecological stability — is a collective good. In fact, it is what philosophers call a **primary good**, without which other goods are very little use. This is important because primary goods must obviously be given priority in making social policies; this may give rise to conflicts with other goals.

Jointness

Collective goods have **jointness** of production and consumption. *Jointness of production* means that goods have to be provided by collective action; they cannot be provided by individuals acting alone. For example, to move a piano, at least two people are needed: one person alone cannot lift it or shift it. Therefore the good of movement, in this case, has jointness of *production*. But it does not necessarily have jointness of consumption: once moved, the piano may benefit only one person — moving men are rarely pianists, and even amateur pianists are rarely movers — and consequently may be consumed individually (i.e. used by only one person, in this case). The crucial point in joint production is *cooperation:* if one of the moving men is sick or on strike, the good cannot be produced. Collective, common, or **public** goods (not all collective goods are public) all require cooperation, and consequently human institutions for cooperation and coordination become very important. Once provided for one person, jointly produced goods can often be extended at relatively low cost to others.

Jointness of consumption means that it is not feasible to reserve a good for one or a group of persons. The good is **indivisible**, or **non-excludable**, or both. Divisibility means the ability to parcel up a good and sell it off or otherwise distribute it. For example, a tray of mangoes or a dozen beers are divisible: they can be split up and distributed among willing buyers or hungry and thirsty consumers. In 'market' economies, virtually the whole commercial mass media culture is devoted to divisible goods, obscuring the lack of real choice offered by the mass production system, and entirely neglecting the fundamental importance of indivisible collective goods.

Clean air, for example, is indivisible: everyone must breathe the same air, and someone polluting it (an industrialist or a smoker, for example) inevitably forces the pollution on their victims. Public roads are a non-excludable good: though it would be technically possible to make everyone pay a toll for each length of road used, it is not really feasible, except for some major arteries on which access can be controlled. It is far cheaper to tax everyone and allow road use free of charge, even if some inequities arise. A radio signal has both indivisibility and non-excludability: if provided for one, it can be provided for all at no further cost; and noone with a suitable radio can be excluded from reception. A pre-technological analogue with even freer access is the town crier, whose job was to 'cry' news and proclamations in the streets. A sad remnant, calling the hours of the night, survives in the *Wizard of Id* comic strip. An *indivisible* but excludable good might be a country club, where all members use the facilities in common, but exclude the general public; a *divisible* but non-excludable good would be a rail service where anyone can buy a ride, but seats are sold piecemeal.

Some goods, though they can be provided privately, can be more economically or more effectively treated as public: in health insurance, for example, universal schemes can save accounting costs as well as ensure that everybody gets at least the same minimum standard of care. Other goods, such as law and order, or clean air, can only be provided publicly.

Fuzziness and Lumpiness

The boundaries between types of goods are often fuzzy. Some goods require more jointness in production than others: moving a piano is more 'private' than building a pyramid, since the one requires two or three workers, whereas the other calls for thousands. The enormous amount of cooperation and coordination required to provide global postal or banking systems, or manufacture silicon chips, means that their production is 'joint' to some degree, even if it is 'private' rather than public, and consumption is strictly individual. Attempts can be made to privatise even the most public of goods: for example, industrialists might use their wealth to live on mountain tops, above the polluted air with which their workers have to put up. (Perhaps it is poetic justice that mountain-dwellers are more susceptible to the thinning ozone layer, thus redressing the balance in favour of jointness!)

Some goods are also 'lumpy', displaying thresholds which have ecological, physical, or technical causes. Floral and faunal species, for example, cannot survive unless they have certain minimum populations; below the minimum level (which can differ widely between species) they die out. If survival of a particular species is a common good, the good is 'lumpy' because each population must be at least the minimum size. Contemporary examples include whales and, more locally, a number of species including the helmeted honeyeater and Leadbeater's possum in Victoria, both acutely endangered, primarily by loss of habitat. Some minerals do not occur at continuously varying densities in the earth's crust: lead ores, for example, are absent at concentrations below about 60% lead, meaning that lead cannot be worked more expensively, at ever-decreasing concentrations, once the best deposits give out. Once those are gone, there is no more lead. Technology can be lumpy, too. Providing a rail service involves building tracks and then running trains. Each track might cost (say) $1,000,000 per kilometre, and have a maximum capacity of 45,000 passengers per hour in 40 trains. If each train also costs $1,000,000, then the first passenger's first kilometre costs the railway $2,000,000, and the fares would be correspondingly high. But each subsequent passenger costs less, so that the *marginal cost* of adding each extra passenger is actually negative — until the capacity of the first train is reached. The 1126th passenger will cost an extra $1,000,000, and the 45,001st will cost $2,000,000.

Lumpiness has two important effects. One is that sometimes a good can be provided in more than one way, and it may pay to do so (the railway, for example, might pay for a taxi or provide a bus for the 1226th and 45,001st and some subsequent passengers, until the traffic justified further investment — or they might just pack 'em in!). The other is that the 'lumps' are often important ecological thresholds with policy significance.

Lumpiness and fuzziness mean that different political systems often have characteristic methods of supplying public goods: the USA, for example, leans to private provision through franchised monopolies, whereas Britain and Australia have tended to use public corporations; in Eastern Europe and the Soviet Union, even divisible goods were until recently publicly provided.

Ecosystem stability, clean air and water, pollution control, preservation of threatened species, and so on, are all problems requiring collective management rather than individual, uncoordinated activity. Furthermore, with problems such as the supply of raw materials requiring longer and longer lead times, planning — by large companies and governments — becomes essential. And with more and more complex technology, the role of research and development in discovering and making technologies available also becomes greater; thus research and development also has to be planned. This need to plan ahead and to manage the human impact on the environment is one of the greatest challenges humanity faces today, and one for which it is ill-equipped. It, too, is a collective good, more and more so in an increasingly interdependent world.

Market Failure and 'Solution by Government'

The **collective action problem** in political economy parallels *market failure* in economics. Markets are often **imperfect**. Even when perfectly competitive, they may fail to allocate resources equitably or efficiently. Scarcities, misallocations, and **externalities** then arise. The 'free market' makes isolated decisions in response to specific stimuli. It cannot plan: it responds only to changes in supply or demand. If some resource suddenly runs out, the market will react only at the point of crisis. It cannot cope with 'stock' pollutants and 'stock' resources; it also handles the threshold effects associated with 'lumpy' goods badly. And markets have no way to value collective goods, because they cannot be parcelled up and sold. This leads to **suboptimal** levels of provision; particularly of public goods.

Some kinds of public goods, such as agreement to act in a particular way, are not subject to market failure because they are never traded in markets: they are purely political. Some of these are very fundamental: one example is **civility**, the propensity to abide by and uphold the law, and to extend courtesy and consideration to other members of society. Civility is one of the primary goods without which market economics is impossible, yet many economic 'rationalists' behave as if they have never heard of it.

CHAPTER 8

Goods such as civility can break down if too many people behave in an uncivil fashion: if the law is broken too often, for example, people may have to break it in self-defence, perhaps by arming themselves, shooting at suspected assailants, and in extreme cases by application of lynching and other sanctions to suspected, but unconvicted, criminals.

Government then has to step in. Political leaders often function like entrepreneurs, 'selling' common goods to the public, and building up **coalitions** of support for the necessary measures. However, their efforts are frequently suboptimal, because the cost of the political effort has to be met from the available resources. In addition, resources are often not equitably distributed, and there are inherent problems in attempting to make social decisions. Rational decisions by *individuals* do not necessarily add up to a rational outcome for society as a whole: a major problem to which considerable time and thought must be devoted.

Common-Pool Resources

It is not always necessary to rely on government. Some collective goods are managed as **common-pool resources** (CPRs), by a group or community which owns the resource in common, with minimal or no government participation. In fact, not only have CPRs frequently proved successful where government has not; they have even been able to function when government has broken down. For example, during the long period of political turmoil and warlordism that racked China from the turn of the century up to 1949, peasant farmers were frequently able to keep up elaborate and extensive irrigation systems, arranging repairs and maintenance and allocating water by agreement among the villages involved.

A feature of CPRs is that they are *jointly* owned, not ownerless. Their boundaries are clearly defined. Access to them is available only to designated persons, typically under quite restrictive conditions. These conditions normally impose some obligations for the maintenance of the CPR, and they also may involve participation in a decision-making process. Examples cited by Ostrom include upland pastures in Switzerland and Japan which are collectively managed by the owning villages, and *huerta* and *zanjera* irrigation systems in Spain and the Philippines. Ostrom stresses that the arrangements in each case are appropriate to the resource concerned, and that an explanation for the durability of the institutions is to be found in their context-sensitivity. In practice, the arrangements are quite variable; in some cases, rights were not transferable, pertaining only to specific individuals. In others — especially with irrigation water — rights were freely transferable within the community; formal markets even arose. The crucial factor was the limitation of overall demand, and frequently the timing of resource availability.

In California, in a modern 'rational-legal' setting, competitive races to pump groundwater developed in more than one basin. This was a typical natural resource conflict situation, in that overpumping would result in a falling watertable, followed by irreversible contamination of the coastal aquifers by salt water from the Pacific Ocean, thus destroying the groundwater resource for all concerned. The problem generated some serious legal tangles involving riparian rights. Where successful solutions were achieved, it was by creating management bodies covering a single groundwater basin ('whole-basin management') which regulated withdrawals, primarily by allocating water rights.

Ostrom also examines a number of joint resources in which management has been unsuccessful or has broken down. A frequent problem is the interference of the central government in local arrangements, with a consequent loss of sensitivity to local needs. Another has been insensitive 'development,' in which expansion of access or introduction of new technology has occurred without consideration of the capability of the resource to support enhanced rates of exploitation. Consequences have included the immiseration of the original users and destruction of the resource, which is particularly easy with 'fugitive' resources such as a fishery.

OVERVIEW: THE CENTRALITY OF COLLECTIVE GOODS

Scarcity frequently results in wasteful and destructive competition, which is especially damaging to ecosystem resources. Many environmental goods are common or collective goods. These must be jointly produced or jointly consumed, or both. Frequently, cooperation or collaboration can produce goods where none previously existed, thus making all concerned better off. Collective goods are often lumpy or fuzzy. Lumpiness indicates the presence of thresholds which may be of ecological importance, as well as having policy implications. Fuzziness refers to the fact that very few goods are purely collective, leading to variations in the ways in which they are provided.

Government is frequently needed to ensure that collective goods are supplied, or to prevent the breakdown of arrangements

Chapter 8

for them. Because the collective goods problem is analogous to 'market failure' in economics, the role of government in regulation is also similar. Coalitions of support for political action depend on the actions of political entrepreneurs. As resources become scarcer, competition for them may grow more fierce, and the regulatory role of government becomes more important.

Collective goods are also found in the form of 'common-pool resources,' or CPRs. These are typically managed by the group owning them, using rules and institutions geared sensitively to the nature of the resources themselves. The institutions have often proved very durable, outlasting governments by many hundreds of years in some cases. Very frequently, insensitive government interference has been responsible for damage to or destruction of such CPRs.

Further Exploration

This chapter moves to a broader, non-institutional way of understanding politics, in which political institutions assume far less importance than patterns of interaction. For a readable and thoroughly competent exposition of the reasons for adopting such an approach, see Laver's *Invitation to Politics*, especially pp. 2–3 of the Introduction.

Chapters 1 and 2 of McLean's *Public Choice* are an excellent introduction to many of the ideas raised in this chapter; alternatively, Laver's *Social Choice and Public Policy*, especially Chapters 2 and 6. Ostrom's *Governing the Commons* is essential reading for anyone wanting to understand more about common-pool resources (CPRs) and their management.

CHAPTER 9

STRATEGIC GAMES

The alternatives to competition, coordination and collaboration, are neither fortuitous nor random. They fall into certain more or less clearly defined patterns, some far more common than others.

Often, the choices facing political actors are restricted by the **payoffs** for particular courses of actions; that is, the gains to be expected from them. Where the payoffs are fairly explicit, and the *structure* of the interaction is clear, these situations are called **games**. A specialised activity called **game theory** is devoted to their dynamics. Even elementary game theory can be very illuminating when considering political interactions.

THREE TWO-PERSON GAMES

The oldest, best known, and most pivotal game is the **'Prisoner's Dilemma'** (PD), which explains why it may be rational for an individual to behave in ways which, socially, are quite irrational and even harmful. PD games crop up with remarkable frequency in connection with environmental problems.

A and B, suspects in a major criminal offence, are caught. The

CHAPTER 9

police have evidence to convict them only for a trivial offence. They invite each criminal to confess, and be let off with a light charge, while the other faces a heavy one. The dilemma for each prisoner is the same: whether to confess, hoping that the other will not, or to remain silent. If both confess, the police will be able to charge both with the major offence, and both will get heavy sentences. If neither confesses, both will be charged with the minor offence, but each will probably spend longer in gaol than he would if he were to confess and the other did not. But if one confessed and the other did not, the one not confessing would probably end up with a heavier sentence than if both confessed. The payoff structure can be represented by the matrices shown below:

Prisoner B

	Not Confess	Confess
Prisoner B Not Confess	Each 1 year	A: 3 months B: 10 years
Prisoner B Conferss	A: 10 years B: 3 months	Each 8 years

Concrete Form

Player B

	Cooperate	Defect
Player B Cooperate	3 / 3	4 / 1
Player B Defect	1 / 4	2 / 2

Abstract Form

152

Once the first individual has **defected** (confessed), the second player's best strategy is to defect as well: that way he gets a bigger payoff, in this case a shorter sentence, or 2 units instead of 1. Thus the individually rational, or **dominant**, strategy for each individual is to defect (confess), hoping that the other does not. But the collectively rational strategy, giving the pair of criminals *taken together* the 'best' outcome, is to **cooperate** with each other by not confessing. The problem is that each criminal must be able to trust his accomplice. If there is 'no honour among thieves' and each fears that the other will confess, then there will be a race to confess first and win a lighter sentence. In that situation, each individual has destroyed the collective good by a self-defensive action.

'Prisoner's Dilemma' problems typify situations in which there are important social benefits to be gained by cooperation, but the pay-offs encourage individual defection. For example, public places look better and as a result everybody is better off if people dispose of their litter in bins or take it home with them; but that imposes a cost — finding a bin or arranging to take the rubbish home — which many people are unwilling (or too careless) to incur. The result is that the cost is transferred into the public domain, or 'socialised' (i.e., it becomes a social cost) in the shape of filthy public places and the expense to local councils of collecting and disposing of the garbage.

'Prisoner's Dilemmas' occur in many familiar contexts. In the original game, the police seek to exploit the mutual mistrust of the prisoners. The British, when they ruled India before 1947, used religious and social divisions to inhibit the growth of a united nationalist movement: the so-called 'divide and rule' policy. It had a terrible legacy: the Partition of India and Pakistan in 1947, politically necessary to accommodate religious tensions, led to the murder of millions and the forced transfer of even more. The biggest 'Prisoner's Dilemma' of all is the problem of political obligation: getting people to obey governments.

The environment bristles with 'Prisoner's Dilemmas'. In Hamburger's 'mutual pollution' game, two cities border a lake, and for each the cheapest way to dispose of their sewage is to dump it in the lake. But if either dumps its sewage, the lake becomes unfit for water supply, recreation, and other desirable activities. Given this payoff structure, the most individually rational solutions to their sewage disposal problems will make both cities worse off.

Closely-related is **Chicken.** Its defining characteristic is that, even if one player defects, the other is still better off to provide the

Chapter 9

good himself. In the classic prototype, in which two hoodlums speed towards each other in junk cars, the one who swerves and averts a collision is 'chicken', but *both* are alive. Had neither swerved, both would be dead. Thus the one who swerves pays the cost (in this case, opprobrium) but provides the good (in this case, life for both). The dominant strategy in Chicken is therefore to commit oneself to defection first, thus forcing the other player to provide the good. The winner in a Chicken game is always a 'free rider' on the loser.

In **Assurance** games the greatest rewards are for cooperation; it is therefore always in everybody's interest to cooperate.

Here are the matrices for Chicken and Assurance:

	COLUMN Player A Cooperate	COLUMN Player A Defect
ROW Player B Cooperate	3, 3	4, 1
ROW Player B Defect	1, 4	0, 0

Chicken

	COLUMN Player A Cooperate	COLUMN Player A Defect
ROW Player B Cooperate	4, 4	1, 2
ROW Player B Defect	2, 1	0, 0

Assurance

In Chicken, the payoff to the second player for defecting is always less than the payoff for cooperating (in PD, by contrast, it is greater). So, while in the PD game the second player must defect, in order to retrieve what little can be salvaged once the first has defected, in Chicken, by contrast, it is worth the second player's while to cooperate *even if* the first has defected. In Assurance, of course, no course of action pays as well as cooperating.

Transformations of Games
It is possible for changes in perceptions to transform mutual mistrust — expressed as strategies appropriate to PD or Chicken games — into trustful cooperation, expressed in a cooperative strategy appropriate to Assurance. Interestingly, this can also transform the payoffs, given suitable circumstances.

This happened in a very striking way with the Mediterranean Action Plan ('Med Plan') of 1975, a cooperative plan for the management of pollution and other problems affecting the Mediterranean. By the early 1970s, pollution was serious:

> The extensive pollution of the Mediterranean is the result of intense coastal population pressures, combined with largely unregulated industrial, municipal, and agricultural emission practices. Constructing sufficient sewage treatment facilities region-wide to handle the wastes generated by up to 200 million summertime tourists and residents would require a regional investment of $10 to $15 billion over a ten-year period. Eighty-five percent of the pollution of the sea comes from land-based sources: agricultural run-offs, industrial wastes, direct emissions from cities lacking sufficient sewerage facilities, and other wastes transmitted by rivers. Eighty to ninety percent of the coastal municipal sewage is discharged into the sea completely untreated. Thus, effective protection required the coordinated efforts of all the coastal states. Common pollution standards had to be adopted for pollutants from tankers, offshore dumping, and a variety of land-based sources. Contending uses of the sea also had to be balanced: for instance, fishermen and tourists require much cleaner waters than do tanker and industrial interests.[1]

Success required the cooperation of numerous states which were in political conflict over other issues: Greece and Turkey, for example, over Cyprus; Israel with Arab states such as Libya and, at one time, Egypt; other littoral states such as Tunisia,

CHAPTER 9

Libya, Algeria and Morocco with ex-colonial powers such as France or Spain. Algeria, suspicious of foreign control, announced that it would not sacrifice its development programme for the environment. Nonetheless, persistence in attempting to establish a system of mutual cooperation eventually overcame resistance. In 1976 protocols were negotiated covering dumping from ships and aircraft, together with an agreement on cooperation over oil spill emergencies. In 1980 a further protocol on control of pollution from land-based sources was agreed, and in 1982 specially protected areas were established. The last Mediterranean nation to join the Barcelona Convention was Albania, in 1985. Research and monitoring projects were well under way; important because of the pivotal role of scientific knowledge in establishing cooperation.

Assurance is interesting, therefore, not because of the peculiarities of the game in its own right, but because of its potential to transform other games, thus defusing tensions and improving the delivery of collective goods.

Which Game is It?
Some situations, depending on the payoffs, might be any one of these games. Take 'mutual pollution'. If the amount of sewage one city dumps into the lake is not enough by itself to be a health hazard, but that contributed by two is, then the matrix is more like Chicken than PD: the second city may have a positive incentive to refrain from dumping (i.e. to 'cooperate') regardless of the first's action, and each has an incentive to dump first so as to make the other refrain. If dumping by any one city is enough to create a health hazard, then the game is either PD or Assurance: in this case, the nature of the game will depend on the relative values assigned to sewage disposal and public health. If the cities are primarily concerned with economy, then they may dump and be damned. If, on the other hand, both value public health highly, they may both opt to treat their sewage. If one values public health more highly than the other, then each may perceive the game differently: one may see the game as PD or Assurance, the other as Chicken. Exactly this sort of 'ideological' difference has marked US-Canadian relations over the management of the Great Lakes, especially Erie or Superior, where the problem resembles Hamburger's 'Mutual Pollution' game more than at Niagara Falls. The matrices might look like this:

156

Strategic Games

Prisoners Dilemma

City A
	C	D
C (City B)	3, 3	4, 1
D (City B)	1, 4	2, 2

Chicken

City A
	C	D
C (City B)	3, 3	4, 2
D (City B)	2, 4	1, 1

Assurance

City A
	C	D
C (City B)	4, 4	2, 3
D (City B)	3, 2	1, 1

For this example, we might call the numbers millions of dollars, and let them stand for the net benefit to the city of each course of action. Thus in the PD case, once one city has fouled the lake, the only thing the other can do is join it; the lake is already a health hazard. In Chicken, there will be a race to foul the lake, because the first to dump will force the other to refrain (assuming that clean water is valued sufficiently highly). In the Assurance case, health is valued more highly by both, so the payoff for cooperating is greater. In real life, if sewage treatment were very expensive, but water purification cheap, the payoffs would tend towards PD; if the reverse were the case, towards Assurance.

Asymmetry

Symmetrical two-person games often fail to capture an asymmetric reality. For example, one reason the US–Canadian situation over Niagara is not strictly a symmetrical Chicken game is that, geographically, the Canadians, being downstream, cannot avoid getting more pollution from the US than the US can from them. The Canadians' inability to retaliate lends the game an asymmetry characteristic of riparian issues. Similarly, in the Murray–Darling system, the South Australians, with no legal redress or means of retaliation, had no option but to put up with the reduced river flows and enhanced salt levels created by New South Wales and Victoria.

Asymmetry often means that no solution exists which is both stable and *just*. It can mean that the situation persists for long periods because the 'winning' players have no incentive to change their posture, in part because the losers cannot retaliate effectively. Especially if there is no legal system or government to appeal to — internationally, or where the legal system is deficient, as in the Murray case — then the only way the worse-off participant can

Chapter 9

induce the 'winner' to desist may be by compensation, thus changing the payoffs. The notion of *compensation* is quite well developed in the economics literature: while highly unjust, it may still be the only *practicable* option.

To gain New South Wales' cooperation in the Salinity and Drainage Strategy of 1988, it was necessary to 'bribe' the State by demonstrating that benefits such as permission for saline discharges could flow from participation in a basin-wide management scheme. A more just solution would no doubt have demanded more of New South Wales and given more to South Australia; but this is to ignore the essential asymmetry of their relationship. What makes Paterson's work for the Murray–Darling Basin Commission in 1986 particularly interesting is that it changed the perceived payoffs — by offering each party a return for tackling its salinity problems — while also making the game less asymmetrical. Changing perceptions in Victoria were important in changing the alignment of the States, as was the fact that New South Wales also began to see itself as affected by salination.[2] But the rest of the transformation was achieved by applying the compensation solution: specifically, increasing the permitted level of saline discharge induced both New South Wales and Victoria to cooperate in other mitigation measures.

N-Person Games and 'Free Riders'

The Murray Valley case also involved more than two players: it was an ***n*-person** game. In *n*-person games, players can readily gang up on minorities, and the payoffs are neither as obvious nor as straightforward. In fact, the distinction between PD and Chicken may be significantly blurred in the *n*-person case. But the analysis of *n*-person games, poorly understood though they are, clarifies issues such as the 'free rider' problem encountered with subsidies to Murray Valley irrigators.

In real life, it is exceptional to find only two actors participating in a game-like situation. *N*-person games are far more common; in them, the payoffs change subtly but significantly. For example, in a three-person 'Prisoner's Dilemma', the defection of one no longer means that the dominant strategy for the remaining two is to defect: it may still be worth their while to cooperate among themselves and enjoy the good. Of three prisoners, for example, one might confess, but the other two stick to a prepared story. If, in the absence of corroborating evidence, the police were unable to secure a conviction, the prisoner who had confessed might get a light sentence, but since the other two are not convicted, they may get an even lighter one. The one who confesses is then a 'free rider,' and gains a benefit to the extent that the others' refusal to confess reduces his sentence.

N-person games rapidly generate 'free riders'. Individuals who avoid contributing but enjoy the good conserve their resources for other purposes; the temptation is strong. People who ride on public transport but do not pay are 'free riding' on those who do.

The 'free rider' problem generates a 'punishment dilemma' which is sensitive to scale: the larger the community, the greater the costs of detection and punishment. If a community of ten, for example, sets out to build a road, and one does not pay their share, there will only be nine-tenths of the money for the road available, and the resulting gap will be rather noticeable. But because the group is small, detection of the defector is easy, and direct pressure to pay can be brought to bear. However, where 1,000 are contributing, then the 'free rider' may escape notice and hence sanctions. In fact, the cost of detection may be so high that it isn't worth trying to catch a single defector.

But 'free riders' cannot be ignored. If there are too many, it may become so costly for the remaining honest ones to provide the good that they too finally defect. Thus, if lawlessness is rife, law and order may collapse, and even honest people may have to resort to lawless means (vigilante groups or lynching, for example) to maintain order. Every complex society has some 'free riders'; the costs of detection and punishment mean that they can never be entirely deterred, except perhaps at unacceptable cost. But no society can tolerate too many, or cooperation, coordination, behavioural rules, law, and even society itself may break down.

Olson showed that the 'free rider' problem also means that large, amorphous interests tend to have more trouble getting together than small, organised ones that have a vested interest. Kellow points out that the subsidy to irrigation might cost the average Australian $20 annually — hardly worth the cost of finding out about it — whereas it may be worth many thousands to irrigation farmers, who thus have a strong incentive to organise and defend their privileges.[3] For this reason, Olson would *expect* the local conservation society to have trouble matching the resources and expertise of the big paper mill, even if they have a better case. This 'public interest' is submerged in many political disputes because no-one has a strong interest, or the resources, to pursue it. Since many environmental problems are about 'public interest' or invoke it as an important consideration, this finding is

CHAPTER 9

highly significant.

The 'free rider' problem also explains compulsory taxes; the 'closed shop' in trades unions; conscription for military service, and other cases where compulsion is used to prevent unscrupulous characters from evading payment of their share. In each case, the organisation providing the benefit — government or a trade union — finds it necessary to suppress 'free riding' among potential beneficiaries. Most government services are provided to all comers: therefore all should pay their taxes. Better working conditions and higher pay benefit all workers in an industry, so trades unions argue that all should join and pay the subscription. Few people are suicidal enough to join the army and help defend the country, *especially* in wartime; government compulsion, in the form of conscription, is what creates cannon fodder.

There are some other curious outcomes of 'free riding': the largest partner in an alliance, for example, pays more. Both the USA and the USSR pay (or paid) the lion's share of the costs of NATO and the Warsaw Pact, respectively. Small nations can change their allegiance, but large ones, especially superpowers, have less room for manoeuvre, being locked into their strategic situation.

Elementary games such as these have clear deficiencies. They can be applied unambiguously only to simple cases. They lack explanatory power when really difficult problems such as asymmetry arise; and such common situations as n-person games are imperfectly understood. Despite this, they are increasingly used to explain a wide range of political behaviour. A surprising number of environmental problems turn out to have the characteristics of PD or Chicken; knowing their structure therefore makes it possible to predict the likely outcomes with some reliability. And even where the 'fit' with reality is poor, games can also make it possible to identify the sources of stress and attempt to recast the payoffs — as Paterson did for the Murray–Darling Basin Ministerial Council.

STRATEGIC GAMES AND COLLECTIVE ACTION

Strategic games are primarily relevant to collective action problems. In particular, they help to explain why 'free riders' emerge, and why collective goods are not automatically provided, even if they are obviously in everyone's interest. They have obvious applications within societies and nations, and also internationally. One major reason why international problems do not get solved, and why nations indulge in 'dog-in-the-manger' behaviour is because there is no central authority to force nations to obey the rules. The international system is *anarchic:* a congeries of would-be free riders all competing for limited resources.

The political collective action 'problem' is to get people to cooperate in the production of necessary public goods, and to make sure that everyone pays — and receives — their fair share. Some argue that the only way is to *force* obedience.

The 'Tragedy of the Commons'
Garrett Hardin takes this line in his 'Tragedy of the Commons', a 'Prisoner's Dilemma' set in a mythical mediæval village with a common just large enough for each family to graze its one cow. Any more, and the grazing will be destroyed. But one covetous villager reasons that a second cow will make no noticeable difference, and puts an extra animal out to graze. However, others then make the same choice, the common is overgrazed, and eventually cannot support any stock. The common resource is destroyed — this is the 'tragedy' — as a result of a series of individually rational decisions.

This tale does not explain why the communal lands of mediæval Europe disappeared. In England, for example, common land was 'enclosed' — taken over — by avaricious landlords in order to run sheep, dispossessing the ordinary villagers in the process. Similar piracy occurred elsewhere. And mediæval society was a subsistence economy, not a market one: there was no demand for the extra produce, and hence no temptation to keep the extra cow. In fact, subsistence economies with communal property systems have proved very durable historically; individual societies have lasted far longer than modern capitalism. Hardin's is a fable of market society. It is relevant, if at all, to modern times, not the Middle Ages.

Hardin argues that, where possible, property must be private, not common, so that each person will be 'responsible' for looking after his own. He pays little attention to misuse of private property by its owners: for example, farmers who 'mine' their soil then move on. He also ignores the long and very successful history of joint management of common property in numerous societies; some of these arrangements — such as jointly-managed pastures in Swiss mountain villages — still flourish. River basin management, as Ostrom shows, has been a particularly fertile field for durable common-property agreements.

Hardin's error is to confuse ownership in common with ownership by none; in the process he assumes that common property will be abused in the same way that property with no owner would be. But where suitable arrangements exist for the management of common property, it is often well maintained and highly beneficial to the joint owners. Where no such arrangements exist, Hardin's arguments have some validity; but the restricted range of options he proposes is radically incomplete.

Where common ownership is unavoidable, Hardin argues for 'mutual coercion, mutually agreed': a crude **social contract** in which individuals endow each other (or a ruler) in advance with powers to enforce the rules. He sees the global environment as just a gigantic common, needing similar solutions to its management problems: in fact, a benevolent dictatorship.

Hardin's analysis does parallel the situation in the Murray Valley. Interstate competition is essentially anarchic, and in this case destructive. The systematic over-use of irrigation water — resulting in the degradation of the common good through salinity — arises from undercharging for a resource which is treated as a 'free good', owned by no-one, and inexhaustible.

Hobbes: the 'State of Warre' and the Covenant

But Hardin was not the first to think of the common goods problem. The Chinese had explored it in the fourth and third centuries BC, and seventeenth century English philosopher Thomas Hobbes, in his book *Leviathan*, offered the first systematic statement of it. He argued that unless there was a strong central power to make and enforce the laws, some people would get away with breaking the rules of society. Others would be forced to follow in self-defence, and a situation of mutual misery would develop, in which none was strong enough to dominate, and all lived in a state of insecurity with no mutual trust: the 'State of Warre', or 'war of all men against all men'. Under these conditions, there would be no

> ...place for Industry; because the fruit thereof is uncertain: and consequently no Culture of the Earth, no Navigation, nor use of the commodities that may be imported by Sea; no commodious Building; no Instruments of moving, and removing such things as require much force; no Knowledge of the face of the Earth; no account of Time; no Arts; no Letters; no Society; and which is worst of all, continuall feare, and danger of violent death; And the life of man, solitary, poore, nasty, brutish, and short.[4]

To escape this unpleasant state, Hobbes argued, men would have to come together and enter into a covenant (i.e., an agreement) to set up a **sovereign,** a ruler to make and enforce laws. This would create a state of **civility** in which men would 'perform their contracts made' and arts, letters, industry, and morality would flourish.

Hobbes created tremendous controversy by arguing that all the law-making and enforcing powers would have to be vested in a single sovereign, in order that the sovereign power could never be paralysed by internal disagreement. But whether Hobbes intended to justify outright dictatorship, or merely a very strong sovereign with a lot of discretion in making and enforcing laws, some modern followers of Hobbes and Hardin have explicitly recommended the setting up of 'ecological dictatorships'.

There are many criticisms of Hobbes' position, but it has been very influential. In fact, most modern justifications of government in the form of the nation-state rest on similar arguments.

The Roots of Cooperation
The core problem addressed by Hobbes and Hardin is the 'Prisoner's Dilemma supergame', in which the basic problem, social, political, or economic, recurs again and again. But neither appreciated the implications either of multiple players or of repetition. Both assumed that the logic of the two-person PD game, in which a single defection could destroy the common good, applied willy-nilly to the n-person case. If it did, the problem would be to prevent the *first* defection, because then further defections would have a domino effect, and the common good would unravel swiftly. This is a very demanding conclusion, implying a high level of social control capable of detecting and preventing (or better, deterring) defections very early. But 'free riders' are actually widely tolerated in human societies; common goods such as civility or even cleanliness do not immediately unravel because of one mugger or one litterbug.

One reason is that repetition permits the development of retaliatory strategies: defectors in one round can be punished in a subsequent round. In a series of 'tourneys' conducted by the US scholar Robert Axelrod, in two-person PD supergames involving the same players, one particular strategy, called 'tit-for-tat', emerged with great frequency. 'Tit-for-tat' involves defecting in the game following a defection by one's opponent; if the opponent cooperates, then the 'reward' is cooperation in the following game. 'Tit-for-tat' is *forgiving,* because it retaliates only once, and

CHAPTER 9

in kind, for each defection, and it is regarded as *robust*, because it is easily understood and seems to emerge spontaneously.

'Tit-for-tat' causes participants in games to learn cooperative behaviour, because it is a strategy of **conditional cooperation**: each player cooperates so long as the other party does so, and defects in retaliation whenever the other does. Other conditional cooperation strategies also exist, and can in principle generate cooperative behaviour in groups.

Thus, in n-person supergames, the mutually self-destructive logic of 'Prisoner's Dilemma' situations can be avoided by adopting strategies of conditional cooperation. Conditional cooperation is very robust — that is, it is frequently adopted and survives challenges from other strategies — and it underlies a lot of social behaviour.

Axelrod's other discovery was that although his tourneys were set up explicitly as 'Prisoner's Dilemma' games, and the participants understood that, most were prepared to invest 40–60% of their resources in attempting to achieve cooperation from an early stage in the game. Players were inclined to start out by cooperating, and to abandon cooperation only where other players were intransigent. This peculiar — 'irrational' — behaviour, it turned out, was based on a conviction that cooperation was superior to defection. The consequent willingness to try to achieve cooperation even at some personal cost is called 'altruism'; frequently it is motivated by a conviction that collective goods are worth having even if there is no obvious benefit to the individual.

Altruism is often based in or reinforced by the adoption of social norms commanding unselfish or cooperative behaviour. 'Do as you would be done by' is one example. **Solidarity** is an important norm in trades unions and the Labour movement. And rules commanding moral behaviour on the ground that, even if you personally do not benefit, society as a whole will, are examples of this kind of norm.

In the Murray Valley, Kellow emphasises the political significance of the incumbency of Labor governments in South Australia, Victoria, New South Wales, and federally during the early 1980s. Cooperation is an important value in the Labour movement, and this was a historic opportunity to achieve agreement on a problem which had festered for 100 years. The Labor governments in question did prove more willing to cooperate, and the Labor federal government more able to coordinate, than any previous combination. There undoubtedly were immediate, short-term losses to some of the parties — especially New South Wales and the Commonwealth, which latter had to provide finance — but the long term *collective* gain was clearly seen to outweigh them.

Rational behaviour theory has often been accused of sanctioning amoral, impersonal, ruthlessly selfish behaviour. Yet it actually leads to a surprising and paradoxical result: despite the rigorously individualistic presuppositions from which it begins, it reaches the conclusion that it is rational for individuals to cooperate, for sound, self-interested (but not *selfish*) reasons. Furthermore, it usefully describes political behaviour as well as suggesting ways of approaching conflict situations.

This does not mean that societies can be self-governing anarchies, except perhaps for very small communities numbering only a few dozen people. But it does mean that most people don't have to be forced to 'do the right thing' most of the time, because they will cooperate voluntarily out of self-interest. That in turn means that governments do not have to be dictatorial; rather, they need **sanctions** to deter would-be 'free riders'. The effect of sanctions is to *change the payoffs*, typically by making defection more expensive, and consequently cooperation more attractive. Sanctions tend to move PD or Chicken games towards Assurance, encouraging cooperative behaviour. Payoffs can also be changed by rewards: good behaviour can be encouraged by various kinds of bonus, at the same time that bad is discouraged by punishment. Farmer cooperation in the Victorian salination mitigation scheme, for example, was secured by various kinds of rewards. Most notably, groups setting up local cooperative schemes gained access to government funding: a clear incentive!

The Importance of Rules
This is why **rules**, **norms**, and *laws* are of key importance. Sometimes, the actual content of a rule is not as important as its existence. Driving on the left, for example, is a useful convention for avoiding collisions; some nations just as happily drive on the right. What matters is that everybody in a given country drives on the same side. Social rules and expectations serve to *coordinate* people's behaviour. As societies grow more and more complex, such coordination becomes ever more necessary; without it, the system would rapidly break down. (The back-up system necessary just to let bank depositors get money from a machine with a plastic card, for example, relies on people the depositors have never met and probably never will.) In some cases — such as air travel — passengers are literally trusting their lives to these unseen 'collaborators'.

Chapter 9

Since any common good by definition offers more utility than the sum of the individual costs which go into it — in a sense, common goods are 'created' by cooperation — there will always be a 'surplus' of utility as long as the good is provided. But enforcement costs, especially in complex societies, mean that some 'free riders' will persist. Their presence means that less of each good will be provided than if everyone contributed: the level of provision is **suboptimal**, i.e., less than is desirable to meet all the needs of the community. Too many 'free riders' will make the provision of public goods impossible by eroding the surplus.

That affects social structure. The use of rewards to encourage cooperation spawns grants, subsidies, honours and prizes. Sanctions to deter those who might drive on the wrong side, or break other rules, require people to apply them, such as police, courts of law, social disapprobation, and all the other ways of keeping people in line. These have to be coordinated, which is one of the things **governments** do.

There also have to be people to make the rules: **legislators**. And there must be rules for them to operate under. Hence there are **constitutions**: sets of rules about making and enforcing rules. These breed **lawyers**, who specialise in interpreting and applying the rules.

Institutions

There is another highly important corollary. Individual decisions in the PD game are **separable**: that is, each player can decide in isolation from the other(s). But in the Assurance game, they are not. Instead, each player needs to know that the other(s) will cooperate, before deciding. Cooperation under these conditions, being conditional, also requires some reassurance of the intentions of others, or a standoff may result, in which no-one will commit themselves, and the good is not provided. As individual expectations that *others* will cooperate rise, so will willingness to cooperate: a mutually reinforcing positive feedback. The problem is to initiate and sustain the feedback.

Runge argues that this reassurance can be provided — and indeed most commonly is — by **institutions**. Appropriately, the analogy is with a friendly society: the members all contribute, and the society assumes the cost of illness, funerals, or whatever else it has agreed to provide. As long as each member knows that the others are continuing to contribute, they will be reassured and continue to contribute.

Institutions are particularly important when change is occurring, because they can provide reassurance that cooperation remains worthwhile. More trivially, they can save the individual from the necessity of recalculating the decision to contribute each time it comes round.

OVERVIEW: GAMES AND GOVERNMENT

Environmental problems are collective problems. They involve goods which are, in whole or part, jointly produced, jointly consumed, or both. These goods cannot be provided by individual action, nor by markets. They require cooperation. Cooperation is often difficult to achieve because the structure of some situations tempts participants to pursue short-term gain rather than collective self-interest. The study of such situations is called 'game theory'; three major games, the 'Prisoner's Dilemma,' 'Chicken,' and 'Assurance' dominate the literature. 'Prisoner's Dilemma' games are especially common in environmental policy. Cooperative solutions of repeated, n-person games can emerge, but are likely to be unstable. Anarchic cooperation, though possible, is limited to societies of very small scale. Consequently, cooperation and coordination must be overseen by government, which must have coercive powers in reserve.

FURTHER EXPLORATION

Until very recently, the literature of public choice and of game theory was forbidding and inaccessible. The work of such scholars as von Neumann, Morgenstern, Tulloch, Buchanan, Sen and even the relatively accessible Olson was widely known, but few had read it and fewer still appreciated its significance. Fortunately, two British authors, McLean and Laver, have recently produced primers which make the whole field vastly more accessible to students and nonspecialists. Furthermore, they neatly summarise much of the rapid and exciting progress which the field has recently been making. The idea that rational social choice theory would lead to a world of coldly asocial behaviour, on the model of the 'economic rationalism' of the political Right, has already given way to an expanded appreciation of the purely rational reasons for cooperation in human societies. This conclusion is very important to social science as an enterprise, for if human behaviour (and especially social cohesion) cannot be explained in rational terms, then social science is obviously a waste of time; at best it could be a kind of sophisticated charity work. For the policy sciences, the

CHAPTER 9

tantalising promise of some very powerful tools for the analysis of political situations is at last being realised.

The basics of game theory are lucidly explained in Chapter 7 of McLean's *Public Choice: An Introduction*; the importance of 'tit-for-tat' in particular is emphasised. He also refers (p. 16) to Laver's amusing discussion of coordinating conventions in *Invitation to Politics*. Chapter 6 of Laver's more recent *Social Choice and Public Policy* also explores the issue of collective goods. Laver's earlier (1981) *Politics of Private Desires* is particularly good reading for those interested in the 'free rider' and cognate games-related problems. But the key work on the 'free rider' remains Olson's *Logic of Collective Action*, an absorbing examination of the impact of the problem on the behaviour of groups in political life. It should, however, be read in conjunction with Taylor's criticisms in his *The Possibility of Cooperation*. And Runge's work elegantly links the decision process in Assurance games to social institutions and their role.

Hardin's 'Tragedy of the Commons' first appeared in *Science* in 1968. His general philosophy is spelt out at greater length in *Exploring New Ethics for Survival*; the 'Tragedy' appears as an appendix. Hobbes' magnum opus, *Leviathan*, was published in 1651 and has gone through innumerable editions; try to get hold of one (like the Everyman edition, quoted here) that preserves the original spelling and grammar, so as to appreciate the flavour of his work. It was so far sighted that many contemporaries regarded him as a 'monster'.

Also at the theoretical level, Taylor's *Anarchy and Cooperation* (1976), recently revised as *The Possibility of Cooperation*, remains the most lucid discussion of the 'cooperation problem' in general. Few pioneering works of political theory are as clear as this one. Taylor's work draws on, and is underpinned by, Axelrod's experiments, summarised in his *The Evolution of Cooperation*. Taylor is explicitly interested in the possibility of anarchistic societies, an issue discussed in both *The Possibility of Cooperation* and *Community, Anarchy, and Liberty*.

The 'Mutual Pollution' game is one of the many examples in Hamburger's *Games as Models of Social Phenomena*, at p. 187.

The story of Med Plan is set out in Haas' *Saving the Mediterranean*, in which he emphasises the importance of the international ecological epistemic community. Some theoretical implications are absorbingly dealt with by Makim in her Honours thesis 'Possibilities for International Ecological Cooperation' (School of Environmental Studies, Griffith University, 1991).

CHAPTER 10

SOCIAL CHOICE

Even if people do decide to cooperate — whether for self-interested or altruistic reasons — and even if there is a government to apply the rules, it is not necessarily easy to come to a clear conclusion about what everybody wants. It is quite possible, as game theory demonstrates, for a series of individually rational decisions to add up to a socially irrational outcome. It can also happen, given certain configurations of individual preferences, that it becomes quite impossible to 'add up' everyone's wants and get a clear 'social sum' which can act as a basis for public policy.

This problem is a serious one. It is the domain of **social choice theory**. It underlies many of the drawbacks of **welfare economics**, which attempts to 'tot up' individual economic well-being as a guide to policy, and the body of related techniques, such as cost-benefit analysis (CBA) and environmental impact assessment (EIA), which depend on the same basic assumptions. It is also the cause of serious difficulties for other 'head-counting' processes such as voting; and this in turn has repercussions for democratic theory.

Collective choice is especially relevant to environmental policy, because so many environmental goods are collective. The problem is to find methods for deciding on and delivering those goods.

Chapter 10

Three Paradoxes

The Paradox of Voting

First a theoretical difficulty: the **Paradox of Voting**, discussed by J-C. de Borda in a paper written as early as 1770, and subsequently by such great names in social theory as Condorcet and E.J. Nansen. It was a favourite of C.L. Dodgson (the real name of Lewis Carroll, author of *Alice in Wonderland* and *Through the Looking-Glass*), who was fascinated by mathematical puzzles.

Imagine a population of three people, each with a definite preference ordering of three possible future social states, x, y, and z. (By a future social state, any future state of affairs that involves more than one person is meant. For example, a group of friends deciding what to do together on Saturday night are choosing among future social states. So are voters making choices between 'capitalism' and 'socialism' at the polls: they are choosing between possible sets of economic relationships among themselves.) These orderings are **complete**: each individual is either indifferent to or favours one or other alternative out of each pair (were the ordering incomplete, the individual might not favour some or all of the possible choices — if, for example, he hadn't thought about it). They are also **transitive**, which is to say that, if someone prefers x to y and y to z, then they necessarily prefer x to z. Now suppose that each person has a different preference ordering:

A x y z = A prefers x to y, y to z, AND (by transitivity) x to z.
B y z x = B prefers y to z, z to x, AND (by transitivity) y to x.
C z x y = C prefers z to x, x to y, AND (by transitivity) z to y.

What do A, B, and C prefer as a group? Comparing each pair of options in succession, A and C both prefer x to y, only B preferring y to x; A and B both prefer y to z, only C preferring z to y. By transitivity, since the group prefers x to y, and y to z, it must prefer x to z. But B and C both prefer z to x! The matrix is **intransitive**: no rational summation of the three preference orderings is possible. Worse, the outcome is sensitive to the order of presentation: if the pairwise comparison process had started with y/z, the outcome would have been different. This property is called **cyclicity**, and the results are **cyclical**. The majorities favouring each alternative change with each pairwise comparison — a characteristic of cyclicity — and there is no obvious way to resolve the paradox.

This is the smallest number of people and of options which will yield the result predicted by *Arrow's theorem*. This says that, given

Social Choice

four simple conditions which are essential if the outcome is to reflect the wishes of the participants (these conditions are also basic to democratic theory, for the same reason), there is no way that a collectively rational outcome can reliably be achieved in every case. Worse still, the larger the number of choices, the greater the probability of an intransitive outcome. But the larger the number of choices and participants, the harder the intransitivity will be to spot.

Thus procedures such as voting do not *necessarily* reflect the wishes of the participants: often the only way to find out if they do is by careful research. To make meaningful choices, political systems will have to be able to *refine* the options down to a relatively few worthwhile ones, without leaving out any important ones.

Ostrogorski's Paradox

This is not easy to do; in fact, many political systems are trapped by paradoxes such as these, and are consequently incapable of the necessary selection and refinement. **Ostrogorski's paradox**, for example, shows how a popularly elected government can find that a majority of its policies are unpopular.

Here two political parties, called P and Q, have policies on three issues, 1–3. Each policy is designated by the lower-case letter corresponding to the party concerned. The voters are represented by the letters A–G. Tabulation of the voters' preferences might produce a matrix like the following:

		\multicolumn{7}{c	}{Voters}	Policy
		A	B	C	D	E	F	G	preferences
	1	p	q	p	p	q	q	q	q
Issues	2	p	p	p	p	q	q	q	p
	3	p	p	q	q	q	p	q	q
Party preference		P	P	P	P	Q	Q	Q	P

In this example, voter A prefers party P's policy on policy issues 1 to 3; but B prefers party Q's on issue 1. Similarly with the other voters. If each voter's *party* preference is summed up by assuming that they will vote for the party with which they most agree, that preference can be written at the foot of each column. Thus A is a one-eyed 'P' and E and G are loyal 'Qs'. As B agrees with two of P's policies, and only one of Q's, a vote for P is

171

CHAPTER 10

assumed. In the above matrix, four voters prefer P and only three prefer Q. So P would be elected in a poll. But the *paradox* is that, if the policy preferences are summed one by one, horizontally, the voters as a whole prefer two of Q's policies, and only one of P's. As a group, should they really vote for Q?

Writ large, this means that a government preferred by a majority of voters could find that a majority of them oppose its policies! And this assumes that everybody is perfectly informed and knows exactly what they want. Obviously, in the real world of uncertainty and imperfect information, choice is much more difficult — for government as well as citizens.

The Logrolling Paradox
In fact, because it may be worth an individual's while to conceal or misrepresent his real preferences in order to gain an advantage, it may be even harder to aggregate preferences in such a way as to produce a true picture. McLean's **logrolling paradox** shows this clearly. Logrolling is a form of bargaining in which individuals or groups support the proposals or programmes of others similarly placed, in return for their support. In essence, it is a network of 'treaties' in which the participants *conditionally* pledge mutual support. But to redeem their part of the bargain, they must misrepresent their own preferences, typically by voting against some measure which they would otherwise support.

On a five-member committee, where there are three issues each of which commands majority support, it may nonetheless be in the interests of the three people, each of whom opposes two of the three issues, to form a logrolling coalition to defeat all three. As there are three, they will be in a majority, and will prevail. Here is a diagram, showing the *real* preferences of each at the outset:

	Issue 1	Issue 2	Issue 3
Voter 1	For	For	For
Voter 2	For	For	For
Voter 3	For	Against	Against
Voter 4	Against	For	Against
Voter 5	Against	Against	For

Each of voters 3, 4 and 5 is against two of the propositions, so a **coalition** with the other two to vote against all three gives each voter two of their three preferred outcomes. Obviously, it is worth their while to form the coalition, by cooperating conditionally. Thus, if each individual were to vote on the basis of true prefer-

Social Choice

ences, there would be a clear majority for each proposition; but a logrolling coalition of voters 3, 4 and 5 would defeat all three propositions.

As in the Paradox of Voting, cyclicity is present — none of the three 'antis' is against all the same things — and it is to eliminate the cyclicity as well as to produce the counter-intuitive outcome of defeating all three issues that voters 3, 4 and 5 are tempted to misrepresent their preferences. Such misrepresentation — in this case, concealing a real preference on one issue in order to gain something on another — is called **strategic voting**.

IMPACT ON VOTING SYSTEMS

Strategic voting happens all the time; in fact, some voting systems make it unavoidable. Consider, for example, the 'first-past-the-post' or **simple plurality** system used for elections in most parts of the world. A 'plurality' simply means the most votes: more than any other candidate. It does not necessarily mean a majority.

Simple Plurality Voting

In this system, votes for each candidate are cast, and the candidate *with the most votes* wins. Thus if Smith and Jones are standing for election, and Smith gets 30 votes while Jones gets 40, Jones is elected. But when there are more than two candidates, this method gives indeterminate or even outright bad results. For example:

Smith	30
Jones	40
Hill	25

Under the simple plurality system, Jones is elected. But more people *did not want* Jones than wanted him: 55 to 40. It is just that they were split between Smith and Hill. There are two alternatives: declare Jones elected, even though more people oppose him than prefer him, or insist that no candidate be elected without a majority (that is, 50% of the votes cast, plus one). Simple plurality systems opt to declare the candidate with the most votes elected, even if that is a minority. This happens quite frequently: the Thatcher Government in Britain never exceeded 40% of the popular vote, but a combination of simple plurality and the 'multiplier effect' gave 'landslide' Parliamentary majorities.

This is why supporters of minority parties, rather than 'waste'

173

CHAPTER 10

their vote, may cast it for a major party they expect to win, and which they see as a lesser evil. For example, if the opinion polls showed the following percentages,

>   Labour   40%
>   Conservative   40%
>   Green   5%

a Green supporter who felt that Labour might do less damage than the Conservatives to the environment might well vote for Labour, given that the Greens have no chance of election. In terms of 'revealed preference' (votes actually cast), this will understate the Greens' support, and overstate Labour's by the same amount.

The Second Ballot

One way round this problem is to insist that any winner must have a majority. This can be done by holding a **second ballot** (or runoff election) in which the candidate(s) gaining the least votes is eliminated, and voting is repeated. Thus, if Hill is eliminated in the three-way example above, then the outcome might be:

>   1st ballot   2nd ballot
>   Smith   30   Smith   50
>   Jones   40   Jones   45
>   Hill   25

Smith would then be declared elected, whereas Jones would have won under simple plurality.

Second ballots are quite widely used; for example, in France. But they have at least two disadvantages. Firstly, the voter turnout is often lower than for the first election, so the candidate elected might not be the second choice of all who participated in the first ballot (note the element of *uncertainty* here: a common problem with counting heads). Secondly, runoffs are expensive.

Preferential Voting

Another alternative is to ask people, at the time of the first poll, to *imagine* that their preferred candidate has lost, and to nominate a second, or even a third or fourth preference. This option, called **preferential voting**, is widely used in Australia. To win a seat, the successful candidate must gain an absolute majority — 50%+1 — of the votes, not a mere plurality. If, after all the first preferences are counted, no candidate has an absolute majority, the candidate with the *least* votes is eliminated, and the second preferences stated on those ballot papers allocated to the remaining candidates. This process continues until one candidate gains a majority and is elected. Preferential voting 'finds majorities', and it has been said that it results, not in the *most wanted* candidate, but the *least disliked* candidate being elected. It can permit a minority party with dispersed support to direct its second preferences to one candidate or another, and thus exercise a significant veto power. The Democratic Labor Party (DLP) successfully denied government to Labor for some 18 years after the split of 1954 by this means.

Landslides and Multipliers

But it is not only in *counting* votes that problems emerge. Difficulties can also arise from the size of electorate and the number of members it elects. The simplest of these problems is the **multiplier effect**.

Depending on the electoral system, either a whole country or a whole region may be treated as a single, gigantic electoral division, or it may be split up into divisions known as **electorates** or **constituencies**. (In the USA, they are known as **wards**.) If all votes are to have the same value — i.e., if some groups are not to be over-represented at the expense of others — each electorate must be the same size. Because of demographic changes, such as migration and regional increases and decreases in population, there must be provision for periodical **redistribution** of electoral boundaries.

Single Members and Multipliers

Each electorate may elect one member, several, or even the whole parliament, as happens under some systems of proportional representation. The oldest system is the single-member electorate, where each district elects a single member to represent it in the national parliament. This is the system adopted for all Australian State Lower Houses, except in Tasmania, and for the Commonwealth Parliament's House of Representatives. The major difficulty with it is that those who supported a losing candidate effectively have no representation.

The multiplier effect is most marked in single-member electoral systems, especially when coupled with simple plurality voting. Under these conditions, a series of narrow wins in particular electorates may be converted into an apparent landslide in parliamentary seats:

Chapter 10

Seat (each seat has 100,000 voters)	Party A	Party B	Winner
1	51,000	49,000	A
2	55,000	45,000	A
3	30,000	70,000	B
4	52,000	48,000	A
5	51,000	49,000	A
6	40,000	60,000	B
7	20,000	80,000	B
8	55,000	45,000	A
9	60,000	40,000	A
10	51,000	49,000	A
Total Vote	46.5%	53.5%	

Thus, in this example, Party A wins 70% of the seats in Parliament — the mass media would call it a landslide — but has actually gained only 46.5% of the votes. Party B, with an actual majority — 53.5% of the votes — gains only three seats. The popular preference is distorted by the presence of electoral boundaries.

Single-member electorates also tend to favour large parties, since small ones with a highly dispersed vote will often be unable to capture them. In Australia, parties with 10–15% of the vote have often gained no seats at all in the House of Representatives for this exact reason.

However, a minority party with a regionally concentrated vote may be able to capture all the seats within its area of strength, and thus be over-represented nationally. For example, in Australia the Country Party regularly held some 16% of the seats in the Federal Parliament during the 1950s and 1960s, with a mere 8% of the total national vote. Thus regional minorities may do very well under a single-member system.

Mitigating the Multiplier

Multi-member electorates can mitigate the multiplier effect, because the number of votes needed to gain a seat drops. If, for example, each electorate elects two members, then instead of 50%+1 votes being necessary to gain election under a majority rule, only 33.33%+1 are needed, since two wins with 33.33%+1 will add up to more than 66.66%, preventing any other candidate from gaining 33.33%. This is one reason why the Greens hold seats only in the Tasmanian Parliament: the **Hare-Clark** system of **proportional representation** (PR) used in Tasmania elects five members from each electorate. The 'quota' for each seat — that is, the minimum number of votes needed to gain a seat — can be found out by adding one, and dividing into 100. Thus $100 \div 6 = 16.66$, so the quota is 16.66%+1. Obviously, this dramatically changes the electoral chances of small parties.

PR systems vary considerably; their only shared characteristic is complexity. Most work by means of a 'list' system, in which each political party nominates a number of candidates, ranking them in a preferred order. (Competition for a 'high' place in the list is therefore intense *within* each political party; jockeying for Senate places on party lists is very common in Australia.) Voters then vote for the *parties* (rather than individual candidates) by marking their preferred list. A quota is calculated for each seat, and as the parties' votes are counted, their candidates are declared elected, in order of their place on the list, as each quota is filled. The remaining votes for that party are then transferred to the next candidate, until there are no longer sufficient to reach a quota. The *transfer value* of each vote is often calculated according to a formula, reducing the value with each transfer. The system used since 1949 to elect the Australian Senate is broadly similar, with the complication that the voter has the option of ticking a list or making a fully preferential vote with every candidate listed.

Under systems of proportional representation in which the whole of the country is the electorate, quotas will obviously be very small, in percentage terms. Many small parties consequently gain seats, producing parliaments which reflect the popular vote with mathematical exactitude, but may be unworkable because no party has a majority. In the Australian Senate, half the ten Senators from each State retire at every election, leaving five vacancies. In recent years, there has been a persistent tendency for two to go to each of the major parties — Labor and Liberal — and one to go to the Democrats or an independent. Ever since its adoption in 1949, PR has resulted in very evenly balanced Senates, in which it has been rare for the party in government to command a majority. The ability to use their potentially decisive vote to defeat government legislation in the Senate has been the basis of the Democrats' pledge to 'keep the bastards honest'.

The advantage of PR lies in its claim to be able to provide mathematical exactness in *representation*, by ensuring that no party is under- or over-represented. It is commonly felt that this is also its disadvantage, since this may result in no party having a parliamentary majority, creating a necessity for coalition government, which is frequently unstable.

The Gentle Art of Gerrymander
Why the concern to avoid over- and under-representation? The short answer is, to avoid bias. Most Australian electoral systems have historically over-represented country areas, especially in state Upper Houses. This directly affects policy. In Queensland, a persistent electoral bias towards rural interests ever since self-government in 1859 — except for the period 1921–1949 — has reinforced an existing heavy dependence on rural support, and so farming, pastoral, and mining interests have been very successful in obtaining assistance from successive State governments. Yet government action on land degradation has been slow and ineffective, despite the Queensland economy's heavy dependence on agriculture and regular public demands for action from community leaders. Successive governments have failed to act strongly, for fear that adverse short term economic effects would trigger an electoral backlash. Although other States do not suffer quite such extreme electoral distortions, they too have been unwilling to anger rural interests: as with Victoria's reluctance to embrace land retirement or clearance controls.

Electoral distortion is a form of election-rigging; a temptation to which governments succumb more often than they should. There are two legal ways to rig, and dozens which are illegal. Really nasty ones include 'stuffing' ballot boxes with false votes — once a common pastime in ALP branch pre-selection ballots, leading to some violent fights — and rigging voting machines, where they are used, to give more votes to a favoured party. Other methods include intimidating voters, laying on free transport to the polls, losing votes, counting them more than once, or 'voting the cemetery': that is, casting votes on behalf of individuals who are dead but whose names still appear on the electoral rolls. Many of the trappings of modern electoral systems, including impartial officials, scrutineers from all interested parties, frequent revision of electoral rolls, and the secret ballot itself, are designed to eliminate abuses of this kind.

Rigging electoral boundaries, the 'legal' approach, is a prerogative of government, unlike the other methods, which are also adapted to private enterprise. Boundaries can be rigged by **gerrymander** or by **malapportionment**. A gerrymander is a situation in which boundaries are drawn in such a way as to maximise one's own electoral advantage by 'locking up' the votes of the opposition so that they are 'wasted,' and ensuring that one's own are so distributed as to give useful wins in as many electorates as possible. Coaldrake cites three examples from the 1985-6 redistribution in Queensland: the excision of the Wujal Wujal Aboriginal settlement from the Barron River electorate; the 'Cunnamulla Nipple'; and the 'Parson's Nose' at Caloundra.[2] The alternative is to vary the size of the electorates — malapportionment — by having fewer voters in yours and many more in theirs. Thus you will get more seats from less votes, and they will get less seats from more. This was the case in Queensland, where, until the reforms of 1991, a vote in the electorate with the largest enrolment was worth only 0.392 of one in that with the smallest.

Systematic malapportionment has been far more common in Australia than outright gerrymander. The 'country' bias of most Australian electoral systems persisted until very recently. The present-day Legislative Councils in Victoria and Tasmania, as well as the Queensland Legislative Assembly, still retain deliberate over-representation of country areas. In this light, the Country Party record as the most successful of all Australia's minority parties, persistently succeeding in gaining a greater percentage of parliamentary seats than of the vote, is less remarkable.

Coalitions

The logrolling paradox demonstrated that individuals will often be tempted to conceal their actual preferences in order to improve their political position; exactly the same problem arises with simple plurality voting. Strategic voting, involving the concealment of preferences, may occur as individual action — as in simple plurality elections — or it may arise in the context of a coalition, in which conditional cooperation among self-interested voters is the central theme.

Coalition formation is basic to politics. Coalitions are *conditional*: they represent compromises between groups of individuals who *do not* share identical interests, but whose interests on some specific issues coincide. That is why they are unstable, and why they often embody mistrust. Most political alliances are coalitions, which is why politics is both volatile and permeated with bargaining behaviour.

Bargaining is quite central to the bulk of political activity. It cannot proceed without recognition by both parties of the existence of a **contract zone**; that is, an area in which some agreement can be reached. Bargaining, and consequently coalition, is frequently

CHAPTER 10

highly fluid because of changing interests and perceptions.

Coalitions are not confined to voting situations, such as elections or parliaments. It is common usage to talk of a 'coalition' of support for a particular political programme or party. Such a coalition will typically consist of various groups in the community who, despite differences among themselves, can agree to support a particular programme or party. Thus the Australian Liberal Party seeks support from big business and small; from doctors, stockbrokers, lawyers, and the 'respectable' professional middle classes generally, as well as from others who may have no identifiable group affiliation. Such groups are neither homogeneous nor unanimous — for example, there is a Doctors' Reform Society which supports Medicare, in contrast to the violent opposition expressed by the conservative Australian Medical Association (AMA) — but very often they display considerable agreement on specific issues. This agreement can be channelled into political support.

OVERVIEW: THE PITFALLS OF COUNTING HEADS

Nineteenth century Liberalism, the political theory which underlies Western democratic systems such as Australia's, assumed that counting heads would reveal the answers to many policy problems. This assumption is not only false, but dangerous. For example, governments winning elections often declare that they have a 'mandate' to implement their policies. But it is not enough to assume that because a government (or other body) has won an election, it necessarily has a mandate to implement whatever policies it chooses. Parties sometimes misrepresent their platforms in order to gain power, as the Liberals under Kennett did in Victoria in 1992; or they may find themselves confronted with circumstances they did not expect. Additionally, electoral systems, either by design or by quirks in the way votes are counted and seats in parliament allotted, may produce what are politely called 'distortions'. The Paradox of Voting, and Arrow's result, suggest that we ought to be suspicious of the voting process; Ostrogorski's paradox that we should suspect the 'mandate' too.

The persistence of irrigation development in the Murray–Darling basin, despite land degradation and poor economics, illustrates the point. Throughout recent history, the basin's land users have voted for governments which provided assistance for the expansion of agriculture in the region. The over-representation of the Country Party has given their views additional weight. The result is that *political* pressure has often prevailed over *ecologically* or *economically* desirable policies: so much so that the established orthodoxy of 'development' has largely excluded such considerations.

The risk of cyclicity, and hence of inability to come to any conclusion, increases with the number of alternatives. This leads to two related consequences. Firstly, a significant part of the political process, and especially political debate, is devoted to attempting to simplify the options, by reducing them to a handful of 'critical' choices. Sometimes this is effective; at others it is a charade. Secondly, because most choice processes — such as general elections — involve a large number of disparate issues, parties and candidates attempt to simplify the problem by offering 'bundles' of issues, rather than single ones. Voters thus find themselves choosing second-best alternatives: the 'bundle' they least dislike, rather than one which embodies all their preferences. This can lead directly to Ostrogorski paradoxes and to logrolling. Because most voters will not support all the policies of their chosen party, governments can never be sure which of their policies have majority support, and which do not.

And this is by no means all. To properly reflect the popular will, voting procedures have to be non-dictatorial: that is, there can be no individual whose will would always prevail *regardless* of the wishes of the rest of society. But, in practice, this condition is often broken. The best government makes decisions on a host of minor issues without consulting public opinion. Sometimes such **imposed** decisions become essential in order to break the Gordian knot, when a policy issue reaches stalemate, whether because of an impossibility or simply because of disagreement. And some persons in authority prefer dictatorship to democracy in any case. A similar form of imposition occurs when 'price-makers' impose their terms in a market situation: logically, voting and markets are identical.

But there are subtler forms of imposition. Some political issues are never discussed: they never reach the public 'political agenda'. This occurs because of social inequality, and especially limited access to channels of communication. In most 'democratic' political systems, the wealthy and powerful control the mass media, as well as being disproportionately represented in official and political positions. Their definitions of what constitute political problems tend to prevail, and many issues of serious concern to less privileged citizens are never raised. This 'agenda control' is dictatorial in effect, since it simply precludes some points of view from

Chapter 10

gaining attention, and distorts popular perceptions of problems.

Thus liberalism can deliver oligarchy, rather than democracy, and the paradoxes of voting and social choice will speed it on its way.

Liberalism also fails utterly to address the issue of distributive justice: that is, the distribution of wealth and power. This is frequently offensive to an ethical commitment to equality, but, as well, it invokes the most intractable of choice paradoxes, often in tandem. In deciding how to divide up a good among three or more individuals or groups, there will be a tendency for social preferences to be cyclical (*unless* all will accept equal shares), leading to temporary coalitions for short-term advantage. Further, there may be no incentive to cooperate, especially if one group can be permanently disadvantaged: a 'Prisoner's Dilemma'. So if people are taught to be greedy and short-sighted, social problems arising from gross maldistributions of wealth will be ignored. In certain cases, as with scarcity of vital resources, this can pose difficult ethical questions.

Where resources are scarce, the enjoyment of a resource by one person can exclude others; in that case, voting (or making other acts of social choice) in a selfish manner can run right into intransitivity problems. In Australia, some resources — good land, for example — have always been short, and always will be. There is every reason to suppose that the question of access to the best resources has often been solved in a highly non-democratic manner: the 'problem' of Aborigines occupying land, for example, was solved quite simply by murder. The moral is that to solve problems arising from shortages of desirable resources, for example, it is necessary to look well ahead, and anticipate the likely difficulties.

Democracy, therefore, while not impossible, is unexpectedly difficult. Worse, because democratic social choice processes are not necessarily ecologically responsible, environmental problems do not always respond well to the compromise approach typical of democratic politics.

Further Exploration

Social choice theory has its roots in the work of Jeremy Bentham, the great British Utilitarian philosopher, and specifically in his idea of a *felicific calculus* — a calculus of happiness — which could sum up the pleasures and pains of everyone in society. This could then be used to choose between alternative courses of social action. Bentham never solved the thorny problems with which the notion of a calculus bristled, though that did not stop some of his followers — the 'economic rationalists' of their time — attempting ruthlessly to implement what they took to be the policy implications of his ideas. A more restricted version emerged towards the end of the nineteenth century as welfare economics, based on essentially the same assumption. But the work of Arrow and Little, from the 1950s on, showed that the underlying assumptions were critically deficient. A series of gifted scholars, most notably A.K. Sen, continued the analysis, leaving many of the assumptions of nineteenth century political philosophy in ruins. Chapters 1 and 8 of McLean's *Public Choice* are by far the most accessible introduction to this field, followed closely by Chapter 7 of Laver's *Social Choice and Public Policy*, which is less theoretical, but is full of useful examples. The Frohlich and Oppenheimer volume *Modern Political Economy* is now a little dated, but may help those requiring a more methodical, less descriptive approach than McLean and Laver offer. For extended reading, Sen's own work, especially the seminal *Collective Choice and Social Welfare*, and the essays in *Choice, Welfare and Measurement*, is indispensable.

Many textbooks deal with electoral systems, but few will yield an understanding of their relationship to social choice theory. McLean's essay 'Mechanisms for Democracy' in the Pollitt and Held book *New Forms of Democracy* is by far the best quick introduction. The Niemi and Riker article 'The Choice of Voting Systems' is also useful.

Australian electoral systems have a long and only partly ignoble history, since Australia was the first country in the world to adopt secret ballot (still known as the 'Australian ballot' in some countries). The other side, of corruption and skulduggery, is brilliantly fictionalised in Hardy's *Power Without Glory*, based loosely but tellingly on Victorian politics at the beginning of this century. Corruption as a theme in Australian politics has had less attention than it deserves, but Pearl's *Wild Men of Sydney*, Cannon's *The Land Boomers*, Kennedy's *The Mungana Affair*, and a recent spate of books on Queensland are useful. Coaldrake's *Working the System* is an excellent introduction to election-rigging as well as a vast range of other scandalous behaviour as practised in Queensland in the 1970s and 1980s; see also Dickie's *The Road to Fitzgerald* and the Prasser, et.al. collection *Corruption and Reform*. There is much less information available on corruption in New South Wales.

Chapter 13 of the Stewart and Ward text *Politics One* has a good short outline of electoral systems and practices in Australia, with

Chapter 10

appropriate attention to built-in biases. Carden has documented the results of a rural electoral bias in his Darling Downs paper, putting forward an interesting line of argument about its effects on policy.

Chapter 11

MAKING DECISIONS

The paradoxes of social action spill over into political decision-making, and even more to planning. The provision of collective goods, such as environmental goods or socially-provided services, requires decision and some degree of foresight. But, because of the inherent paradoxes of social choice, government is frequently obliged to impose decisions arbitrarily.

N-person games and paradoxes of voting may seem rather abstract; but the institutional arrangements of governments are often intended, *inter alia*, as solutions to precisely the problems they identify (though they may not always be successful). A *constitution*, for example, is a set of **meta-rules**, that is, rules about rules. It provides a framework for arguing about what the rules ought to be, separate from specific cases and the passions they arouse. Getting people to vote for their legislators not only eliminates power struggles; it also makes them feel that the resulting laws are 'theirs' in some sense. This is good for government legitimacy, and reduces the costs of compliance.

Once established, however, governments have to do three kinds of things: they must plan, coordinate, and innovate; produce goods and services; and regulate. To do these things, they have to make decisions.

Chapter 11

Political Decision-making

Because Bentham's idea of counting heads often results in paradox, most social and political decisions are taken by a subset of the members of society. This is called 'leadership', 'statesmanship', 'dictatorship', or 'fascism' depending on the observer's political bias; whatever the label, it definitely involves imposition, which is a breach of Arrow's nondictatorship condition. The decision makers work with partial information about the preferences of society; biases of their own; and technical and scientific information which may range from comprehensive to woeful. Often they are unable to interpret the latter or to appreciate its significance, and so decisions frequently reflect ideology or prejudice rather than enlightened self-interest, altruism, or sound judgement.

Some decisions are taken long before the event, some on the spur of the moment and in a tearing hurry. Decision–making in any complex society is necessarily diffused among large numbers of individuals, not all of them formally charged with the responsibility. Politicians do make decisions, but so do public servants, members of the public, and all sorts of other people. Coordination is therefore a serious problem in any complex system.

Policies not only have to be made, but also implemented; this can be costly, and sometimes, if they are unpopular, those charged with carrying them out will drag their feet, reinterpret, or sabotage them. Farmers, for example, will often resist policies regulating salination or land clearance — let alone land retirement! — unless they can see clear advantages for themselves. Similarly, pollution laws require rigorous enforcement.

Influence; Resources; Bargaining

Influence, the capacity to get people to do things they would not otherwise have done, is the currency of politics. Like regular money, it can be accumulated, but it is subject to even more rapid inflation: it must be continually renewed to be useful. It is also positional, so that 'lame duck' politicians nearing the end of their terms may find their influence rapidly eroded. **Resources** convertible into influence include money, votes, friendship, obligation, and even the threat of coercion. Votes are very important, because demonstrated support increases influence, especially if uncoerced.

Public policy-making has to compromise differing interests; 'perfect' solutions are rare, and most policy decisions are 'second-best' in some way. Strategies for bargaining become important. But bargaining can have undesirable side-effects. Logrolling, for example, can result in distributional conflict, in which short-term gain prevails over long-term self-interest, but everybody is ultimately worse off. Bargaining is very often unequal; the more influential can command higher prices for their support and are more likely to have a veto over specific policies. Bargaining creates difficulties where environmental goods are concerned, since the room for compromise is highly restricted. A decision to preserve half a habitat, for example, will not guarantee the survival of the relevant species.

Decision-makers also face conflict between 'public interest', the needs of government as an entity, and **sectional** pressures. Optimal solutions require perfect information and are time-consuming; so often a remedial approach is adopted: 'putting right' the most pressing problems. But the process of taking the first satisfactory solution to arise — technically known as **satisficing** — may ignore eminently desirable policies. By its very logic, bargaining is especially likely to tempt the decision-maker into satisficing.

Making Decisions

What is a decision? Common parlance assumes that everyone comes, at some point in time, to a situation where choices must be made among various alternatives, whether it is a trivial matter such as what to eat that evening, or a serious one such as buying a house or car. But this tends to assume that there is such a thing as a 'decision', that it occurs at a discrete point in time, and that — in principle at least — it can be 'detected': i.e., that the decision maker can reach back in memory and say 'Yes, on the sixteenth of March last, at 2:44 pm, I made up my mind to invest $20,000 in Central Landscape Destruction'. The problem is that the person concerned may have been thinking about the move for some time before, and have been waiting (for example) for a favourable share price or the arrival of some loose cash from another source; or the decision may have been contingent, in the sense that no other worthwhile investment was available at the time. In political systems, the 'decision' process can be quite stately, with long investigations, repeated meetings, and major delays. While this may reflect a genuine desire to arrive at a fully informed, deliberate decision, it more frequently means that the government's mind has long been made up, but it has encountered serious and determined opposition within its own ranks. Enthusiasm for 'fast-tracking' allegedly important projects reflects frustration with this exact problem.

However, despite the difficulty of identifying decisions, orthodox thought about decision-making long assumed that decisions were sufficiently discrete to be used as a building block in political analysis, and, further, that the pattern of decision-making could itself be analysed.

For many years, the most influential line of such thought focused on the notion of **decision strategies**. A decision strategy is a generalised procedure for the defining of problems, the gathering and analysis of data, and its interpretation in arriving at a decision. Three major strategies have been discussed at length in the literature:

1. the synoptic, rational-comprehensive, or 'root' method;
2. the incremental or 'branch' method;
3. 'mixed scanning', a hybrid of the other two.

All have weaknesses, the common factor being a tendency to formalise the decision process beyond the limits to which it goes in real life.

The Synoptic Decision Strategy

The synoptic, or rational-comprehensive decision-making model, assumes that the decision maker will establish goals, then look for suitable means to achieve them ('policies'), and will not decide finally on the means until comprehensive information has been collected on the problem. Lindblom sums it up thus:

1. Clarification of values or objectives distinct from and usually prerequisite to empirical analysis of alternative policies.

2. Policy formation is therefore approached through means–end analysis: first the ends are isolated, then the means to achieve them are sought.

3. The test of a 'good' policy is that it can be shown to be the most appropriate means to desired ends.

4. Analysis is comprehensive; every important relevant factor is taken into account.

5. Theory is often heavily relied upon.[1]

This is a **maximising** strategy: it seeks to isolate the end or ends to be served by the policy and secure as much of it/them as possible. It obviously requires precision in the specification of goals, as well as much factual information, before ends and means can be related and policy made. It is also the prototype of the strategy advocated in much of the planning literature, and is the inspiration for such 'once-and-for-all' decision techniques as environmental impact assessment (EIA) and cost-benefit analysis (CBA). Its shortcomings as well as its strengths are of great importance.

But Lindblom and Braybrooke claim that 'the synoptic ideal is not adapted' to

1. 'man's limited problem-solving capacities'. Problems are often so complex that to work through them in careful detail is impossible — ways have to be found to 'cut corners'.

2. 'inadequacy of information'. Complete information is often unavailable or costly, yet the synoptic strategy requires it.

3. 'the costliness of analysis'. There are time and resource constraints on information analysis, especially for complex problems.

4. 'failures in constructing a satisfactory evaluative method'. We have no all-purpose evaluative technique and probably never will.

5. 'the closeness of observed relationships between fact and value in policy-making'. Particular policies will favour some values over others; they cannot be considered in isolation. Policies offer a 'mix' of values, which must be optimised, in typical cases. Values are unstable over time and compete with each other.

6. 'the openness of the system of variables with which it contends'. It is easier to solve a problem if it can be isolated, as required by the synoptic strategy, and if all the variables can be readily identified and their interrelationships understood. But real-world problems often have wide-ranging implications, and their variables have unknown relationships to others remote from them.

7. 'the analyst's need for strategic sequences of analytical moves'. In other words, it doesn't tell the analyst how to proceed nor how to identify suitable solutions once they have been found.

8. 'the diverse forms in which policy problems actually arise'. Not only do problems tend to interlock and have interlocking solutions, but policies are often trade-offs among elements of the cluster rather than a maximising of predetermined goals.[2]

The synoptic strategy, then, often fails because it is very difficult to gather together, analyse, and translate all the necessary information into decisions. Information is costly, time often short, and human capacity for exhaustive analysis restricted. The strategy therefore has limited utility as a technique for the decision maker, and is often inappropriate. Consequently 'short-cuts' which subvert comprehensiveness will often be taken: these can be dangerous, especially where it is implicitly assumed that a problem has been taken into account. For example, electrical equipment which has been designed without thought for safety can deal out nasty shocks; environmental impact assessments which conceal their use of inadequate data are misleading at best. Advocacy of the synoptic method can therefore be dangerous.

Nor is the synoptic model very common in practice. It is therefore inadequate as a description of how decisions are actually made. Lindblom proposes that the incremental model is at once a better strategy and a better description of actual policy-making processes.

The Incremental Decision Strategy
Of the 'short cuts' the most important is **disjointed incrementalism**. When political scientists began to study budgeting and decisions in US Congressional Committees during the 1950s, they were surprised to find that the synoptic method was not used. Instead, to avoid rethinking policies from scratch, politicians would often ask 'what did we do last year?' and 'did it work?' Rather than trying to rethink each policy from scratch, the assumption was that it was reasonably satisfactory but would need some adjustment. They thus proceeded by **incremental** adjustments; means and ends would be considered simultaneously, and satisficing was the norm. As a lot of political decisions, especially budgeting ones, are **iterative** — regularly repeated — this is quite rational.

Lindblom argues that the incremental or 'successive limited comparisons' method is better because it takes account of the impossibility of complete analysis, it is adapted to work with poor information, and it is capable of adjusting to multiple goals rather than maximising a single one. It permits the examination of means and ends to go on simultaneously, thus allowing the implications of particular policies to be followed through.

The characteristics of the strategy are:

1. Selection of value goals and empirical analysis of the needed action are not distinct from one another but are closely intertwined.

2. Since means and ends are not distinct, means–end analysis is often inappropriate or limited.

3. The test of a 'good' policy is typically that various analysts find themselves directly agreeing on a policy. (Without their agreeing that it is the most appropriate means to an agreed objective.)

4. Analysis is drastically limited:
   i) Important possible outcomes are neglected.
   ii) Important alternative potential policies are neglected.
   iii) Important affected values are neglected.

5. A succession of comparisons greatly reduces or eliminates reliance on theory.[3]

This is a **satisficing** approach with a **meliorative** orientation. It works by building directly on existing policy and available information. As public policy is very rarely made completely from scratch, for obvious reasons, it provides a series of clear steps for the policy maker by directing attention to the 'worst' problems first. Satisficing permits termination of the analysis as soon as the first *satisfactory* policy emerges, thus saving time and resources. By so doing, it reduces the problems of attempting to optimise or maximise — which may involve massive informational requirements — to more manageable proportions. Incrementalism, being iterative, is well-adapted to repetition — as in annual budgeting — and is at least in theory receptive to feedback from the actual operation of policy. Thus gains can be consolidated, and mistakes abolished, before too much damage is done.

Lindblom argues that decision-makers rarely have time to work through a formalised means–end analysis. Their information is incomplete and is limited by acquisition costs and time constraints. Their analysis is partial, often consisting of comparisons between marginally different policies, which is a primitive way of 'controlling' variables. Not all policies are examined: 'reasonable' limitations on possible options are accepted on *prima facie* grounds. The need to consider multiple goals means that trade-offs rather than maximisation are the rule. (A practical advantage is that 'winner take all' can produce disgruntled losers; something for everyone is better politics.) Lindblom and Braybrooke claim that this is close to an ideal strategy, especially in 'open' societies.

Because incrementalism assesses means and ends simultaneously, and because its test of a 'good' policy is that those involved can agree on it — regardless of their own ultimate ends — it is well-adapted to the bargaining situations that permeate everyday politics.

But incrementalism has been criticised for the arbitrariness with which possible alternatives are excluded; this limits sharply its interest in and receptiveness to information. The resulting narrowing of the decision maker's field of view, critics have argued, makes incrementalism unduly conservative. Lindblom has demurred, arguing that incrementalism is compatible with quite rapid change. But he and Braybrooke both argue, in support of incrementalism, that in fact rapid social change is very difficult in any case; thus they appear to be attempting to have their cake and eat it as well. The argument that arbitrary exclusions of policy directions will occur is not met by Lindblom, and remains a powerful criticism.

Additionally, incrementalism is unlikely to be sensitive to problems involving limits or thresholds. This is for two reasons: firstly, most incrementalist strategies implicitly assume that policy processes are reversible. Thus if a mistake is made, it can quickly be reversed, once the ill-effects are known. This does not sit well with ecological processes displaying irreversibility, since irrevocable damage — such as extinction of potentially useful species — may have been done by the time feedback is available. Secondly, incrementalism assumes that policy responses to particular levels of effort are roughly linear. Thus, if this year's budget applies $10,000 to a problem, and some improvement is seen, then $100,000 ought to do ten times as much good. This may not happen, because the amount of money or effort may reach saturation, at which point returns diminish, and no additional amount can effect further improvement; or alternatively, incrementalism may fail to detect a threshold. If, for example, a minimum of $100,000 is needed to make a noticeable impact on a problem, then an experimental allocation of $10,000 will produce no discernible improvement, and lead to no further allocations. In short, incrementalism is likely to cope poorly with lumpy goods, the more so because, in interpreting feedback, simplistic assumptions about the processes involved are made.

Dror argues that incrementalism is designed to maximise security in making policy, by attempting to offset the effects of poor information and uncertainty. But three essential conditions must be met:

1. the results of the present policies must be in the main satisfactory (to the policy makers and to the social strata on

which they depend), so that marginal changes are sufficient for achieving an acceptable rate of improvement in policy results;

2. there must be a high degree of continuity in the nature of the problems;

3. there must be a high degree of continuity in the available means for dealing with the problems.[4]

In other words, incrementalism is likely to work well in times of economic and social well-being, when major change is not perceived as necessary, but gives no guidance at other times. This weakness is critical: Dror argues that 'when the results of past policies are undesirable, it is often preferable to take the risks involved in radical new departures'. Under stable, predictable conditions, outcomes of policies are likely to be better understood than under unstable ones, and incrementalism is consequently more reliable.

Lindblom grants this objection, but does not see it as serious. Dror, counterattacking, argues that the real cause of conservatism in incrementalism lies not in its restriction of attention, but in its endorsement of inertia in policy-making agencies. He argues that they are often indolent and do not explore viable possibilities; the synoptic model at least demands that they bestir themselves, while the incremental excuses inertia, leading to far more 'muddling through' than is necessary or desirable.

Dror's position has been powerfully endorsed by Robert Goodin, in his book *Political Theory and Public Policy*, in which he argues that incrementalist attitudes are often an excuse for inaction, and underpin claims that unfavoured options are impossible. In particular, he argues that (contrary to some critics of Utilitarianism) it is quite often possible to foresee the outcomes of policies, and that a wilful refusal by policy makers to take account of those outcomes constitutes a powerful reason for rejecting incrementalism as a method for policy-making.

'Mixed Scanning': a Hybrid Strategy

Dror argues that good policy-making requires an attempt to increase the rationality content, through explication of goals, search for new alternatives, exploration of expectations and outcomes, and clearer formulations of criteria for decision. Extra-rational processes such as intuition and knowledge of the situation 'on the ground' should not be ignored; and use both of spe-

Chapter 11

cialised knowledge and of 'brain-storming' may improve the process. Since the synoptic method is too comprehensive, and the incremental too blinkered, Dror suggests a 'mixed strategy' in which an attempt is made to determine whether a policy of minimal risk or one of innovation is desirable, and a strategy adopted to match.

Dror's mentor, Etzioni, proposes such a strategy under the name of 'mixed scanning', resting on a distinction between **contextuating** (or fundamental) and **bit** (or item) decisions. Contextuating decisions are goal-related overviews which omit detail data in order to gain breadth. Bit decisions follow them, and are incremental within the context thus defined. Etzioni argues that the decision maker's need to survey the whole area of interest is occasional; much of the time only particular issues and problems need be in focus. Thus in making *new* policy, a preliminary survey of the whole 'policy space' would locate interesting policy directions, at the same time eliminating the obviously unfeasible. Each 'promising' approach would then be examined, progressively eliminating those to which crippling objections become evident. Finally a choice is made between two or three possible policies. Each of a hierarchy of contextuating decisions generates bit decisions, which in turn become the contextuating decisions for the next level (or stage) of the process. For example, middle-level decisions may be contextuating for those below and bit in relation to those above: a nested hierarchy, in short.

Etzioni uses the illustration of a satellite remote sensing device. This has a wide-angle lens to survey the ground below it; the pictures are transmitted back to earth, and the results scanned by the analysts for anything of importance. If some phenomenon of interest is spotted, the satellite can be ordered to switch to a telephoto lens, in order to take a photo of a much smaller area, but at considerably higher resolution. Etzioni compares this with the process of scanning a room just entered, or a soldier entering a battlefield making a quick survey in order to locate danger spots.

'Mixed scanning' therefore works by making strategic occasions an opportunity for a wide review of alternatives, including those which at first sight seem unfeasible. Those displaying crippling objections are eliminated and attention focused on more promising areas which are examined in progressively more detail. The process is repeated until one alternative emerges. This policy is then broken down into its components, about which a series of 'bit' decisions are made, and implemented in steps, facilitating feedback on effects. Decisions to continue can be made conditional on this feedback. Review of the policy and its effects continues while implementation is in progress.

This approach is well adapted to feedback, because decisions can be implemented step-by-step — permitting flexible response to implementation problems — and in addition, 'bit' decisions, essentially incremental within the framework of their contextuating decision, are themselves small enough in scale to permit very thorough detail examination. Etzioni also argues for comprehensive reviews either at regular intervals or when certain kinds of problems — which might make the existing contextuating decisions doubtful — come up. In other words, 'trip wires' can be built into the system to trigger review at certain critical junctures; provided that they can be foreseen. Thus, in combining the synoptic and the incremental approaches hierarchically, the 'mixed scanning' method also addresses the problem of *relevance:* how to choose only appropriate information and to eliminate the irrelevant.

At the same time, Etzioni rejects Lindblom's argument that decisions must always deal with multiple goals. He argues that many social decisions do have a single major goal — generating power or providing pensions, for example — and that hence a more synoptic method, where maximising is the aim, is well adapted to these. Secondly, the ability of the mixed scanning strategy to discriminate between the *kinds* of decisions needed at different junctures is an obvious advantage, likely to result in considerable economies of effort. In addition, the method is not closely tied to a particular social philosophy or style of social interaction, with consequent limitation in scope.

A difficulty with any hybrid approach is the need to avoid the pitfalls and maximise the virtues of its elements. Further, it needs to specify when the techniques appropriate to each should be used. Seizing on this point, Davis, Wanna, Warhurst and Weller have criticised 'mixed scanning' on the ground that it does not offer a guide to which decisions should be incremental and which synoptic; that it is vague about the boundaries between the general and the particular; and, in short, that it has the disadvantages of both its precursor systems.

This harsh view may betray a misunderstanding of Etzioni: firstly, because it is obviously a matter of judgement in each specific case exactly where each strategy may be introduced; and secondly, because Etzioni's notion of a hierarchy of decisions is a genuine attempt to come to grips with the structure of decision problems, implying that the decision and its implementation should mimic the problem structure. It might be that that itself would inhibit novelty, though this would be true only if novel approaches were never adopted in the overview phase.

A more telling criticism is that 'mixed scanning' does not really overcome the problem of failure of comprehensiveness in its synoptic phase. Etzioni's implicit assumption is that *some* comprehensiveness can be achieved, without significant distortion due to its own inherent failings. While he does attempt to deal with one of the central problems — finding a principle of selection, to eliminate unneeded data — Etzioni fails to resolve the question unambiguously. The danger is that if comprehensive overviews do fail to exclude data systematically and without compromising relevance, 'mixed scanning' will replicate the arbitrary exclusion tendency of incrementalism. This, of course, is detrimental to its claims to be a 'third way'.

DECISION STRATEGIES AND SOCIAL STUCTURE

Etzioni also perceives an association between decision strategies and political systems. For example, dictatorships are prone to reject their past and attempt rapid and fundamental policy changes. Comprehensive, perhaps — but far short of rational, as incomplete and misinterpreted data often lead to glaring policy deficiencies, while at best dictatorships tend to overplan. Democracies, by contrast, tend to display many of the worst features of incrementalism, being conservative and too slow to act, especially on social problems and on urgent foreign affairs issues such as stopping dictators before they have murdered large numbers of people. Democratic politics tends to involve more bargaining, trade-offs, and logrolling, in direct proportion to its responsiveness to popular demands. On the other hand, policies in 'open' societies will tend more often to be consensual, and hence acceptable, thus reducing 'decision costs'.

Incrementalism is also associated with **pluralism**, a doctrine that argues that all political systems are split up into a myriad different interest groups, all of which make regular demands on government. Government adjusts policies to these shifting claims, and a process of mutual accommodation at the policy-making level is seen as making for a harmonious system in which everybody gets at least some of what they want. Lindblom has formalised the process in his notion of **partisan mutual adjustment**. Standard criticisms of this view point to enormous disparities in the size and resources of interested groups; to the disenfranchisement of some groups, such as the poor or ethnic minorities, who belong to no powerful, organised groupings; and to the fact that pluralism largely ignores the interests and ideological affiliations of politicians. Like incrementalism itself, pluralism suffers from an arbitrarily restricted world-view, in which nothing is fundamentally wrong.

This is Etzioni's point of departure in proposing his 'active society', based largely on a 'mixed scanning' approach to policy decisions. This is obviously intended as a way out of the problems of both the synoptic and the incrementalist strategies by combining their best features. Implied is a similar 'halfway house' at the political level, not so much between democracy and dictatorship, nor planning and pluralism, but between purposeful policy-making and outright drift. The notion of the 'active society' therefore depends on problem-focused decision processes rather than partisan ones; and public decision-making for public benefit, not private profit.

Decision-making is also a function of political intelligence and will. Democratic political systems are often accused of decision-making by drift, an incremental process of marginal change. This, it is argued, leads to an inability to face challenges, especially environmental and economic change, and external pressures. On the other hand, autocratic single-mindedness can also ignore viable policy options. Australian public policy-making is often marred by inadequate discussion of the issues, and a consequent failure to identify desirable policies.

ARE DECISION STRATEGIES RELEVANT?

The discussion so far has assumed for convenience that decisions do actually get made, that there is an element of choice in them, and that decision makers do actually seek strategies to aid them.

No decision maker operates in a vacuum, however.

The 'Myth of Decisionality'

Some scholars criticise 'decision strategy' approaches for creating a 'myth of decisionality', which obscures the importance of the context of decisions, by imposing a spurious voluntarism. Decision makers often report feeling that they have no or very little choice in many typical 'decision-making' situations, and feel compelled strongly by the circumstances to adopt a particular policy. Thus the decision context may be as important as the strategy adopted or the data available. The standoff between the

CHAPTER 11

Australian States delayed the transition from distributive to regulatory policies in the Murray–Darling basin because of the inherent potential for conflict.

Policy options are often severely restricted by available resources, expected responses, for example from an electorate or a foreign power, or by the available knowledge and the understanding (or lack of it) of the policy makers. Ignorance of scientific principles is a major current problem.

Even where the policy process is open and above board, the 'myth of decisionality' may prevent analysts from understanding the processes under way. The implication is that if decision makers have limited autonomy, external constraints, such as the political economy and the structural constraints of specific political systems must be studied if the policy process is to make sense.

Decisions and Non-decisions

In a brilliant paper published in 1962, Bachrach and Baratz, then of Temple University in Philadelphia, argued that, in a society in which there existed considerable concentrations of power, a focus on decisions *actually made* would be unable to detect such power. Consequently it would shed no light on the 'real' reasons for existing policy. Not only do centres of power and wealth make decisions affecting those at the periphery, but, even more important, one of the most important powers wielded by the ruling group is its power to block or prevent change. This power, they argued, expresses itself as the ability to prevent public discussion, to ignore proposals for change, and to present those in power as sane, thoughtful and capable, while their challengers are — virtually by definition — wild-eyed fanatics. The irony of such a situation becomes especially poignant during nuclear arms races and other such manifestations of leadership madness. The crucial concept was that of the non-decision: a situation in which the powerful could prevent an issue even from reaching a decision.

Bachrach and Baratz later published a study in which they applied their analysis to the city of Baltimore. They showed that for a wide range of issues, the city's rulers could prevent or inhibit public discussion — especially through their control of the mass media — thus disenfranchising the less powerful and subverting the democratic political process. There was no 'free market' of ideas and proposals; the public was spoon-fed from an agenda determined by the powerful ruling clique. The rulers' capacity to block change and censor criticism in this way was an essential underpinning of their power. The existence of **agenda control** of this kind supports a view of social decision processes which underlines the importance of **power**, and locates that power in classes and elites: groups enjoying a disproportionate say in public affairs by virtue of their wealth or standing.

Garbage Cans

Bachrach and Baratz were explicitly critical of pluralism. They rejected the view that power in the community was widely dispersed among a group of decision makers who represented various community interests, as well as its corollary, that the decision-making process was sensitive to individual needs and wants. They were critical of pluralist atomism, which, while recognising that political systems are made up of many different groups which bargain for their needs, tends to assume away differences in power and difficulties in gaining access to government.

A further problem, indeed, is that in 'weak' pluralist states, the agenda for policy may not be set by government so much as by powerful forces within the community. These may include business, the media, various important ethnic and religious groups, and (least likely of all) academics and other knowledgeable sources. Government then reacts to these pressures, and selects among the policy issues thus placed on its agenda rather than raising and stimulating debate on the issues themselves. Occasionally, such hijacking of political debate is quite spectacular, as happened in Australia when the gnomes of John Stone's federal Treasury managed to reorient the political agenda in line with economic 'rationalist' prescriptions.

This is where the metaphor of the garbage can enters, by analogy with derelicts fossicking in rubbish bins. Government, devoid of a programmatic policy direction, selects policies from among those offered to it by the various groups pressuring it, rather than devising its own. The danger with such a process, of course, is that the policies in question come with a 'hidden agenda' attached, generally favouring the sponsoring interest group. This can lead either to policy-making by drift, in which the risk of grounding on a reef of ignored or repressed problems becomes great, or to a surrender, partial or complete, to one or more of the interests involved. The Hawke and Keating government's media strategies seem to be an example of the latter.

'Garbage can' decision-making is likely to be associated with a lack of planning and overall policy direction; it is likely to be reactive, rather than thoughtful and forward-looking; and it is likely, in the short to medium term, to lead to problems of cooperation and coordination, due to the lack of overall thought given to the implications of the policies adopted. This may be particularly undesirable where collective goods are at stake, and where policy coordination is consequently of central importance.

Room for Manoeuvre
A further objection to decision strategies and the associated paraphernalia of formalised processes has been raised by B.B. Schaffer. He points to their rôle in stifling creativity. Not only may incrementalism aid and abet 'tunnel vision', but the pressures of formalised processes in general may limit and constrain decision makers. Schaffer particularly stresses the creative element in decision-making, and suggests that 'room for manoeuvre' is essential. That is, the 'decision space' must not be too constrained, by circumstance, perception or procedure if the best and most creative decisions are to emerge. Decision-making needs to be highly sensitive to those affected, as well as to its own environment; but one of its purposes is transformatory. It must change the very environment.

OVERVIEW: DECISION-MAKING AND POLITICAL POWER

The processes by which political systems make decisions are complex. Decisions may or may not be discrete events, in which a specific problem is isolated and rationally resolved.

For many years, decision theory has relied on the idea of decision strategies, in which decision makers identify problems and follow defined algorithms to resolve them. Argument has raged over the merits of the rational-comprehensive, incremental and possible 'mixed' decision strategies. But in recent times the very notion of discrete decisions has come under challenge as a distortion of reality. Instead, decision contexts are emphasised.

One serious deficiency of the decision strategy approach is that the power to prevent decisions is critical. Control of the media and the public purse often confer power to prevent issues from reaching the agenda of public discussion. Ruling elites can 'block out' discussion of important issues and thus reduce or eliminate public awareness. In turn, useful alternatives will then not be discussed, and the formation of public policy is likely to be biased. This capability to reinforce the status quo relies on **agenda control**: the ability to determine what issues will and will not be discussed publicly, for which control of the press and media is essential.

Another weakness is that over-reliance on a 'cut-and-dried'

procedural model may stunt decision makers' capabilities, either by blinding them to important opportunities or by eliminating their 'decision space'.

Political systems contain many centres of power, which settle important questions by bargaining. Some pluralist political systems may permit their policy agenda to be set for them, adopting 'garbage can' solutions. This may be inimical to collective goods and especially to policy coordination, essential in environmental issues.

FURTHER EXPLORATION

The literature of public policy was for many years dominated by the 'rational-comprehensive' school, of which H.A. Simon was the most famous. The challenge from incrementalism emerged in the early 1960s, one of the most important events being the publication of Braybrooke and Lindblom's *A Strategy of Decision*, with its strong advocacy of incrementalism. The literature for the next two decades was largely devoted to argument over the relative strengths of alternative strategies, punctuated by occasional dissent.

Essentially, the critique of incrementalism took two forms. Firstly, there were technical criticisms, such as those of Dror, Etzioni, and many others who argued that incrementalism did not live up to its promise, and that it encouraged conservatism and torpor. The most sophisticated of these is Goodin's, in *Political Theory and Public Policy*. The other line was the radical, power-based critique, of which Bachrach and Baratz' is the best known; their article of 1962 was later incorporated into *Power and Poverty*, still a very influential statement of their position.

The consensus that there are serious deficiencies with the decision strategy approach has emerged much more recently, in part as appreciation of the implications of social choice theory grows. The argument is stated very succinctly by B.B. Schaffer in 'Towards Responsibility: Public Policy in Concept and Practice' in Clay and Schaffer.

Davis, Wanna, Warhurst, and Weller offer a useful and concise overview of all these perspectives in Chapter 7 of their *Public Policy in Australia*.

Part 4

Political Economy and Public Policy

Governments make policy decisions within a framework of constraints, some of which are outside their control. Game theory shows how the pattern of interaction can be constrained by the available options, expressed as 'payoffs', as well as the structure of the interactions themselves. Just as their mode of adaptation has important impacts on how humans relate to their natural environment, so the nature of state societies imposes limits and priorities on their actions.

Part IV deals with *political economy*. It shows how all state societies must necessarily give priority to survival and maintenance of internal order, and how these priorities — especially since the Industrial Revolution — flow on into other major policy areas, such as 'economic transformation'.

The result is a pattern of major convergences of interest between government, industry, and (frequently) the military: the 'military-industrial complex'. This distorts the pattern of resource allocation on a national scale and creates various 'no-go' areas of policy, which are particularly likely to be defended by 'non-decision-making' and agenda-control techniques. In Australia, the most spectacular of these is the 'settled policy' of development, which has dominated politics for nearly two centuries.

Australia's political economy has been built around a pattern of state-sponsored 'development', in which the rapid exploitation and sale abroad of natural resources has loomed exceptionally large. Government involvement in 'development', largely through overseas borrowing to provide 'infrastructure' for industry and commerce, has led to a distinctive pattern of governmental participation in the economy. This in turn has bred numerous environmental problems.

CHAPTER 12

THE STATE IN ENVIRONMENTAL MANAGEMENT

Why is government so frequently unresponsive to freely available scientific data? Why does it proceed with programmes even when the weight of advice suggests that they will not be successful? For a short-term answer it is often sufficient to look at the political pressures of the time and see what relief the offending programme offers. Thus the promotion of irrigation in the 1860s–1880s was a powerful answer to the pressing political problem of 'closer settlement', in the context of declining activity in gold-mining.

But the environmental insensitivity of government goes much deeper; it is a well-documented fact, extending far back into history. To understand it, it becomes necessary to look at the problems facing government *as a form of social organisation*: that is, state society. (In this context, the term 'state' is not used to mean a small component of a larger nation, as in the Australian States, but rather to denote a venerable form of social organisation, with distinctive characteristics.) The study of the pressures and constraints on state societies, and their reactions to them, is part of **political economy**. This, the study of the political, social and economic pressures on government, is essentially complementary to the more embracive notion of *political ecology,* previously discussed.

Australia's problems in this area are not unique: environmental problems have a long history, and some have triggered off major crises. Many of the earliest known human-induced environmental disasters are associated with state societies. States have existed for at least 9,000 years, and have evolved at different times and different places, apparently in response to similar ecological stimuli. They radically transformed nearby societies and were a major pressure for growth, both of population and of economic activity.

The large populations of modern times make coordination essential, leading to a dependence on government that was less necessary in earlier, less 'developed' societies. This dependence has ecological importance, because in order to support the huge populations they have encouraged, governments have to exploit the environment. This leads to a conflict of rôles, because in general governments are also the only authorities powerful enough to secure common goods, including environmental protection.

In fact, the nation-state's rôle involves a central conflict: long selective pressure has made it short-sighted and expansionist, with a commitment to development and growth, yet government is the only authoritative body capable of coping with longer-term, collective problems such as environmental management. The 'rationality' of state systems often leads to their pushing aside well-established, ecologically sound practices in favour of short-term goals. Resource extraction for political and military purposes increases pressure on natural ecosystems.

From their inception, states have treated environmental management questions as secondary to the more pressing issues of survival and interstate competition. The resulting attitudes and ideologies persist; the predicament of the state is a direct result of the pressures on it.

STATE SOCIETIES

Governments are very old and pervasive human institutions. Nowadays nearly everybody in the world is directly affected, every day, by the power of some government. Governments penetrate further into everyday life and are more powerful than at any previous time in human history.

Governments, organised as **states**, everywhere emerged as a result of **ecological imbalance** or stress, and their structure and organisation was strongly influenced by the selective pressures of

CHAPTER 12

interstate competition. This meant that they had to mobilise resources in order to survive, which implies exploitation of the natural environment. Thus the conflict between exploitation and protection of the natural environment is deeply embedded in the very structure of state societies themselves.

Pressures on Pre-state Societies
Though states — social systems with differentiated, specialised political institutions — emerged as a response to ecological stresses encountered by pre-state societies, their emergence led to an amplification of those stresses.

The earliest human societies were 'hunter-gatherer' societies in which no cultivation of crops or animal husbandry took place. People hunted animals and gathered fruits and other edible vegetable matter. Such societies are now rare, though the !Kung bushmen of southern Africa still follow this lifestyle, and Australian Aboriginals were hunter-gatherers prior to European settlement. Even here, there was some *division of labour*: that is, particular persons performed different tasks, in order to produce more for the community, with less overall effort. And, despite the lack of any formal government, functions such as adjudication of disputes and collective decision-making existed; but they were diffused, unattached to any particular person or office, and performed in an *ad hoc* manner, when and as required.

But by neolithic times the adoption of agriculture had led to considerable specialisation. Sowing and harvesting took place at different times of year, and new tasks, such as preparing, grinding, and cooking grains, appeared. Once shifting agriculture had been abandoned, and settled agricultural villages became the norm, extensive and minute specialisation became possible. The need for coordination meant an expanded leadership rôle: some people specialised in giving directions and making decisions, for example about when to plant and harvest. Furthermore, some agricultural societies began to show social stratification: differences in rank and status. 'Important' people, who gave orders to the 'unimportant', increasingly emerged. This was a critical development, because the first state societies developed from agricultural societies with established settlements and some social stratification, often with technologies such as irrigation.

Social stratification is thought by most scholars to be an important precondition for the emergence of states proper. It is often associated with scarcity: by 'cornering' essential resources, powerful people within society can deny them to those who do not 'toe the line'. Scarcity, and the associated ecological imbalance, themselves underly the emergence of social and, later, political power.

Emergence of the State
To explain the emergence of state societies, modern scholars stress the interaction of ecological pressures with social structure, culture, and other factors. The *ecological pressures* are typically mismatches of resources with demand: for example, shortages of food or other staples. These are frequently caused by population growth, though it is important to note that other factors such as climatic change can have similar effects. It is *ecological imbalance* that is critical here, not population growth as such. Some imbalances can be temporary — for example, as a result of poor harvests following bad seasons — while others may be permanent, as when declining soil fertility or soil salination reduce yields. Societies facing such imbalances must find strategies for rectifying them, or, if temporary, smoothing them out.

In a region of ecological diversity — where, for example, bad harvests might affect one region, but never all — there may be considerable advantages in pooling resources, so as to average the seasonal variations; that way, all else being equal, a larger population could be supported by a cooperative grouping than by a single unit. However, achieving such cooperation may be quite difficult, especially if there is a history of mutual mistrust or conflict. Alternatively, a society which had experienced population growth, and was consequently experiencing a shortage of resources, might cast covetous eyes on the resources of neighbours who had been more frugal. The problem then becomes how to lay their hands on the resources.

This is particularly so because pre-state societies typically did not have large accumulations of tradeable goods with which they could purchase food or other needs when in short supply. They therefore lacked the means to stimulate temporary flows of resources. They could not get credit against the promise of future payment.

An obvious response to this predicament is warfare. Even in pre-state societies there appear to have been strong links between warfare, male supremacy, and population pressure. A society with a surplus population and a shortage of resources will have manpower available for mobilisation, and superior numbers to ensure success in combat. It is unlikely to resist the temptation to conquer a less numerous, frugal neighbour with resources to spare. But in doing so, it creates an immediate paradox: the more 'ecologically

Chapter 12

balanced' and 'responsible' neighbour is overrun, and the warlike, expansionist society predominates. In all probability, it will again get into ecological imbalance and need once again to expand. Virtue is not necessarily rewarded; and competition between societies results in the most powerful, not the 'best' or most 'moral', winning.

Naturally, there is considerable variation in the various mechanisms which have been observed — or, more typically, reconstructed from available archeological evidence. Consequently, historical anthropologists all have their favourite theories about the conversion of social power to political. Harris stresses the growth of **competitive redistribution**, in which 'big men' seeking social status by accumulating and distributing wealth eventually acquire real power, and convert this to social control. This process went on in the Pacific northwest of North America, through the custom of **potlatch**. Webb, by contrast, emphasises chiefly power in highly organised, **circumscribed societies** already engaged in extensive trade. He sees a relatively high level of economic development, accompanied by intensive exploitation of the circumscribed resource base, as important: established trade followed by military struggle over resources triggers transition from chiefly proto-states to state organisation proper.

Circumscription requires a steep **ecological gradient** between the resources most central for production, and the outside environment, thus preventing defection by the conquered and the recalcitrant. Examples include Egypt, a fertile river valley surrounded by desert, where a series of separate societies progressively merged to form, first states, and then an empire; and the river valleys of Peru, surrounded by barren mountains, where the amalgamation process was lateral, linking adjacent valleys. But Webster rates **ecological diversity** more important: availability of resources in a nearby society at times of scarcity can trigger competition via warfare, leading to looting or actual takeover of the society concerned. All, however, agree on the significance of ecological diversity in combination with circumscription.

Hockett stresses another factor of central importance: takeover rather than looting. By contrast with looting, which requires no very high level of organisation, takeover of another socio-political system will require an ongoing administrative structure, leading to institutional innovation. The kinds of activities implied are exactly those which developed in early states.

There is every reason to suppose that all of these factors have been important in the emergence of states: differing material circumstances can easily lead to differing development paths. The crucial point on which all scholars are agreed is that, under ecological stress, a novel form of social organisation emerged, in which, for the first time, the functions of government became part of the regular division of labour in society. Pre-state societies had chiefs; state societies kings, Pharaohs, and emperors.

Even the early **pristine** or primary states were so strong that nearby societies either had to reorganise themselves in the same way or submit to conquest. The reorganised societies are called **secondary** states. Thus the existence of states requires a **state system**: in regions within the reach of a state, other states must form, or their societies will be absorbed. Intersocietal competition, which certainly existed long before state society, is therefore sharpened by the emergence of states. The state system institutionalises competition and makes it more regular, as well as more deadly. A greater share of resources is devoted to competition, primarily by means of warfare.

State Structure and Function

This history of struggle over resources, territory, and populations shapes the structure and functions of the state. Its conversion of low-level, sporadic conflict between societies into organised, structured conflict between states — **interstate competition** — exerts a selective pressure on states. They must look to their own survival first; just to ensure survival in the short term, they must organise militarily and put the needs of the military ahead of other considerations. This requires resources, which must be extracted from the rest of the population, by way of taxes, forced labour, and so on.

There are also two structural consequences: growth of administration, and increased social stratification. Any state will require continuing administration: as a minimum, maintenance of control, allocation of resources, and collection of taxes. Where populations have been taken over, they will have to be prevented from revolting and seizing back control of their own resources; where territory has been conquered so as to alleviate a shortage of staples (foodstuffs, for example), a system will have to be devised, both to redistribute the staples, and to keep track of them. The earliest known writing is no more than a series of tallies, aids in keeping track of stocks of some kind.

But as writing developed, its possibilities for recording, not only information about stocks of materials, but also data about citizens and the myths and legends of the community, also grew. In pre-lit-

CHAPTER 12

erate societies, myths and legends often recorded especially important information about environmental conditions and techniques, wrapped up in mnemonic envelopes such as rhyme and song. The emergence of literacy was an important shift, because it made possible 'science', in the sense of written, technical data without mythic trappings. In ancient Egypt the yearly flooding of the Nile, which fertilised the fields, was a critical event. Villagers had to be ready for it, both to avoid drowning and to plant their crops at the right time. Thus not only did the Nile become central to Egyptian mythology and religion, but the priesthood, seeking to make themselves indispensable, cunningly added calendar-making to their activities; this required both writing and arithmetic.

Administration increases social stratification. In addition to the actual rulers, states develop a functionary class, consisting of the officials who keep the records, maintain the calendars, and teach writing and arithmetic. Often, though not always, these are linked directly with the priesthood, thus augmenting structural power by religious sanctions. The material needs of these classes are a further pressure for growth and enhancement of extractive power: to free labour for administration, the productivity of peasant labour has to be increased.

Additionally, ruling classes typically seek to increase their share of the wealth, or to provide for a growing population. The problem boils down to a simple dilemma: in order to increase the resources available to rulers and functionaries, the state must either tax more peasants, or tax the peasants more. Both have limitations. Military expansion is limited by increasing costs as well as by technology. The larger the territory, the larger the populations to be fed, housed — and administered. On the other hand, taxes cannot be increased to the point that there is not enough left for the peasants to survive. Often states do both: they encourage population growth while seeking to expand territory, and tax as heavily as they dare.

They may also intensify the productive system, through irrigation or other major works, which make possible greater production per unit of area or of labour. This requires the mobilisation of considerable manpower, giving the state an entrepreneurial rôle. State encouragement of population growth can easily create a vicious cycle of intensification, in which utilisation of both land and labour becomes increasingly heavy. In some societies (modern Java is an example) almost every inch of land will be exploited. If novel technologies or new resources are not found, intensification can lead ultimately to crisis.

Specialisation of function in economic activity is greatly

increased by the greater scope of state society as well as its need for novel activities. The consequent efficiencies produce rapid expansion in population, the economy, in social differentiation, and political activity. The emergence of a 'better-off' ruling class meant a demand for consumption and luxury goods which encouraged more and wider trade. However, the hunger of the wealthy classes of early states for consumption goods, such as gems, precious metals, exotic textiles, spices, and much else, caused resource overuse and expansionism for plunder. There was little reinvestment of wealth, which meant that the explosive growth of early states often dwindled into stagnation. This was reinforced by diminishing returns to military effort, especially in multi-state systems.

Internally, states repress dissent and opposition; those who were conquered must be prevented from organising or even expressing their dissatisfaction. Other criticism of the operation of the system must be contained or put down. The result is expansion of the internal repression function, shared by the military, 'police', and networks of spies and informers. The state rapidly becomes 'stronger than society', moulding and constraining the social sphere.

The greater 'rationality' — goal-direction — of state organisation may be a positive disadvantage in ecological terms: it replaces culture with policy, and may discount practices previously conducive to ecological stability, thus destroying 'automatic' balance. Taboos on particular foods or activities, for example, may be discarded as 'superstition', resulting in the over-exploitation of particular resources.

Unnatural Selection

Because, once formed, states found themselves competing for territory and resources, interstate competition became a continuing preoccupation, especially where growth was through territorial expansion. This competition became a kind of 'unnatural selection': some states survived, others became extinct. Competitive pressure, however, meant that states had to put short-term goals — such as mobilising resources for war — first simply in order to survive. This often meant that they failed in the long run, through ecological collapse. Weiskel instances Mesopotamia, Phoenicia, Palestine, Egypt, Greece and Rome in the Mediterranean alone. Often the technology adopted was to blame, as in Mesopotamia, where salination eventually overcame a whole civilisation based on irrigation agriculture.

Chapter 12

Expansionist states thus enjoy a selective advantage over static societies, introducing a paradox. Expansionism intensifies environmental stresses, and may ultimately doom the state to extinction. Thus interstate competition may work directly against long-term stability. Like natural selection, unnatural selection is not a process which necessarily results in a happy outcome. It is non-teleological — that is, there is no set goal toward which it trends — and blind. Outcomes are contingent — the result of the interaction of circumstance — not designed or foreordained.

Empires

In premodern political systems the bulk of the population were poor, engaging in subsistence farming, or acting as slaves or servants to the few wealthy citizens. The cash economy was small, and politics impinged on most people only when they were conscripted or saw their livelihood devastated in their rulers' wars. Many governments did not even collect taxes directly, instead 'farming out' the work to local notables. This led to considerable variation in tax rates, as well as extortionate attitudes on the part of the tax farmers, for very often, once they had remitted a fixed amount to the government, the remainder was theirs to keep. Poor peasants saw little or nothing back in the way of services for their taxes; instead, they were often called on to fight for their remote rulers.

Empires of antiquity, though limited in physical size by transport technology, and in administrative capabilities by both communication technology and organisational capacity, allocated large shares of their budgets to military purposes, absorbed a larger share of the productive surplus than other societies in the process, and were most successful at mass mobilisation. Their major ecological disasters included deforestation and desertification in North Africa and Asia Minor.

Only a few — notably the Roman Empire — ever succeeded in mobilising more than a tiny fraction of their wealth and population. Only the Chinese empire developed an extensive administrative bureaucracy, recruited through competitive examination, and offering some — though very limited — opportunities for social mobility.

Poor communications and slow transport combined to make the physical extension of states beyond certain limits quite impossible. But some states did expand to the limits of their technological and organisational capacity. This could involve great distances. The Roman Empire was so large that troops took two weeks to move from one end to the other; the Chinese extended to boundaries a week from the capital by the fastest means of transport then available. Both empires placed emphasis on transport and communications, the Chinese via a system of roads and posting stations, where messengers could get replacement horses; the Romans by a system of well-engineered roads designed for rapid movement of troops on foot.

Empires concentrated wealth and privilege; to do so, they tended to plunder and loot the extremities. Often taxation levels varied sharply between the centre and the periphery, the outlying areas being loosely ruled but heavily taxed. Often, too, peripheral areas were subjected to near-continuous warfare, and as a result were extensively depopulated.

The social burdens often found their expression in ecological stress as well. Forced to intensify their exploitation of the resource base, subject populations were frequently forced to adopt practices they themselves knew to be inimical to long-term productivity and stability. An example is North Africa: in the days of the conflict between Carthage and Rome, this was a rich agricultural area, which became the granary of the Roman Empire. Now, 2000 years later, most of the Mediterranean coast of North Africa is desert.

Ancient states were limited in their extent and power by technology. Communications were poor, transport expensive, and warfare costly; even those empires which most energetically improved communications were limited in their geographical scope. Ancient states also had primitive administrative capabilities: even their control over territory near their capitals was limited. They were also hostile to technological innovation, which was generally seen as threatening to established political power and social order.

European Expansion

The exception was Europe. Plunged into chaos by the collapse of the Roman Empire in the fifth century AD, Europe became a welter of small, fragmented political units, all vying for power and territory. The political centres of gravity, as well as the centres of culture, were at this time further afield: Europe was frontier territory. Intense military competition led to rapid innovation, mostly in the technologies of war. Some inventions, such as stirrups, the horse-collar and the use of horses for traction (replacing oxen, which were much slower), had direct civilian spinoffs; others, such as gunpowder, did not.

Chapter 12

But even by the Middle Ages, conditions in Europe were exceptional: energy use was greater, technological innovation more rapid, and social change, though slow by modern standards, was proceeding apace. One result was extensive foreign trade, and the early emergence of institutions — such as banking — which were to be essential to the later emergence of capitalism.

In the late Middle Ages, with the rise of Arab political power and of central Asian empires which threatened the ancient trade routes to the Far East, Europe faced a shortage of trade goods, including some which were vital to daily life. Spices, for example, were essential for the preservation of food, and could not be grown in Europe. In consequence, coastal states such as Portugal and Spain began to develop improved navigation, with the aim of reaching the Spice Islands (modern Indonesia) and other desirable destinations directly. Coincidentally, they were aided in this by the Chinese Empire's near-total withdrawal from seaborne trade. European navigational technology was sufficiently good by the fifteenth century to permit very long (though extremely hazardous) journeys by sea.

Thus, from about 1450, the states of Europe exploited their superior military technology and organisational capacity to spread right across the world, conquering 'primitive' — as well as quite advanced — societies and plundering their resources. European expansionism was fuelled by growing populations and resource shortages. It nurtured the most striking development of all: the Industrial Revolution. Financed by the spoils of empire, it vastly enhanced economic productivity, but at the cost of enormously increased environmental impact. By tapping fossil resources — especially coal and oil — industrialisation simultaneously made possible expanding populations and improved standards of living. But this expansion depended on continual economic growth, which in turn depended on limitless supplies of resources, and an ability for the biosphere to absorb constantly expanding human impacts.

Overview: the Pre-Modern State

Pre-modern states emerged, not in one place but in many, as a result of ecological stresses caused by human environmental exploitation. Biophysical preconditions for their emergence include circumscription, ecological diversity, and in general, a steep ecological gradient between the proto-state and the outside environment. Social preconditions include social stratification, a sedentary lifestyle, trade, and some military capability.

Needs for enhanced administrative skills and improved control over subject populations led to the development of greater specialisation, including military and administrative capabilities. The development of writing and arithmetical skills followed. This in turn produced greater social stratification, leading to the classic dilemma of primitive statecraft, of attempting to balance taxation of the producing classes — predominantly peasants — with the needs and desires of the consuming strata of society, such as rulers, functionaries, and military.

States replaced the low-intensity intersocietal competition with a greatly sharpened interstate competition, in which warfare played an enlarged part. The imperatives of survival in a competitive state system led to a strong emphasis among states on military policy. This 'unnatural selection' tended to prefer expansionist states because these could survive in the short term, generating the paradox that short-term survival frequently bred long-term collapse. Ecological impacts were sharpened by the emergence of state societies, leading to yet another paradox as states encouraged intensified exploitation of their natural resource base while remaining the only social institutions powerful enough to compel cooperation in environmental preservation.

Historically, state encouragement of population growth meant that improvements in productivity generally resulted in a rise in population rather than a rise in standard of living. Living standards for the bulk of humanity probably declined rather than improved.

State expansionism also produced empires, which succeeded in mobilising unusually large amounts of resources for military and political purposes. Most were limited in extent by poor communications and transport technology as well as by administrative inefficiency, but a few managed to push their borders to the technical limits. All are associated with accelerated ecological damage, and often with some of the most spectacular ecological disasters of antiquity.

The vast European expansion from the fifteenth century onwards exploited superior military and transport technologies as well as administrative skills. The resulting system of mercantile capitalism penetrated most areas of the world, bringing them for the first time into a global trade system. The wealth so accumulated was reinvested in industrial development in Europe, leading to the Industrial Revolution and the massive qualitative changes in the human condition that characterise modern times.

Chapter 12

Further Exploration

There is an extensive and exciting literature on the emergence of state societies, and keen controversy is fuelled by continuing archeological discoveries of great interest. Among the best introductions for new readers is Hockett's *Man's Place in Nature*, especially Chapters 33–38; an alternative account, throwing light on legitimate differences in interpretation, is Harris' *Culture, People, Nature*, in particular Chapters 10–13 and 17–20. Webster, like Divale and Harris, is explicitly concerned with the role of warfare in state formation; Webb, on the other hand, emphasises the importance of trade. Fried's *The Evolution of Political Society* can also be recommended; though dated, Carneiro's 'A Theory of the Origin of the State' remains one of the seminal articles in the field. Finally, attention should be drawn to the Cohen and Service anthology *Origins of the State*.

The literature linking the history of the state with environmental changes and problems is still very thin on the ground. In 1989 Wieskel offered a condensed overview in his *Ecologist* article; much more detail is to be found in Hughes' *Ecology in Ancient Civilisations* and Bilsky's *Historical Ecology*. The author of this book attempted to spell out some of the more general problems in his paper 'The State in Environmental Management'; histories with a much broader sweep, such as Ponting's *A Green History of the World*, are now beginning to appear.

The technological history of Europe, especially in the Middle Ages, is quite fascinating. It is still best explored through the work of White, in particular his *Mediaeval Technology and Social Change*. O'Neill's *The Pursuit of Power* is a very detailed history of militarism in European history, with a considerable emphasis on the links between military action and technological change. Tilly's *Coercion, Capital, and European States AD 990–1992* links ancient and modern with a useful and highly analytic approach.

CHAPTER **13**

THE MODERN STATE

Acquisition of colonial empires after 1500 gave the European nations access to a vast range of raw materials and trade goods, forestalling resource scarcity and declining living standards. Instead, it created vast pools of wealth, most of it the ill-gotten gains of what Marx calls 'primitive accumulation': more precisely, piracy. By the eighteenth century, when the Industrial Revolution began, Europe already possessed a well-developed system of **mercantile capitalism**, and was evolving the necessary legal framework and financial institutions to support it. This had in part been made possible by the earlier emergence of **absolutist** states, far stronger than the loosely-articulated, weak Mediæval states. Early examples included Tudor England under Henry VIII and Elizabeth; later came the absolute monarchy in France, and Bismarck's Germany. These régimes both codified and enforced law, while encouraging mercantilism for the sake of the trade and wealth it brought.

Mercantile capitalism, abundant raw materials, and a new attitude to technological innovation, seen now as increasing possibilities for profit rather than as destabilising, were the essential preconditions for the Industrial Revolution. The combination of industry and modern statecraft was literally to revolutionise the world.

The Industrial Revolution made possible very substantial increases in human capabilities by applying novel technologies, such as water and steam power. The latter initiated the exploitation of non-renewable natural resources, initially coal and iron ore, later oil and a further range of metals. This made available vast amounts of energy, at the cost of rapid depletion of natural resources, including materials such as metals. It also tended to treat non-industrial natural resources, such as soil and timber, as inexhaustible, 'mining' them at ever greater rates. The consequences are the problems of pollution and depletion familiar in the modern world.

The Population Explosion
This vast explosion of productive capacity did not change the ingrained tendency to use additional resources to expand human population. Global population was about 200,000,000 in 0 AD, reached only 500,000,000 by 1650, but exploded to 1,000,000,000 in 1850, 2,000,000,000 in 1930, and 4,000,000,000 in 1975.[1] In the year 2000, it will be over 6,000,000,000. But the rate of growth in productive capacity during both the nineteenth and the twentieth centuries has been so great that — at least for the fortunate few in the richer countries — an unprecedented expansion of population has gone hand in hand with a vast improvement in the standard of living. Many of these improvements come not from consumer electronics or 'miracle' new products, but such simple, fundamental improvements as decent sanitation and effective public health measures: clean water, sewage systems, antisepsis, immunisation, and so on.

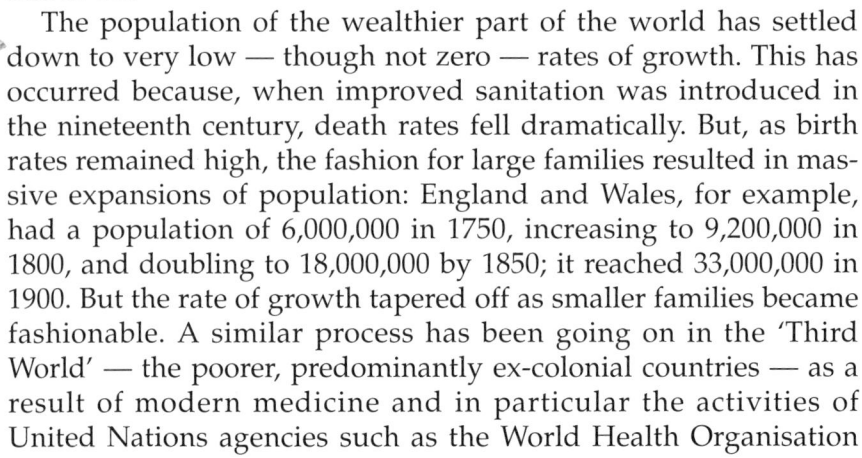

The population of the wealthier part of the world has settled down to very low — though not zero — rates of growth. This has occurred because, when improved sanitation was introduced in the nineteenth century, death rates fell dramatically. But, as birth rates remained high, the fashion for large families resulted in massive expansions of population: England and Wales, for example, had a population of 6,000,000 in 1750, increasing to 9,200,000 in 1800, and doubling to 18,000,000 by 1850; it reached 33,000,000 in 1900. But the rate of growth tapered off as smaller families became fashionable. A similar process has been going on in the 'Third World' — the poorer, predominantly ex-colonial countries — as a result of modern medicine and in particular the activities of United Nations agencies such as the World Health Organisation

(WHO). As some basic public hygiene measures are still lacking, further population expansion can be anticipated, at least in the absence of 'positive' checks such as disease or famine. The 'demographic transition' — the tapering off of population growth due to smaller families — is in the future for most poor countries.

The critical factor in triggering a demographic transition is life expectancy at birth. It has been shown that desired family size in most countries is similar: parents want two children to survive them to adulthood. This is especially important in countries with no social welfare system, since children are an investment for security in old age. But in India, for example, infant mortality rates are high: more than 50% of children die before the age of 15. Under such conditions, it becomes necessary to have more children to allow for the risk of death, and still finish up with two or three reaching adulthood. In practice parents 'overcompensate,' with families of six to seven, yielding population growth rates in the range 2–3%. This means that India alone adds the population of Australia to itself *every year*.

In the face of such rapid rates of population growth, scholars such as Georg Borgstrom and Lester Brown have asked whether there will be enough food for all these people. The answer is that, unless a demographic transition or some other limiting factor comes into play very soon, there probably will not. What is worse is that present-day patterns of consumption, of food as well as other resources, favour the wealthier nations at the expense of the poorer ones. This ensures that American pets are healthy at the same time that Asians and Africans starve. The *global policy problem* — which has been posed but not tackled — is whether to wait for 'natural' forces, such as famine, disease, and war, to reduce populations, or whether enlightened and civilised policy measures should be introduced to make the adjustment rather less painful.

Population growth is not the only problem. With such large numbers, human impacts on ecosystems are greatly magnified: some estimates suggest that as much as 40% of all biological production is now diverted to human use! Overloading of natural systems previously able to supply ample resources or absorb substantial amounts of waste now regularly occurs. Competition between individuals, social groups, and nations for food, natural resources, and even space for waste disposal is a side-effect of the congestion problem that this represents.

A global system of competing, squabbling states has so far proved ineffective in coping with such problems.

Chapter 13

State Power

Since the Industrial Revolution, the number, variety, and extent of state functions has vastly increased. Modern states reach far more deeply into the everyday lives of their citizens, and their decisions affect more people than ever before. The power of modern states depends directly on rapid communications, industrial productivity, and control of information. This is particularly true of some of the more frightful dictatorships, which could not maintain themselves at all without coercive technology.

Giddens identifies four 'institutional clusters' associated with modernity: 'heightened surveillance, capitalistic enterprise, industrial production, and the consolidation of centralised control of the means of violence'.[2] The cumulative effect of these is to make modern states qualitatively different from pre-modern ones. The crucial change is that the 'segmented' social structure of pre-modern state societies, which placed strict, technologically derived limits on the rulers' power, is replaced by a far more homogeneous social structure in which state power is immeasurably stronger.

Heightened surveillance depends critically on the ability to store and manipulate vast amounts of *information* about citizens, through government bureaucracy and record-keeping, and nowadays through 'information technology' such as computerised databases.

Capitalistic enterprise has included the development of skills in large-scale management, plus novel institutional forms especially for the transfer of funds, the management of credit, and the promotion of trade in general.

The impact of *industrial production* has been, firstly, to greatly increase the quantity and quality of goods available for the use of ordinary people, as well as introducing new consumption goods unheard of in pre-industrial times. Through mass-marketing, these goods have become extremely common. Examples include photography, electrical and electronic consumer goods, and motor vehicles. Services, such as transport, medical care, municipal waste disposal, and many others, have also come to be taken for granted in most 'Western' societies at least. The second effect of industrial productivity has been the release of large numbers from direct production tasks. Most of these have been absorbed into information-related work of one kind or another; with the displacement of many simple tasks of calculation and recording by computers and other advanced technology, a further major displacement is currently under way.

The third, unavoidable characteristic of industry is its close linkage to war. Military technology both aided and benefited from the Industrial Revolution. For example, the cylinders of early steam locomotives were bored using techniques first developed for cannon. Improvements in metallurgy and the invention of machine tools and mass production vastly increased the output of military weaponry. The industrialisation of warfare is one of the most striking changes of modern times. The direct effects are easily observed in countries such as Somalia, Iraq, and Bosnia, seething with modern weaponry, where its misuse for the creation of misery is most evident. But the industrialisation of warfare has also created many environmental problems — both the USA and the USSR irresponsibly dumped huge amounts of nuclear waste from their military facilities, for example — as well as fuelling interstate competition in technology development.

Industrialised warfare has been the key to *consolidation of centralised control of the means of violence.* Coupled with better information technology, it has led to the conversion of *frontiers* — remote border areas poorly controlled by central government — into *boundaries*, clearly demarcated and subject to direct control. This has been both a cause and a consequence of the discouragement of small-scale military enterprise — warlordism, banditry, and piracy — in favour of a monopoly by the state. Dependence on relatively few sources of supply, especially for the most sophisticated military technology, has also been a factor.

These four crucial changes have enabled states to develop novel forms, unknown in the past, such as military dictatorships, parliamentary democracies, and totalitarian states; all, in their various ways, dependent on better information and communications and backed by the monopoly of force. Even modern democracies frequently deploy paramilitary police units with firepower undreamed of as little as 50 years ago, and keep remarkably detailed records on their citizens.

Despite this qualitative change, however, many of the goals and policies of states continue to be determined by the same factors. Every state is a part of a larger, highly competitive and unstable interstate system, in which survival is a critical priority. The priority of national survival means that states continue to allocate large shares of their resources to military purposes, and that arms races and warfare — local and global — have the potential to break out without notice.

CHAPTER 13

The Modern Industrial State

Industrialisation and the concurrent development of the capitalist economic system saw massive urbanisation, in part through technological change and displacement. The changed relationship between city and country led to intensified pressure for resource exploitation. Urbanisation led to a pervasive monetarisation of the lives of ordinary people, who no longer held land or engaged in subsistence farming. This left them at the mercy of economic forces, and reduced most to selling their labour to gain a meagre living, scandalising Karl Marx and his followers.

This change transformed not only society but social control as well. Mediæval history, as well as the recorded history of other societies, shows that peasant revolts were common and often violent. Because they were largely self-sufficient, peasants could assert their independence of the state. Resentful of its imposts, they frequently rebelled. They were invariably put down by violence; when, rarely, rebellion was successful, the leaders rapidly adopted the statecraft of those they had deposed. The mechanism of social control in segmented societies was violence, directly applied through military means. By contrast, the primary source of discipline in modern industrial societies is the need to earn a living: in a monetarised economy, landless workers are not autonomous, depending instead on the facility to sell their labour on a continuous basis. This dull and undramatic form of social control is nevertheless powerful.

Government was also drawn increasingly into the provision of services, in the form of urban infrastructure (water supply, sanitation, later electricity supply and transport), mass education, regulation of economic activity (pure food and drug Acts, regulation of safety standards, working conditions, contracts and stock markets) and finally, in the twentieth century, into the conscious management of the economy. This has transformed politics. Even newly freed ex-colonies strive for industrialisation and the enhanced material life-style they associate with it.

However, the dependence of modern large-scale agriculture on fossil fuels arouses considerable concern for the future, given global dependence on agricultural surpluses from the USA, Canada and Australia.

One curious consequence of urbanisation and industrialisation is the notion of the independence of economics. Not until Adam Smith invented modern economics in the 1750s did the notion of an autonomous sphere of exchange-related activity gain currency. Prior to that, economic activity was thought of as just one more aspect of social activity, to be regulated by the state, often through the granting of monopolies, confiscation of wealth, extortion, taxation, and so on. Economic laws had their consequences: there is ample evidence of bad management leading to inflation of currency, collapse of trade, and numerous other consequences. It is just that they were neither understood nor even conceptualised as a separate domain of human activity. The modern New Right would have been literally unthinkable before 1750.

However, as Giddens points out, this notion of an autonomous economic sphere of activity is itself confusing, because politics impinged far less on the everyday lives of most people in pre-modern states, where production was carried out independently of politics. Nowadays, by contrast, the two are intimately intertwined, with economic management an important political campaign issue.

Policy Priorities

The enormous enhancement of the state's power through technology has not greatly modified its environmental behaviour. The policy priorities of modern states remain strikingly similar to those of the past. In a study of national policies in 32 selected, predominantly European, states from 1849 to 1972, Rose found three distinct classes of activity: 'defining', 'resource mobilising', and 'social welfare':

Defining functions are the things the state must do to exist at all. They include:

1. Defence and foreign affairs: there will be an army, and ministries with attached departments for both functions;

2. Maintenance of internal order: a legal system, police, and a ministry or ministries to supervise them;

3. Securing finance: the raising of taxes and the issuing of currency.

Resource mobilisation includes '...building canals, roads and railways, or creating a postal and telegraph service'.[3] It amounts to the provision of infrastructure, from tax revenues, in order to provide a framework for economic activity within society. European states in the nineteenth century regularly encouraged the development of 'national capital', by making available grants, subsidies and loans, offering preferential tax régimes, and creating investment opportunities. This tendency cut across the previously quite cosmopolitan and internationalist character of capitalism, inherited from its mercantile era.

Social welfare includes pensions, health services, education, etc. It can be thought of as investment in 'human capital'.

States attend first to their defining functions; resource mobilisation follows, with social welfare last, showing considerable divergence in levels of provision. Only 20 of Rose's 32 states had 'at least one ministry concerned with questions of land use and/or with the protection or exploitation of natural resources'.[4] In other words, 12 did not concern themselves as a *matter of policy* either with conservation or with exploitation! Presumably activity in those areas was covered piecemeal by other ministries.

The industrialisation of warfare creates a technological link between Rose's first and second categories: increasingly, the ability of the state to maintain its 'defining' activities (and hence survive) is dependent on its ability to mobilise resources. The primary resources concerned are those needed for military technology, but these inevitably impact on the industrial system at large. Since the Industrial Revolution, a primary goal of statecraft has been industrialisation. Industrial capacity means self-sufficiency in armaments, and consequently a greater ability to 'project' the state's power when needed. This is why economic management, and particularly the health of the 'military-industrial' sector, becomes of pressing importance to all modern states.

All major industrialised countries consequently have a 'military-industrial complex', which closely links government and big business. Because of its centrality to the 'defining' functions, governments typically go to great trouble to ensure the 'health' of this sector of their economies. This includes, if necessary, propping up the profits of inefficient firms, either by overpayment for goods and services, or by direct subsidies. In addition, most countries — including Australia, in recent years — attempt to make their own weapons production more economical by generating overseas sales. The biggest sellers of arms, globally, are those nations with the biggest military-industrial complexes: the USA, the former USSR, Britain, and France.

The consequences of such activity in a competitive global state system, which has become very unstable since the collapse of USSR, can be ascertained by direct observation. Not only are the human consequences dire, but there are significant ecological effects as well.

The State and the Economy

In democratic countries, economic management as a political goal makes a 'politics of distribution' almost inevitable. Political parties gain and maintain support by offering handouts to favoured groups. This activity is justified by Keynesian economics, which stresses the use of government expenditure to stimulate the economy. But Keynesian economics requires 'counter-cyclical' policies at all times: governments should run deficits during slumps, but budget for a surplus in times of boom, thus 'dampening' the cycle. However, there was no budgetary restraint in the post-1945 economic boom. This is because public pressure and political resistance makes the distributive expenditures of depression difficult to eliminate when booms occur. The outcome is a 'semi-Keynesian' pattern of deficit expenditure at all times, with inflationary consequences: the 'Keynesian ratchet'. Governments generally rely on increased economic growth to finance their way out of this problem, but this often worsens it instead.

It is not clear whether all economic growth is bad for the environment; certainly much of it is. The state's close intertwining with corporate interests, as well as the structural consequences of its own economic management strategies, limit its freedom to determine policy. 'Economic transformation' — primarily by promoting capital accumulation — locks states into indiscriminate economic growth. Additionally, small states may find their options acutely restricted when they are heavily dependent on one or two crops. Options are also limited when powerful multinational corporations occupy strategic and powerful positions in their economies. This restricts their freedom to attend to ecological or other long-term issues. Add to this the need to maintain a military-industrial complex, and the range of options open to government shrinks dramatically.

Minorities

There is also another very serious problem for nation-states. Most nations consist of more than one community; often there are substantial minorities, some speaking different or variant languages, possessing distinct cultural identities and practices, or adhering to differing religions or sects. The loosely segmented states of pre-modern times co-existed unhappily with such minorities, unable to absorb or fully control them, and often acutely aware that their allegiance was unstable. But, with the rise of the modern state and the extension of administration, the problem is aggravated. Some states have attempted to absorb them, deliberately proscribing their languages, interfering with their religious practices, and suppressing their culture. In extreme cases, such as Hitler's Germany, Stalin's Russia, or Karadzic's (one hopes very short-lived) Bosnian Serb 'Republic', minorities have been murdered, tortured, robbed, and deported on a massive scale: the crime of genocide. Other states have tolerated minorities to a greater or lesser degree. It is striking, however, that even the most 'democratic' of states have significant insurgent minorities, some of them very stubborn. Britain and Ireland have their IRA and Ulster Unionist terrorists, France the Corsicans, Spain the Basques, Greece the Macedonians, and so on. Many newly independent ex-colonial nations have boundaries drawn by some remote European bureaucrat or diplomat with no knowledge of local conditions. These often divide communities, or create 'artificial' minorities. Indonesia controls East Timor and Western New Guinea (Irian Jaya), both areas with which it has little or no affinity in ethnic or cultural terms. Papua New Guinea controls Bougainville Island, more properly a part of the Solomons, and home to an intractable independence movement. India controls tribal territory in Assam and adjoining areas. There are many more examples, some amounting to internal colonial repression, as in the case of Brazil's persecution and extermination of its Indians.

Conflicts of this kind artificially complicate environmental management, which is caught up in, and often falls victim to, disputes which should never have arisen in the first place, and which often could and should be settled by negotiation and grants of suitable autonomy. This does not happen because states are not evolved to include out-groups or their ideas. Typically, the dominant group exercises agenda control by closing communication channels, and defining the problems involved out of existence. This, of course, only displaces them in time, frequently intensifying their ultimate impact.

Adaptively speaking, the failure of states to come to terms with their minorities is a major weakness. Ecologically speaking, it is irrational for two quite separate reasons. The first is simply that war, conflict, and exploitation tend to increase environmental impact, and consequent degradation. The second is that large amounts of creative energy, which could otherwise be used for solving ongoing environmental problems, is instead taken up with the processes of repression, rebellion, and so on. The adverse impact of military conflict and its surrogates on human welfare is well known; nor should its environmental impact be ignored.

THE GLOBAL ORDER

The extractive pressures of the powerful industrialised world have been so great that they have seriously disrupted traditional, better adapted productive systems, especially in agriculture. In modern times, this has resulted directly in famine and malnourishment on a vast scale. In part, these have been a consequence of overpopulation; but the situation has been made markedly worse by imperialism.

Imperialism
Imperialism is the maintenance of control over the politics and economy of one country by another. Direct control via colonial possessions is now rare; economic ties are more common. 'Neocolonies', sometimes in the hands of 'puppet' governments, and rarely governed democratically, supply raw materials to the industrial nations. Sometimes, they are stripped of important natural resources as a result of corruption among ruling elites; the military in Burma and Thailand, for example, have been heavily involved in deforestation.

Imperialism results in an imbalance in the *use* of materials and energy: the USA, with 5% of the world's population, consumes 60% of global mineral production. Better prices for manufactured rather than primary goods mean imbalanced terms of trade. This imbalance leads to impoverishment of primary producing nations. A major example of this imbalance is the so called 'protein swindle'. Poor nations often export high-quality protein products in return for low-protein grains, causing serious malnutrition among their own poor. The result is serious global maldistribution of food, materials and energy. 'Neocolonies' also have other roles as markets for finished products, sites for investment, and sources of cheap labour, which are growing in importance.

Attempts to redress some of these problems have been hampered by a number of factors. The primary source of finance for 'developing' countries is the World Bank (or International Bank for Reconstruction and Development), which has tended to insist on market deregulation, low tariff barriers, and a number of other measures which often have the effect of increasing unemployment, poverty, and rural misery. Some countries are so deeply in debt that there is no prospect at all of their ever paying off the debt. In the worst cases, such as Brazil, the interest alone exceeds all the nation's foreign currency earnings. The effect has been greatly increased pressure on natural resources such as the

Amazon forest and Brazil's iron ore reserves, which can be quickly exploited and sold abroad to raise foreign exchange. The effects, which include massive deforestation, serious ecological damage including spectacular rates of species loss, and the deprivation of many poor Brazilians of their livelihood, have aroused serious concern.

'Ghost' Resources and Stretched Feedback Loops

Maldistribution of materials and energy increases environmental pressure: the rich squander resources, while the poor are forced to adopt ecologically bad practices to survive. In addition, waste of resources for military purposes is significant worldwide, but is most grotesque in countries such as Ethiopia or Somalia.

Catton has shown how the increasing dependence of the Western countries on 'ghost' resources and 'ghost' acreage has significantly worsened the ecological problem, lengthening feedback loops and insulating populations from the consequences of their actions. The rich nations of western Europe and North America have exhausted most of their better mineral resources and run their populations up to levels that cannot be supported from domestic food production alone. This makes them dependent on overseas sources, many of them located in colonies or neo-colonies. These are 'ghost' resources because they are offshore and remote from those who consume them. Cutting them off can cause immediate hardship — the Dutch suffered badly in 1973, when their oil was embargoed by Gulf nations — and their use represents a net transfer of wealth as well as of nutrients. Reliance on 'ghost' resources, Catton argues, lengthens the feedback loop from environment to user, who may have literally no idea where the necessary resources are coming from, nor what ecological damage results from their extraction. This leads to an enhanced risk of *overshoot*: that is, that the resources might run out or an irreversible ecological decline set in long before the wealthy consumers become aware of it. By the time they are, it is too late, and nothing can be done to retrieve the situation.

Imperialism, therefore, results not simply in a maldistribution of wealth: it also increases the risks of environmental catastrophe because it lengthens the feedback loop so drastically.

Global Problems

Worse still, conflict can seriously exacerbate global 'commons' problems. One important global common is the world's oceans and the mineral reserves on the sea floor. The possibility of mining the ocean beds for rich manganese nodules stimulated the recent series of Law of the Sea conferences. The less developed nations demanded an international regime to manage such resources for the common good, meaning greater shares for them. However, countries like the USA, whose multinational corporations already had the technology to exploit the deposits, would not agree to this arrangement, and no decision was reached.

Increasing global interdependence will require more cooperation, and eventually global institutions, including world government. But the difficulties are many. A global order would conflict with many national and corporate interests. It would also become a target for control by these interests, for example by one or two 'great' powers. As a result the world body might be given too little (or too much) power. The problems in devising a system of fair representation for all would be immense. The existing network of international cooperation, even though more extensive than ever before, is but a small beginning.

Overview: Political Economy in Global Perspective

Interstate competition is an ancient, deeply embedded feature of state societies. Most often, it manifests itself as war. It deeply influences policy priorities and social structures, and the sustaining mythologies become part of the very culture. Other interests of ruling elites influence both social structure and public policy; the social inequalities historically inseparable from state societies continue to generate strong pressure for overexploitation of resources. In modern times, the institutionalised pressure for continued economic growth expresses not only the demands of the military-industrial complex, stressing growth and innovation; but also the mutually reinforcing interests of ruling elites and industrial corporations. In a conflict between 'hard economic realities', particularly the effects of population growth, and ecologically-derived desiderata, the former dominate.

Conflict over resources, especially land, between the governments of different nation-states leads to interstate competition, a powerful selective process. Like natural selection in evolution, it is unpredictable. Often, through conquest or ecological failure, whole civilizations have become extinct.

Because of this competition, states need a military capability to survive; this depends on the availability of physical resources. To satisfy this demand, nations increasingly need to access resources efficiently. Comprehensive economic management is a vital com-

Chapter 13

ponent of this process. The competition between states of the modern world has led to an emphasis on 'national' capital; state support of capital; and close linking of capital and military. Powerful 'military-industrial complexes' have arisen from this convergence of state and corporate interests.

Environmental stresses become visible in a number of ways: an acute disjunction between contemporary political culture and scientific knowledge; ignorance and inadequacy of public knowledge and attitudes; neglect of issues such as reform of population policy or assessment of technological capability. All are surface phenomena, indicating the presence of a deeper problem. This is the domination of public policy by the forces emerging from the internal dynamics and external competition of state societies. It is graphically illustrated in the relations between states and their minorities, which shows how political factors endogenous to a specific system can impede resolution of urgent environmental issues.

The logic of interstate competition poises it permanently on the edge of outright conflict, with the attached risks of environmental damage. In the insecure, post-USSR world, this is a serious problem. It may become significantly more difficult to create ongoing cooperation, let alone supranational institutions for environmental regulation and management.

This is why environmental management, and the risk of ecological crisis, are a major adaptive challenge to modern states. The best scientific and technical knowledge suggests that growth in gross national product (GNP) and military activity must be curbed. More appropriate and temperate patterns of resource allocation should be sought. But these objectives confront deeply embedded pressures, which are likely to resist change.

Further Exploration

The general thesis spelt out in this and the previous chapter is to be found in the author's paper 'The State in Environmental Management'. Broadly, the themes resemble, and are informed by, Giddens' *The Nation-State and Violence*, which is required reading for anyone wishing to understand modern state societies. However, Giddens is not greatly interested in the ecological impact of the state, and as a result, although he recognises the important continuities between ancient and modern states, tends to underemphasise them.

Geographers in particular are now paying great attention to

The Modern State

human environmental impacts; Thomas' great 1956 symposium *Man's Role in Changing the Face of the Earth* can now be supplemented by entry-level texts such as Goudie's *The Human Impact on the Natural Environment* or Simmons' *Changing the Face of the Earth*. Both usefully summarise some theories of cultural development which are touched on only lightly in this and preceding chapters, as peripheral to the major themes, as well as providing useful overviews of current environmental impacts.

The figures on population growth in England and Wales are from Ponting's *A Green History of the World*, p. 241. Ponting's Chapter 12 is full of useful comparative figures, as well as a discussion of population changes in the nineteenth century.

Georg Borgstrom's works on food include *Too Many* and *The Hungry Planet*. He emphasises not only food shortages, but also the global maldistribution of food resources, arguing that time to deal with overpopulation and its related problems could be gained by fairer distribution of food. Lester Brown is well known as Director of Earthwatch, a foundation which issues regular State of the Earth reports and supports research into, and dissemination of information about, environmental problems. His books include *World Without Borders*, in which he argues that the rich nations cannot escape the consequences of overpopulation, food shortage, and poverty, and *The Twenty-Ninth Day*, a dramatic discussion of the implications of exponential growth.

CHAPTER 14

DEVELOPMENT AND ENVIRONMENT IN AUSTRALIA

Australia was and is a part of the colonial system created by the European expansion. It was settled late, mainly because it was remote and offered no obvious lures to explorers or adventurers. Its development came very quickly to depend on overseas markets and on overseas resources.

All of the Europeans' colonies proved to be rich sources of plunder. But some adapted relatively easily to imposition of European technology and culture, though none succumbed without some human resistance and ecological damage. Australia, as one of the remotest and most forbidding, was among the least easily settled. Though the problems were similar to those encountered elsewhere, the specifics were unique: a remarkable ecology and a human population remote culturally and adaptively from European concerns.

The Australian Environment

Australia has been geographically isolated for nearly 45 million years, although for the last 15 million it was close enough to Asia for some migrations to occur. Consequently, Australian flora and fauna have evolved largely independently of the rest of the world, to fit a singularly harsh environment.

A number of factors were of especial importance in making European settlement difficult. Australian soils are less fertile than those of Europe. The continent is an old and stable landmass, with ancient and severely weathered soils, typically deficient in phosphorus, sulphur, carbon and nitrogen. All are essential for plant growth; European crops and livestock require them. Good, relatively fertile soils are limited to the coast and eastern highlands. Until an understanding of soil deficiencies and the need for trace elements developed, a European pattern of agriculture faced declining productivity.

And Australia is the driest continent in the world. Two-thirds of it receives less than 500 millimetres of rain annually. Rainfall is seasonal, with most of it falling in the summer months in the north and in the winter months in the south. But it is unreliable, and drought is a common occurrence: water is the major factor limiting European-style 'development'. Additionally, extreme temperatures, common in many areas, can directly impede plant growth and exacerbate water shortages by increasing evaporation rates. Even in the Murray–Darling Basin, where water is relatively plentiful, conditions are harsh.

The native flora and fauna are well adapted to these conditions but European crops and livestock suffer both from water stress and from local diseases and pests. In such a forbiddingly dry environment it is not surprising that water became a central theme in Australian politics.

Flora and Fauna

Australia's flora consist not of one assemblage but many. All are unique, ranging from the temperate rainforest of Tasmania to the unusual desert environments of the Centre. Many Australian plants turn out to be of considerable scientific importance, and have substantial commercial potential, often displaying adaptations of considerable interest.

Faunal adaptations are also of great scientific interest, shedding light on evolutionary processes. Because of Australia's isolation, lineages known only from fossils elsewhere in the world survive. The best known of these are the monotremes (platypus and echidna) and marsupials. Many display highly unusual adaptations: apart from obvious examples such as platypus and marsupial gliders, there are ground parrots, small desert marsupials, and many others.

Aboriginal Adaptations

The adaptation of the Aborigines to this environment is still the subject of some dispute. It was long thought that Aborigines had lived in Australia for a relatively short period — perhaps 10–20,000 years — and that their culture had undergone little development. But some modern archæological discoveries threaten to push back this time frame by a factor of as much as 10, and

CHAPTER 14

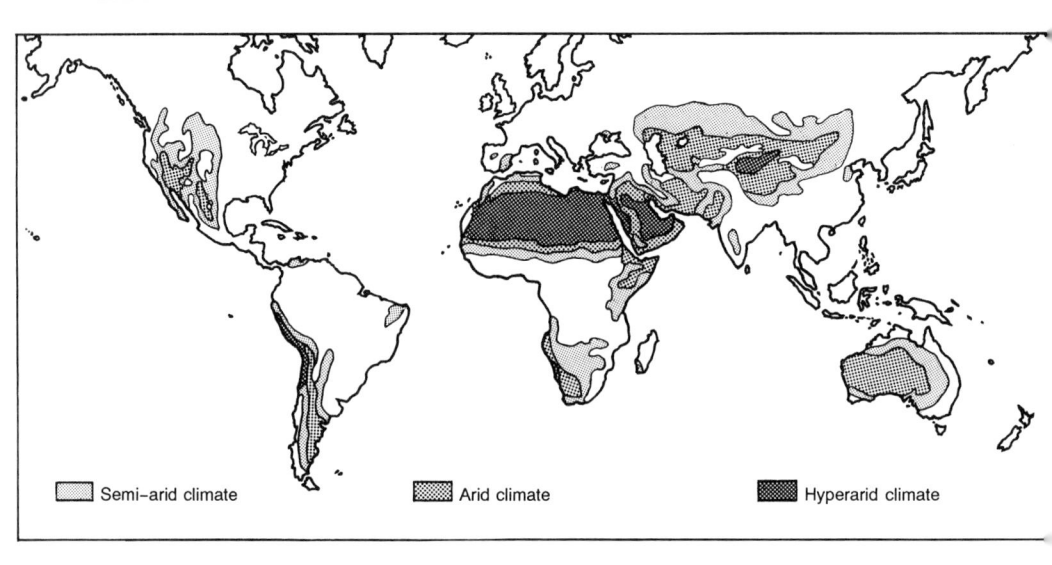

Fig.14.1
Global distribution of arid lands. A larger proportion of Australia is arid than of any other continent.
(Source: Dregne, H.E. *Desertification of Arid Lands*, p.6.)

Aboriginal artefacts have been securely dated to periods 38,000 years and more ago. This makes human occupation contemporary with some now-extinct species such as the diprotodon (a huge relative of the wombat) and the giant kangaroo.

Such a long time-frame has intensified speculation about Aboriginal environmental impacts. In particular, one school of thought suggests that modern, highly conservationist Aboriginal environmental attitudes may be a reaction to an earlier, far more profligate approach which significantly depleted available resources. Certainly, Aboriginal environmental modification by fire is well documented, and the dominance in modern forests of fire-resistant and fire-dependent flora such as *Eucalyptus*, *Acacia*, grass trees, and the like, has been attributed to the extensive use of fire.

What is often not appreciated is the subtlety of this adaptation, and the very substantial knowledge base on which it was built. Modern scientists and anthropologists are coming to recognise the importance of this knowledge, and are attempting, with the help of surviving Aboriginal sources, to reconstruct it.

EARLY EUROPEAN ATTITUDES TO THE ENVIRONMENT

No such attempt was made when the First Fleet arrived at Botany Bay in 1788. The lack of established agriculture was only one aspect of a technology so unfamiliar that the first settlers could use none of it. Their contact with the local Aborigines was mostly antagonistic, primarily because they sought to drive them off their land. As a result little transfer of cultural information occurred, and thus the settlers could not draw on Aboriginal knowledge. They did not know which organisms were safe to harvest for food or commercial exploitation. The colony at Sydney Cove nearly starved to death at first; despite experiments in cultivation, it was almost totally dependent on Britain for the first ten to twenty years. Despite minimal assistance from England, the settlers ignored native resources. They led miserable lives in harsh conditions until farms were established using traditional European techniques and crops. Development of techniques for local conditions was slow and painful. It was not until the Blue Mountains had been crossed, and land suitable for sheep pasture (and later grain-growing) opened up, that the colony became self-sufficient.

Australian flora and fauna did not fit in with the nearly universal rationalist attitude, at the end of the eighteenth century, that nature could be 'made over' into something more comfortable and acceptable to humans. In this respect, Australia compared unfavourably with Britain from almost every standpoint. The local ecology was rejected as inferior, intractable, and unsuitable as raw material for transformation into a new Europe. So, although a handful of naturalists saw Australia as a wonderland of strange species, 'practical' men saw it as infertile bush. Their imperatives were: ringbark, clear, supplant.

Colonial Exploitation

Australia was geographically very isolated. The early settlements were far from the main shipping routes; transport costs were high and transit times lengthy. The economic historian Geoffrey Blainey has called this the 'tyranny of distance'; it retarded economic development and reduced the range of goods that could be sold on international markets. Distance also led to a pervasive feeling of insecurity, summed up a century later in A.D. Hope's image of

> ... a vast parasite robber-state
> Where second-hand Europeans pullulate
> Timidly on the edge of alien shores.
>
> ('Australia'.)

The lack of fertile land in the interior, and the consequent confinement of development to the coastal fringe, no doubt reinforced this insecurity.

In reaction, European Australians continued to identify with Britain as 'Home'. The dream of recreating a European society in Australia became a persistent feature of their culture. Explorers, on the basis of a few days' observation, described lush, fertile landscapes, and peopled

Chapter 14

them in their imaginations with tidy European villages every few miles. Other nineteenth-century ideas, such as Progress, also influenced attitudes: if cast iron and faith could do anything, the dream of total mastery over the environment could be realised even in Australia.

These Victorian attitudes were corrosive, resulting in careless attitudes to the land. Native species — and frequently Aborigines too — were treated as as 'vermin' to be destroyed. A persistent unwillingness to recognise Australia's limited carrying capacity lingers on in politicians and businessmen to this day. Attempt after attempt has been made — the Brigalow scheme as recently as the 1960s — to promote 'closer settlement', through clearance schemes, soldier settlement, homestead Acts, etc. These schemes have often failed: a spectacular example was the Ord River scheme in Western Australia.

But the swift and devastating environmental impact of British settlement did not depend simply on grandiose development schemes. Rather, it was a process of systematic plunder, careless of ecological impacts and informed by a 'get-rich-quick' mentality. If Australia was a temporary berth, and the real object was to return Home with a fortune, the Australian environment, natural or social, could not matter.

Goyder's Line: Ecological Constraints

The major limits to development in Australia have always been ecological; but settlers responded by ignoring and actively attempting to modify them.

Australia's early growth was mostly in the coastal 'fertile crescent' of the southeast. The effective limits of settlement were defined for the most part by rainfall, and the result was a series of major port cities, each serving an agricultural hinterland. The paucity of overland trade and the barrenness of the interior discouraged the development of inland centres.

In the east of the continent, the Great Dividing Range proved a formidable barrier, not only to settlement, but also to transport. For example, the road over the Blue Mountains, completed by convict labour in 1815, opened up the Central West of New South Wales for pastoral exploitation, but only wool was valuable enough to absorb the transport costs by ox-cart. The railway, which followed in the late 1860s, slashed cartage costs by a factor of five, but the difficult terrain imposed two zig-zag reversing stations on the builders, making the line difficult and expensive to work.[1] Though the railway made wheat-growing economic west of the Blue Mountains, South Australia's wheatlands were by then well established.

South Australia's export wheat trade had been facilitated by the proximity of the wheatlands to Spencer and St. Vincent's Gulfs, with-

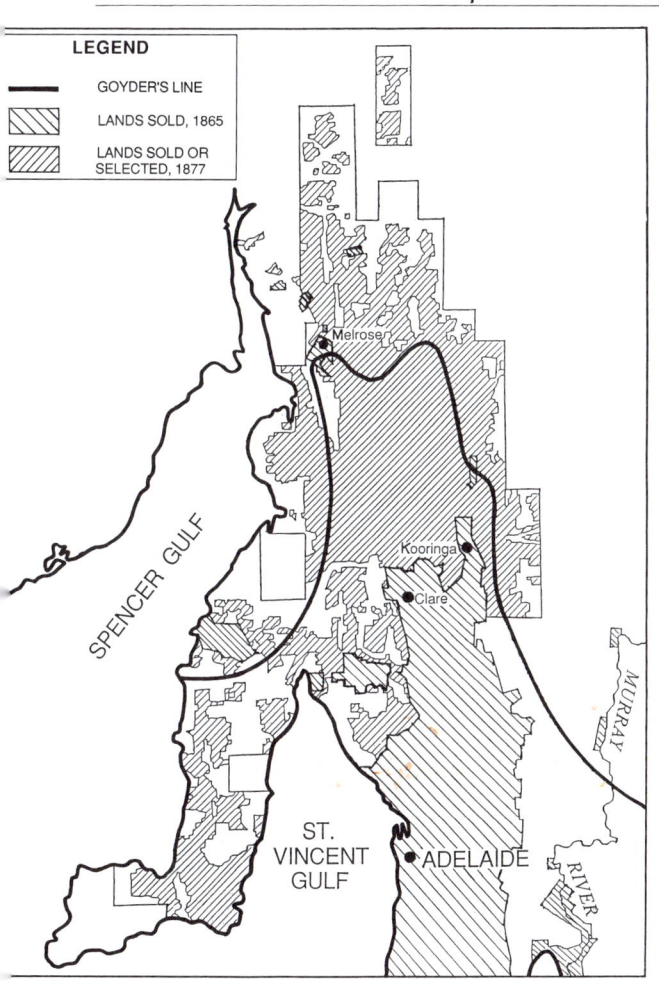

Fig. 14.2 Goyder's line. Land sold before 1865 was the basis of the early success of wheat in South Australia. That sold between 1865 and 1877 was largely in response to political pressure to 'open up' more land. Most of these later sales were 'marginal' land, capable of supporting a crop only in wet years.

(Source: D.W. Meinig, *On the Margins of the Good Earth*, Map 8, p. 44, and Map 11, p. 68.)

out an intervening mountain chain. The crop was cheaply transported the short distances to a chain of small ports by light, horse-drawn tramways. The success of wheat led quickly to pressure to clear and plant further land in the same general area. But the then Surveyor-General of South Australia, G.W. Goyder, conducted a careful survey of the climate and vegetation following the drought of 1864-5. He reported that perennial cultivation of wheat outside a line corresponding roughly to the ten inch isohyet would be impossible. This line became known as Goyder's Line. Its history illustrates the problem involved in gaining recognition for ecological constraints.

Goyder was not taken seriously. The success of wheat in the lands adjoining Spencer Gulf led to land hunger, in the form of demands for more land in contiguous locations. A series of unusually wet years in the late 1860s and early 1870s led to a rejection of his observations. Myths such as 'rain follows the plough' were used to justify extensive clearing and cultivation in areas as far north as Hawker and Lake Torrens. Goyder was ridiculed, and the successful results of the first few years' farming in the newly cleared land used to justify its opening up.

But between 1879 and 1902 there was an almost unbroken run of drought, the worst in Australia's short agricultural history. Many of the new farms of the previous decade were abandoned, never to be cultivated again. Travellers between Quorn and Hawker, for example, can see the ruins to this day. A wide swathe of the land opened up between 1865 and 1877 was reluctantly recognised as 'marginal': at best, it could be cropped only in wet years. Goyder's views gained belated respectability, and have been basic to the teaching of geography in Australia ever since.

This was an early example of government and farmers thinking they knew better than the experts. It was to become a common problem in Australia, and would lead to extensive environmental devastation.

Acclimatisation

The response to Goyder's line was a part of the broader attempt to 'make Australia over' in the European image. What the settlers failed to appreciate was that the attempt was doomed. Rain did not follow the plough; in fact the effect of clearing was often to modify both climate and soil fertility, thereby increasing aridity. Unsuitable crops and introduced species might offer initial gains in production, but they were bound to lead to long-term problems.

This can been seen very clearly in the two major agricultural sectors, pastoralism and cropping. In pastoralism the introduction of sheep and cattle displaced native fauna, with immediate and drastic environmental impact. Australian soils and flora were adapted to soft-footed fauna, such as marsupials. Sheep and cattle, with their hard hooves, cut into vegetation and exposed soil, which rapidly eroded. The effects were especially marked around watercourses, where sheep and cattle frequently fouled and degraded formerly clean and beautiful creeks.[2] The introduced species also have poor drought resistance, creating a persisting problem. Failure to destock during drought led to damage to vegetation, as starving stock depleted all accessible greenstuff by gnawing the bark from trees and shrubs, ringbarking them. Pastoralists made the situation worse by cutting shrubs to feed stock, accelerating degradation of pastures in drought years. Because sheep in particular are selective feeders, preferring young, succulent shoots where available, the pastures frequently did not recover, remaining permanently degraded.

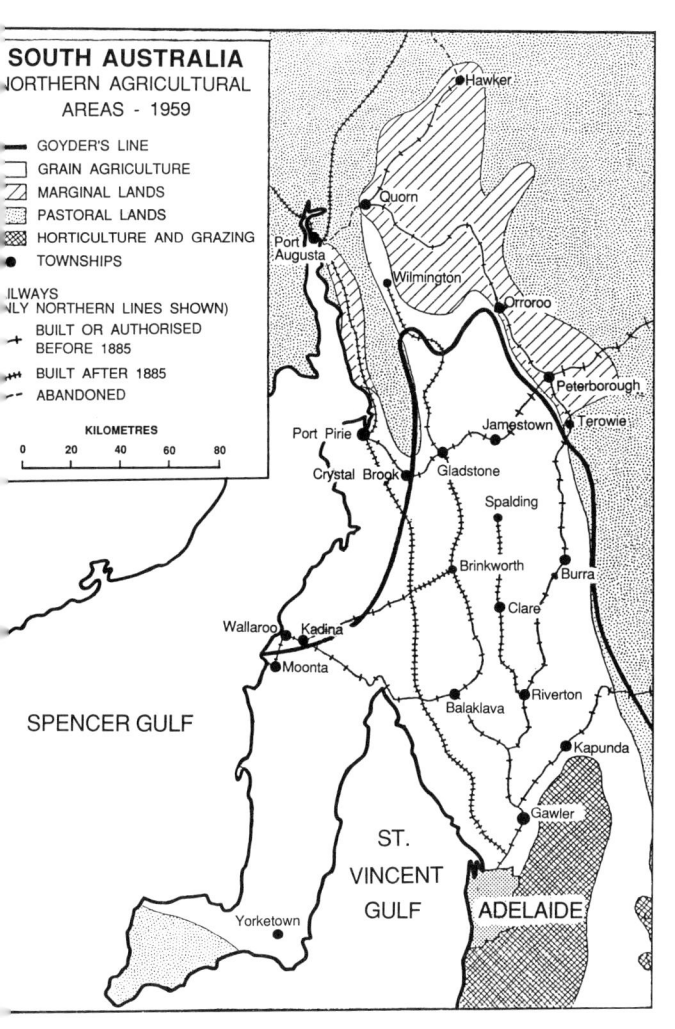

Fig. 14.3 Marginal lands in 1959. By 1959, grain cultivation north of Adelaide was largely confined to the area within Goyder's Line. Much of the land subdivided between 1865 and 1877 is now recognised as 'marginal', and is cultivated — if at all — only in wet years.

(Source: Meinig, D.W., op. cit., p. 208.)

Reluctance to destock was in part due to the cost of restocking, a consequence of the poor adaptation of sheep and cattle to conditions of frequent drought. The soft-footed marsupials, such as kangaroos, have a unique capability to 'freeze' the development of already-fertilised embryos when conditions are harsh. The embryos develop only when conditions improve, leading to the often-remarked capability of marsupials to expand their populations explosively as soon as feed is abundant. Soft feet and the ability to adjust population to drought are obviously desirable traits in pastoral species for a drought-prone country; both traits are lacking in sheep and cattle.

Unsuitable stocking habits were worsened by the political and financial systems. Despite their origin as illegal squatters, most pastoralists held (and still hold) their land under leases from the Crown (in the shape of the State government). Rents for the leases, by and

Chapter 14

large, do not represent the true value of the land for alternative uses (including wilderness) and in consequence do not return to the community the full costs of pastoralism. At the same time, most pastoral leases are overcapitalised and economically precarious. This leads to overstocking, with the damage to land noted above. Pastoralists are locked into this situation, however. Their political power has made it possible for them to award themselves large quantities of 'drought relief' in the form of outright grants and low-interest loans from government. This ignores the need, economically and ecologically, to provide for drought as a normal phenomenon, treating it instead as a completely unforeseen natural disaster. Drought 'relief' — in practice a programme of government-subsidised ecological destruction — is then built into the economics of pastoralism. Banks will lend to graziers in the expectation that government subsidy will keep them afloat, thus ensuring that they cannot afford to destock — with adverse environmental consequences already noted — and their property is overvalued. Government policy — both low rents and drought relief — inflates the value.

The same is true for crops. The European varieties introduced to Australia were almost all annual. This inevitably meant bad yields in dry years; when coupled with cultural methods insensitive to the risks of soil erosion, severe land degradation followed. Before 1900, there was general ignorance of the need for trace elements in the soil, and varieties of wheat resistant to endemic diseases such as rust had not been developed. A direct consequence was falling yields, as nutrients were exhausted and pests became established. These factors were exacerbated by the adoption of traditional European open-field systems, which required the removal of all trees and scrub. This destroyed the habitat of many benign native species, such as birds, which are useful for pest control. It also significantly reduced soil fertility, since many species of native flora return nutrients important for crops to the soil.

Cropping, too, was affected by short-sighted government policies. Many selectors were burdened with conditions, including a requirement to 'improve' the land. 'Improvement' meant the erection of farm buildings, fencing, etc.; but it also meant clearing. Tax concessions and subsidies were offered for clearing, often meaning that farmers would clear at times when activity in cropping was low — such as during droughts — specifically in order to attract the grants and subsidies. The resulting erosion was so dramatic that modern studies have found Australian soil particles, transported by wind, as far afield as Africa! And, just as water charges had the effect of inflating the market price of irrigation land, so subsidised 'improvements' and the expectation of drought relief enhanced the 'value' of farming land. What it could not do was save it from economic marginality and steady ecological decline.

Introductions

Deliberate attempts were also made to modify the Australian environment by introduction of European species, often in the hope that they would successfully drive out Australian ones. Crops and domesticated animals were among the first, but many introductions were made to afford sport, or for purely æsthetic reasons. During the nineteenth century, 'acclimatisation societies', dedicated to the introduction of exotic species, flourished in every state.

Their best-known introductions include rabbits, which spread everywhere despite the construction of rabbit-proof fences at great public cost. They displaced many native species, and significantly reduced the productivity of pastures everywhere. Foxes, introduced for the hunting pleasure of the would-be upper classes, have spread more slowly, but they are efficient predators, highly damaging to native fauna. Goats and pigs have also become feral, doing considerable damage to flora. And domestic cats, frequently dumped in the bush by tender owners unwilling to kill them, are a major pest, especially because measures to eradicate them are politically unpopular.

Introduced bird species include sparrows, starlings, thrushes, blackbirds, and many others. These have frequently competed with local birds, rising to pest proportions in some cases.

As dramatic has been the impact of European and other exotic flora. Some, such as blackberries and prickly pear, have gained the status of noxious weeds, proving difficult and expensive to control. Others have merely supplanted native vegetation, resulting in loss of important and representative flora. Native grasses, for example, are confined to a relatively few areas, in most of which they have survived by accident. As their properties come to be appreciated, scientific effort is being devoted to attempts to ascertain their full extent, and to preserve representative areas of grassland.

Despite the importance of a handful of introduced species to agriculture, the bulk of the introductions have been harmful, some directly and visibly so. Very few have had a neutral impact.

Environmental Impact

The full impact of the European settlement is unknown, and may never be known. But Australia has experienced the world's highest rate of extinctions over the last 200 years, and the scale of other problems, such as soil degradation, is commensurate.

CHAPTER 14

Because of the paucity of scientific effort, large numbers of Australian species, plant, animal and insect, have not yet been described. The possibility that some have been lost beyond hope of retrieval due to the dismissive attitudes of the colonists is very real. Of those species that are known, at least 15 mammals and one bird species have become extinct; 26 mammals, 22 birds, 9 reptiles and 6 amphibians are known to be endangered, while there is little or no information available about fish and invertebrates; and no less than 2206 plant species are rare or threatened.[3]

Survival by Export

Australia's economy, for the first 150 years of its existence, was closely tied to Britain's. The UK was the source of capital, of equipment, manufactured goods, and much else; at the same time it was the primary market for Australia's produce, the bulk of it either pastoral, agricultural, or mineral.

Australia's first exports came from the exploitation of the seal colonies of Bass Strait and abundant migratory whales. Bass Strait was so heavily exploited from 1792–1810 that sealers moved to Macquarie Island and other remote spots. Even these were exhausted by 1820. Whale populations dropped so sharply that by 1845 hunting was uneconomic; Southern Right Whale stocks had not recovered in the 1970s.

The other major source of export income was timber. Sandalwood from New Caledonia and the New Hebrides fetched high prices in China, and many Pacific islands were stripped bare between the 1830s and 1860s. Red cedar (*Toona australis*), a fine timber, abounded on the east coast from Sydney north. It was first exported to India in 1795. Cedar-getters moved progressively north from Sydney, cutting even small trees, and felling many giants 100 years old. The rate of depletion was rapid: all the cedar on the Clarence was gone by 1850, and Moreton Bay was exhausted by 1847. Today no commercial stands of cedar remain.

This pattern of rapid, unsustainable resource depletion necessarily gave way to more stable sources of exports.

Wheat and Sheep

In pastoralism and agriculture, wool and wheat became the primary sources of Australia's foreign currency earnings. Britain, already importing raw materials for its textile industries, plus food for a population grown too large to feed from purely domestic sources, was a ready market. Wool and wheat were easily transported and relatively non-perishable. They were recognisable and hence acceptable to a

European market, and they could be raised in Australia. They were also familiar to the settlers.

This led initially to extensive dry land agriculture. Clearing was extensive, but other improvements, such as fences, bores, or improved pastures, were kept to a minimum. Farms were predominantly monocultures or specialised in only two or three crops. Clearing meant displacement of native species; native grasses suffered particularly badly. Though most fauna, especially small mammals, suffered catastrophic decline, clearing improved opportunities for some. Koalas, for example, have suffered steady habitat loss, but those species of kangaroo or wallaby adapted to open plains have multiplied.

Settlement patterns, despite many attempts to impose order, largely eluded government control. Frequently title was granted after occupation of the land, in a reactive response to a *fait accompli*. The rapid establishment of large grazing holdings quickly established the political power and economic dominance of graziers. This made attempts at 'closer settlement' more difficult when demand for land grew stronger.

As indicated in Chapter 5, this is one of the reasons why irrigation emerged in the Murray valley. It, too, was geared to exports, notably of dried fruits. With the development of chilling and later of refrigeration for ships, the UK market was opened up for meat, and later dairy products.

Mining and Industry

Also export-driven, mining was a 'rush that never ended', in Blainey's telling phrase. Gold was mined from the 1850s. Less valuable minerals were uneconomic, due to high transport costs, until railways were available. Thus silver, copper, base metals, and iron ore were gradually added, the richest and most accessible deposits, near the coast or handy to rail transport, exploited first. Slowly the realisation came that Australia was very well endowed with minerals; major discoveries such as Broken Hill, Cloncurry, Mount Morgan and Mount Lyell punctuated the late nineteenth and early twentieth centuries. But often new techniques had to be developed for mineral extraction; as with flotation, first used at Broken Hill for its unusual silver-lead-zinc ores. Mining followed a spectacular boom-and-bust pattern, littering the outback with once prosperous 'ghost' towns. Some, like Chillagoe–Mungana in north Queensland, were the foci of major scandals over misappropriation of public funds.

Mining has always had spectacular environmental impacts; moon landscapes such as the denuded hills around Mount Lyell became tourist attractions in their own right. Recently, extensive open-cut mining of coal and bauxite, beach-sand mining, and awareness of issues such as toxic waste pollution and habitat destruction have led to such impacts becoming less acceptable. Denser populations make mining more competitive with other uses, such as recreation, urban development, conservation, or agriculture, sharpening land use conflicts. The Aboriginal land rights debate is one side effect. Like other primary industries, mining remains predominantly export-oriented.

Secondary industry came late to Australia, and has never thrived. It always suffered from the small size of the domestic market, making it difficult to achieve economies of scale and hence low prices. High wages and small scale meant high production costs. There was severe import competition. Because of its structure, Australian industry has always been vulnerable, and it is at present in a period of absolute contraction in size due to foreign competition and a shift of investment interest to mining. The crippling dependence of industry on foreign investment and ownership has often meant that it was essentially 'branch plant' in character, controlled and directed from elsewhere.

A New Britannia?

One direct result of the climatic and ecological restrictions on settlement was a rural-urban dichotomy from the very earliest times. Rural people often felt exploited by the cities, especially when conditions were difficult.

Rural standing was based on crop and wealth. Farms tended to fall into two classes: 'stations' owned by wealthy squatters, and small family farms, often economically marginal. The wealthier squatters were almost always graziers; they ran sheep or cattle, and employed large numbers of itinerant labourers such as stockmen and shearers. They frequently had interlocks with the urban rich, by way of common schooling, intermarriage, and shared social occasions such as the Races and the Show.

But while squatters might spend Cup Day at Flemington or Show Day at Randwick, and buy pianos and expensive furniture with the wool cheque, the small selector had no such pleasures. Life for small settlers, such as wheat and dairy farmers (the original 'cow-cockies') was hard. There were many disputes between squatters and 'free selectors'; and a mythology of 'closer settlement' grew up. Poorer farmers were more isolated, and politically far more radical. Whereas graziers supported conservative political parties, small farmers were an early source of support for Labor, and later a moving force in the growth of the Country Parties. Rural workers were very numerous until comparatively recently. Many, such as shearers and cane-cutters, were highly itinerant; especially in Queensland, they were a major force in the growth of the Labour movement (the Australian Workers' Union [AWU] is still a key to Queensland politics) and a source of continued union strength until recently. This was due most notably to the importance of the wool crop, good communications due to their itinerant movement pattern, and possession of an egalitarian ideology forged as early as Eureka and strengthened in the Depression and the Shearers' Strike of the 1890s.

The urban class structure was divided between a working and a middle class. The latter engaged in import–export trade and later manufacturing, plus professions such as medicine and law. The middle classes were noticeably Scottish and Presbyterian, while the workers had a strong Irish and Roman Catholic tinge. Issues such as Irish Home Rule thus aroused violent passions. Australian politics was largely Conservative/Liberal in flavour in the nineteenth century, but settled down after full adult franchise to a Labour/non-Labour orientation, with later emergence of a Country block in the 1920s. The fact that the Australian middle class have always been 'clients' of the foreign interests with whom they deal has had a significant influence on attitudes to foreign capital and foreign power: for example in defence and foreign investment.

Australian politics in the nineteenth century seemed to be very similar to that of Britain, but the similarly-named political parties concealed substantial differences in the policy issues involved. Development strategies assumed far more importance, and with them control of the 'pork barrel' of public funding, rents for Crown land, and so on.

Overview: Colonial Hangovers

Australia's nineteenth-century dependence on exportable dry land produce and minerals led to a reciprocal dependence on imports of manufactured goods, and continuing identification with Britain. Radical environmental modification followed the clearing of land and the introduction of exotic species. While the culture resembled that of Europe, it was a derivative transplant. Committed to a European pattern of development, it assumed that what was Australian was automatically inferior: the 'cultural cringe'. A lack of patience with the peculiarities of the Australian natural environment resulted. These attitudes are still very common among politicians, economists, and ordinary people.

Chapter 14

The wholesale expropriation of the indigenous population led to a situation in which the bulk of the continent was unalienated, 'Crown' land, and the colonies became major landlords, often by default, since they were unable to prevent or control the spread of pastoral settlement. The strong stress on 'development' which arose from the cultural and economic imperatives led to active government involvement in the promotion of economic growth.

Further Exploration

The most radical account of the Aboriginal impact on the Australian environment is Pyne's *Burning Bush*, in which he argues for the central importance of anthropogenic fire in creating the modern Australian landscape. Blainey's *Triumph of the Nomads* offers a more conventional view of Aboriginal impact on the land. Part Three of Smith's *The Unique Continent* deals both with fire and with the Aboriginal population and the European impact on it.

Blainey's *The Tyranny of Distance*, still controversial, explores the reasons for the settlement of Australia in the first place, and the effects of distance, both locally and from Europe. A brief critical overview of the controversy, by Martin, is in *Uncertain Beginnings*. Bolton's *Spoils and Spoilers* is by now a standard history of environment in Australia, contrasting with Lines' new *Taming the Great South Land*, a passionate exposé of the genocide and environmental destruction associated with English colonisation. Rolls' *They All Ran Wild* and Marshall's *The Great Extermination* document the direct impacts on animal species. Early Australian attitudes are documented by Gilbert in 'The State and Nature'. Parts 4 and 5 of Smith's collection *The Unique Continent* also deal with these issues.

Blainey's *The Rush that Never Ended*, though published in 1963 and consequently now rather dated, remains an invaluable history of mining in Australia. Kennedy's *The Mungana Affair* deals with that shattering political scandal. McQueen's *A New Britannia* humorously and irreverently dissects the social patterns of Australian society; see, for example, his chapter on the social significance of pianos! More orthodox is Duncan Waterson's *Squatter, Storekeeper, and Selector*, a history of the Darling Downs in Queensland. For a flavour of the times, it can be teamed with Steele Rudd's *On Our Selection*, and the other 'Dad and Dave' stories. One of the most amusing of these collections is *Dad in Politics*, a hilarious history of Dad's participation in the Brisbane Parliament in the days before Federation, drawn from Rudd's own experiences (under his real name, Arthur Hoey Davis) as an officer in Queensland's courts and Parliament.

CHAPTER 15

THE POLITICAL ECONOMY OF 'DEVELOPMENT' IN AUSTRALIA

The most striking feature of Australia's history has been the persistence of the myth of 'development'. The pursuit of this chimæra has led to some distinctive political institutions and policies, which have persisted for well over a century.

All capitalist economies are essentially systems of 'state-subsidised profit-taking', in which a substantial part of the infrastructure for private industry is provided out of the public purse; education, roads, law and order, communications, are examples. Australia, dependent on and derivative of Europe, was no exception. But the problem lay in the 'tyranny of distance': populations were sparse, the cost of providing infrastructure was so great, and the returns so low, that no significant private investment occurred. Even government provision was scanty: the major cities are coastal precisely because navigation was by far the cheapest form of communication; reaching the interior was costly, and there was little incentive to go because of its low productivity.

'Development' in Australia was politically driven from a very early date, and a distinctive pattern of government economic intervention grew up. The twentieth-century economy is a prisoner of the nineteenth: the latter's ingrained attitudes and ideas continue to dominate visions of the 'correct' way to develop.

Chapter 15

Myth and Mainspring

After the initial excitement of the gold rushes had subsided, the mineral industry settled down to fitful growth, passing wool in importance as an export only in the 1980s. In the nineteenth century, Australia's one really viable development path was dry land agriculture and pastoral industry: wheat and wool. But a persistent myth that Australia's potential was 'unlimited' meant ongoing flirtation with 'closer settlement' and diversification of agriculture, though strictly within the European mould. Lack of a serious and sustained attempt to explore more indigenous development paths meant continued rejection of the potential of native species, floral or faunal, in favour of the exotic, 'European' species. Only one plant of Australian origin — the macadamia nut — entered commerce, and its recognition came as a result of adoption as a crop overseas. Despite frequent warnings from those 'in the know', the environmental impacts were largely ignored.

This pattern of exploitation generated slow economic growth, punctuated by major economic depressions such as those of the 1890s and 1930s. These were aggravated by Australia's continuing dependence on world markets, and consequent vulnerability to global market fluctuations. Government intervened — often without success — in attempts at stabilisation and amelioration of their effects.

By nineteenth-century standards, the 'development' ethos in Australia led to unheard-of levels of government intervention in the economy. For enormous riches to reach the pockets of the aspiring middle classes, they had to come from somewhere; when mining languished, and land booms collapsed, the 'obvious' place to turn was government. In the absence of hungry capitalists eager to establish the needed infrastructure, governments invested in communication, transport, public utilities, and even marketing for primary producers. The single-minded focus on 'development' meant neglect of environmental and even social problems, seen as irrelevant to nation-building. Diseconomies and environmental damage were ignored; by the turn of the century, fear of the 'Yellow Peril' led to the 'populate or perish' slogan and the White Australia policy.

Political Economy

Australia closely resembles other 'dominion capitalist' economies such as Canada, New Zealand, Argentina, and Brazil. All have similar histories of colonisation from Europe, involving the pushing aside of the indigenes and their institutions, followed by attempts to 'develop' along 'European' lines. All are relatively undeveloped in secondary industry, and markedly dependent on their rural economies. The relevant characteristics are:

- dependence on exports of primary or minimally processed products;
- dominance of foreign capital in manufacturing investment; and
- a relatively large service sector.

Prices for primary products on world markets are much lower than those for manufactured products with a high level of 'added value'. Historically, prices for primary products, including minerals, have steadily declined. At the same time, global markets are dominated by manufacturing nations, which set prices and create policies that frequently do not favour countries dependent primarily on unprocessed exports. Financial dependence — evidenced by reliance on overseas capital — also means technological dependence and a 'branch-plant' economy, controlled by remote overseas owners. This means that important decisions with major local implications — such as the opening or closing of a plant — are made, not by those whose living depends on it, but by owners concerned only with profit.

Statist Developmentalism
Governments have generally tried to escape from this situation by pursuing economic growth, attempting to attract industries by offering various kinds of inducements, such as tax relief, services at low prices, free land, and so on. They have also attempted to take the initiative where private investors did not see a profit. Thus state governments have created marketing boards and engaged in the promotion of Australian products abroad, often paying for these activities from taxes. They have become involved in the direct marketing of services such as transport, energy, and communications. This has created a common interest both in government and in private industry in generating 'development'.

This interlocking of industrial and governmental interests was expressed in a development pattern that Butlin, Barnard, and Pincus dubbed 'colonial socialism'. Essentially, the colonial governments borrowed from Britain against the security of their tax revenues, and used the money to build 'infrastructure' — railways, roads, telegraph lines, and other facilities — in order to attract or stimulate private investment. Rates of growth were

Chapter 15

thereby pushed up, beyond what a purely private market would have achieved. But in the process, not only did governments become investors, they also became involved in marketing, and, because of their control of Crown land, were 'the landlords of most of the Australian continent'.[1] Their revenues came from land rents, customs and excise and, more unusually, from their various enterprises, such as railways. Proceeds were often used to cross-subsidise uneconomic services, or reinvested in further infrastructure. The irrigation schemes of the Murray basin are examples.

This pattern continued into the twentieth century. Since 1945, electricity supply has replaced railways as the major focus for investment, but the basic pattern remains the same. However, the name 'colonial socialism' was and is inappropriate. Colonial it certainly was; the term socialism was chosen, no doubt, to reflect the strong element of government ownership. But this had nothing to do with traditional socialist objectives of income equality, social welfare, and a 'fair go' for working people. The 'colonial socialist' policy pattern emerged well before the rise of a strong Labour movement, and long before Labor as a political party had any significant say in policy. Its motivation was developmentalist, not socialist or welfare-statist. A better term would be 'statist developmentalism', emphasising the strong role of the state, but also underlining the developmentalist component.

Structural Consequences

Statist developmentalism resulted both in government involvement in the economy and in a number of characteristic political institutions. The pressure for development emphasised 'projects' to 'open up' the country and thus provide opportunities for private economic activity. Budgetary and electoral considerations tended to give these projects a strongly practical, 'bricks-and-mortar' flavour, making them 'visible' in both a political and a physical sense.

Communications, especially railways, dominated government investment from the 1850s to the 1920s. Only the 3½-mile Melbourne to Sandridge (Port Melbourne) line of 1854 was profitable under private ownership; most of the rest had to be taken over by government even before completion. Australia's railways were the first to be nationalised anywhere in the world; but they quickly became a focus of corruption and political influence. MPs attempted to secure railway access to their own districts, jobs for individuals to whom they owed favours, and concessional rates for traffics in which they had pecuniary interests. Railways were built for political reasons rather than economic returns or as part of soundly based developmental strategies, and were frequently staffed — at least in part — with incompetents.

Their management problems led in the 1880s to the emergence of the modern statutory corporation, a government-owned but managerially independent organisation, such as now run all railways, Qantas (until recently), and even the Australian National (shipping) Line. Similar semi-independent bodies, often called boards, commissions, or trusts, have been placed in charge of such functions as water supply and irrigation, power, urban transport, grain elevators, and marketing. Examples include the Electricity Commissions in each State; the South Australian Housing Trust — an important prop of the Playford Liberal government's industrialisation policies, which it supported by cheap housing and price controls — until recently the Commonwealth Bank; the (Victorian) Grain Elevators Board; and of course the River Murray and Murray–Darling Basin Commissions. Frequently these organisations were vehicles for indirect subsidies, such as those to farmers via preferential rail rates. Rapid growth of subsidies, tax concessions, inducements, and publicly provided services resulted. Australia was one of the first countries in the world to develop bodies of this kind, and still makes unusually extensive use of them. One of their purposes is to depoliticise routine administrative activities; but, as the River Murray Commission's experience showed, they frequently have the effect of 'freezing' public policy in the status quo of the time at which they were set up.

Ill-effects

Unfortunately, the effects of extensive government infrastructure provision were not always as anticipated. Frequently, the taxpayers saw no significant return, even indirectly, on their investment. Sometimes, they were actually 'taken for a ride' by private investors. In Tasmania, for example, the long-established 'hydroindustrialisation' policy, adopted early in this century, sought to attract industry to Tasmania by providing cheap power. To implement this policy, the Tasmanian Government created the Hydro-Electric Commission (HEC), a near-autonomous body dedicated to the development of hydro-electric power from Tasmania's many lakes and rivers. Unfortunately, the HEC's tariffs (the prices charged for electricity) discriminated from the start against small-scale, local industry. So instead of developing an extensive, highly diversified small-scale industrial economy, all enjoying the fruits of public investment, the HEC managed to attract only a handful of major clients. The biggest of these were 'footloose'; that is, they were industries which processed raw materials from outside Tasmania, for sale to markets which were also outside Tasmania. The only reason for their presence was cheap power. Relatively capital-intensive, they created fewer jobs per kilowatt than small-scale local industry would have, but their power demands grew increasingly large and costly to service: by 1968, eight major consumers accounted for 68% of consumption. Worse still, their cheap power was subsidised by domestic consumers, small business, and the Australian taxpayer. Comalco's announcement in July 1993 that it intended to close down the elderly and inefficient Bell Bay smelter, the largest of these footloose consumers, was not the first failure of this policy.

In 1981, after years of State subsidies and attempts to keep the enterprise afloat, the Electrona carbide works, one of the oldest of the bulk consumers, was closed by its British owners at a net loss to the State of Tasmania of over $14,500,000. The Tasmanian Government then attempted to revive it as a silicon smelter. The considerable local opposition was based on the unsuitable topography and climate of the area, which would tend to magnify noise and trap pollution. The local council, anxious to enhance the residential and aesthetic qualities of the area, zoned for light industry. It was overridden by the State Government, which also intervened in the assessment process in a blatant and heavyhanded manner. The silicon works was established with a large State subsidy, but despite technical success, the plant created considerable noise and pollution, generating concern about health issues among local residents. A mere 75 jobs were directly created. The plant operated for only four years before closing down in 1991. The State Government refused to publish details, but the further loss to taxpayers was reputed to have been $22,000,000.

Electrona is a recent example of a problem which arises when an entrenched policy is pursued regardless of its real worth. Australian governments have a long record of subsidising economic activities which have brought no long term net benefit to the community; additionally, they have frequently persisted with policies long after they have lost their relevance. Some, like hydroindustrialisation, have never delivered the claimed benefits.

Pressures on States

Statist developmentalism made the Australian colonies dependent on their revenue from land sales and leases, customs and excise charges, and profits from public enterprises, notably railways, urban public transport systems such as tramways, and electric power supply. After Federation, they lost customs and excise to the new Commonwealth, and their share of the tax 'take' began a long period of relative decline. The Financial Agreement of 1927, which placed limits on their capacity to borrow, formalised this decline in power. The loss of the power to collect income tax, during and after the Second World War, was a further disadvantage.

And from the late 1920s onwards the States began to lose their 'natural monopolies' in transport and urban services. Motor trucks began to compete with the railways, and motor cars and diesel buses — the latter often privately operated — adversely affected the finances of urban tramway and commuter rail systems. The resulting loss of revenue, especially when taken together with a falling capacity to tax, made state governments more dependent on their remaining sources of revenue.

The most important of these were from natural resources such as minerals. States collect royalties, calculated on the volume of ore mined, and in addition may make profits from the provision of services. Queensland Railways, for example, makes its only profits in the Central Division, hauling coal from mines to ports. Obviously, the more that is mined, the more royalties the State collects, and the more profit it makes from ancillary investments such as railways. States therefore become enthusiastic about the extraction of minerals, and other projects which will increase their revenues. For example, the Premier of Queensland, Mr Goss, said of the new Century zinc prospect in the Gulf country that it was 'too important to let anything stand in its way'.[2] Similarly, States are reluctant to abandon any taxes which remain to them — such as payroll tax — even when there are clear economic gains to be made from their removal.

Both the loss of taxing power and the decline of the natural monopolies locked the States into interventionist 'development' activity even more surely than the active promotion of private capital and its interests found elsewhere in the world. To service growing populations, States saw it as essential to expand their revenues. This, increasingly, could only be done by encouraging natural resource extraction and by attracting industries to the State. The resulting taxes, direct and indirect, would enable the provision of additional services and the expansion of existing ones.

Since the 1970s, the progressive abandonment of tariffs on imported goods has devastated Australian manufacturing, adding substantially to the burden on social services at the same time as cutting revenues. In the post-1980s climate of deregulation and financial stringency, the Federal government has been forcing the States to bear a disproportionate part of the burden of cuts in government services, primarily by limiting their annual grants. This has led to squabbling and intensified rivalry among the States. Against a backgound of steady long-term decline in the real price of mineral and primary produce exports, one effect of this strategy was to increase the environmental impact of State government activity. Increasingly, 'development', whether it took the form of new mines or coastal tourist resorts, has had highly visible, sometimes devastating, environmental effects.

Government and Business
Government prominence has never excluded private business. The purpose of infrastructure provision, as well as of States entering business themselves, was to attract private investment and stimulate economic activity. 'Statist developmentalism' depends on private investment to complement it, making government highly dependent on business, and vulnerable to threats to withdraw or simply not to invest. The relationship of government and business is therefore one of mutual dependence: business depends on government to create opportunities for investment, but government depends on business to invest once it has provided infrastructure.

This pronounced interlocking of Australian business, government, and bureaucracy reflects the convergence of government and private interest in 'development'. Politicians exploit the term because they think it is electorally popular. 'Development' therefore becomes a stalking-horse for the interests of business: it permits politicians to justify attention to business, and the privileged position it enjoys, by claiming that it is for the welfare of everyone in the community.

Like other capitalist governments, Australia's develop distributive policies: handouts and tax concessions to various groups in the community. This results in a fiscal 'ratchet' of expenditure growth: governments pursue inflationary policies in times of low economic activity, but cannot bring themselves to deflate in better times, because this would involve making cuts which would certainly be protested by those affected. This sabotages Keynesian counter-cyclical policies, creating steady inflation accompanied by economic growth pressure. But the development ethic in Australia further increases the pressure: more government subsidies can be demanded sooner, and, better still, selflessly. Developers can claim that their demands for subsidy are not for their own enrichment, but for the benefit of the country as a whole. The rhetoric of the mining and woodchip industries is an example.

Development Ideology

The development ideology has stressed 'more and bigger is better' and the belief is still prevalent that Australia's potential is 'unlimited'. Perhaps part of the appeal of 'bricks and mortar' is that it gives politicians something to open and name after themselves, often even before they are decently dead.

The language of development also conveniently excludes some social and environmental concerns, identifying economic growth with the public good. If development is patriotic, those who are against it are un-Australian 'knockers'; this straitjackets public debate. Yet the capital hunger associated with 'statist developmentalism' helps to explain the massive increases in foreign indebtedness, and chronic debt problems for State governments.

CONSEQUENCES OF STATIST DEVELOPMENTALISM

Statist developmentalism is a *distributive* approach to public policy, geared to big 'projects', often with substantial government investment. All have involved committing large lump sums, the beneficiaries being contractors engaged in construction work, and their employees, or recipients of subsidy, such as farmers benefiting from cheap water or fertiliser. Frequently, for political reasons, State governments in particular would become committed to grandiose schemes for which they had few resources. Often, like the Burdekin scheme in Queensland, these would be undertaken in dribs and drabs, a little capital being allotted in each year's budget. In consequence, States became committed to paying interest on loans for schemes which would not yield revenue until far in the future, if at all. That in turn increased the tax burden and tied up funds that might have been devoted to other, more useful purposes.

The other consequence has been large-scale corruption, which has been remarkably common, especially in New South Wales and Queensland. Because government controlled the granting of contracts and, frequently, appointment to jobs as well, it was easy to trade favours.

Statist developmentalism begins to become a liability when resources are no longer abundant and cheap, and consequently there is competition, not only for the resources themselves, but

Chapter 15

from alternative uses, ranging from other, 'cleaner' industries to recreation and wilderness. The environment can no longer be freely exploited; the end of the formerly inexhaustible resources is in sight. Investment no longer expands employment; instead, the biggest, most capital-intensive projects create the most disruption and the fewest jobs. As the effects of development increasingly impose diseconomies on the population, the benefits are no longer 'self-evident', and the associated policies fall dramatically in popularity. In other words, when 'ecological scarcity' begins to bite, and the resource régime has to change from distributive to regulatory, statist developmentalist policies become increasingly irrelevant and even harmful.

The obsolescence of statist developmentalism in turn leads to policy dilemmas for government. In order to break out from the old pattern, new and radical ways of looking at economic management are needed. No Australian government, state or federal, has yet seriously attempted the transition. The sole comprehensive exploration of the problem, the 'Greenprint' for Tasmania, was ignored and shelved by the very government that had commissioned it. Distribution, not regulation, remains the dominant theme in modern Australian politics.

This is in part because statist developmentalism is as much a state of mind as a development strategy. It embodies a series of assumptions, among which the most important are that development (i) is imperative, (ii) is popular, and (iii) has self-evident advantages. The Brisbane *Courier-Mail*, for example, still advocates large-scale 'development' through massive government investment in arid country areas. This mind-set ignores evidence that development damages ecologies and diminishes amenity for the population at large. Instead, its assumption that ecologically rational policies will be costly and will eliminate jobs sets up an unnecessary and damaging conflict. The outcome is frequently that policy makers are unwilling to consider perfectly viable options. Coupled with deeply entrenched prejudices and interests in government departments, such attitudes explain the long persistence of ecologically, economically, and even politically irrational policies.

Australia's Problems

Many of Australia's current problems are consequences of statist developmentalism. High levels of State government debt reflect falling revenues and the persistence of overseas borrowing for infrastructure. In the 1980s, the situation was aggravated by poorly conceived and ineptly regulated attempts to engage in entrepreneurial activity, also for 'development'. Several spectacular bank crashes, most notably in Western Australia, Victoria, and South Australia, ensued. In Victoria's case, the situation was further worsened by the collapse of an ill-regulated building society in Geelong, which led to the electoral defeat of the Cain–Kirner Labor government.

Australia's modest adaptation to its international environment was based on efficient primary production for export. The only break in this pattern was from 1945 to the 1960s, when attempts were made to build up manufacturing. It was assumed that 'comparative advantage' could protect manufacturing: high-wage, high-cost economies could compete with low-wage economies because, being more developed, they had better technology, and could thus offset their high wages with greater productivity. Faith in comparative advantage was so weak that high tariff barriers protected Australian industry until very recently. Comparative advantage is now rapidly disappearing as multinational firms transfer modern technology to low-wage economies such as Taiwan, China or Indonesia. These same countries keep wages low by military force, thus harming Australian workers, who simply cannot compete with wage rates of 21¢ per hour.

The decline in Australian manufactures cannot entirely be made up by exporting primary resources, especially minerals. The real prices paid for primary produce and minerals in international trade have fallen steadily throughout the twentieth century, leading to the present rural crisis. In recent years, despite increased volumes of exports, Australia's share of the world export market has fallen from 2.6% in the early 1950s to about 1.2% in the 1980s; its ranking as an exporter has fallen from 8th to 23rd. Globally, the fastest growing export markets are in manufactured goods with high 'value added', an area in which Australia does not shine. The recent opening up of domestic markets to more international competition has probably made the situation worse, not better. In particular, it leads to government preoccupation with short-term market fluctuations, and makes it more difficult to plan for stable, long-term, ecologically rational management.

The paradox, as Bell has it, is that Australia 'has increasingly attempted to base a first world social structure on a third world export profile'.[3] The implications for environmental management are serious, since economic stringency may lead to short-sighted, *ad hoc* decisions with serious and irreversible consequences.

Chapter 15

Overview: Development as Constraint

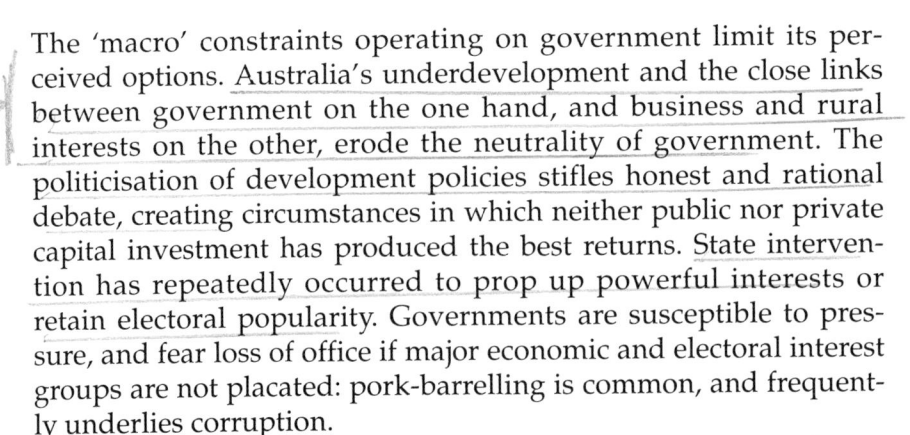

The 'macro' constraints operating on government limit its perceived options. Australia's underdevelopment and the close links between government on the one hand, and business and rural interests on the other, erode the neutrality of government. The politicisation of development policies stifles honest and rational debate, creating circumstances in which neither public nor private capital investment has produced the best returns. State intervention has repeatedly occurred to prop up powerful interests or retain electoral popularity. Governments are susceptible to pressure, and fear loss of office if major economic and electoral interest groups are not placated: pork-barrelling is common, and frequently underlies corruption.

The lack of manufacturing industries creates a heavy reliance on rural industries: mining, agriculture, pastoralism. The unbalanced nature of the Australian economy makes it both dependent on and vulnerable to outside economic forces. This has meant that rural industries have obtained greater government assistance than otherwise would have been the case. This has favoured the persistence of the 'development' myth, which helps to explain why governments continued to support irrigation development in the Murray–Darling basin when there was no economic justification.

The dominant pattern of Australia's development over two centuries has been what Butlin, Barnard and Pincus have called 'colonial socialism', but which, as explained above, is better characterised as 'statist developmentalism'. Its characteristic feature is the procurement of capital on overseas money markets by governments, which then invest the capital to create infrastructure, which is understood in turn to attract private investment. Because governments also controlled land and were willing to engage in commercial activities for profit, they came to play a very large part in commercial life and in the economy. In practice, statist developmentalism has often failed to attract sufficient private investment, leading to direct government participation in the economy in response to pressure for 'development'. It has also led to a politicising of the many economic activities of government, so that investment decisions were made according to political, not economic, criteria.

State intervention to attract industry and investment in competition with other nations leads to direct subsidy as well as 'infrastructure' provision on a massive scale. Under contemporary conditions, it can also lead to increasing dependency, as balance of payments problems progressively worsen.

The Political Economy of 'Development' in Australia

Statist developmentalism was essentially distributive. It was geared to government investment in big 'projects', in transport, communications, land development, water storage, and so on. These all involved committing large lump sums which would then be disbursed to the various beneficiaries, whether contractors engaged in construction work or recipients of subsidy, such as those supplied with cheap water or fertiliser. This permitted corruption on a grand scale, since granting of contracts and appointment to jobs were all in the 'gift' of government. That in turn led to novel institutional arrangements — most notably the development of the statutory corporation — to distance the government's businesses from political jobbery. In the twentieth century, as economic management became an important goal of government, this pattern expressed itself as an explicit 'politics of distribution', with well-documented inflationary consequences.

Statist developmentalism was also a state of mind. It tended to assume that the advantages of development were self-evident, while ignoring evidence that development damaged ecologies and diminished amenity. This mind-set was deeply built into the various government departments, State and Commonwealth, which administered the (mostly very fragmented) functions which government oversaw. In some cases, as with the Hydro-Electric Commission in Tasmania, this led to virtual autonomy, resulting in savage battles if its power or judgement were questioned.

A major source of conflict during the various environmental battles of the 1970s and 1980s has been the entrenched assumption that environmental management was a 'frill', subordinate to the stern business of making a living. Since making a living in Australia almost always means depleting natural resources, this conflict has been savage. Worse, it has led to a false dichotomy between 'jobs' and the environment, the assumption being that 'conservation' always results in job losses.

FURTHER EXPLORATION

The broad outlines of Australia's development pattern, as well as many of the details of its present-day predicament, are discussed in a number of works. Head's books *State and Economy in Australia*, and *The Politics of Development in Australia* both contain some relevant essays as does the recent book by Bell and Wanna. Butlin, Barnard, and Pincus' *Government and Capitalism* reinterprets Australian history according to the precepts of the (then) fashionable economic 'rationalism'. An insightful work, it should be read with an awareness of its biases, and consequent caution about its conclusions.

The 'Keynesian ratchet', which affects many Western democracies, is described in Crouch's chapter 'The State, Capital, and Liberal Democracy'.

The failure of the hydro-industrialisation policy in Tasmania is dealt with by Tighe, and more recently by Crowley, in her excellent short account of the Electrona smelter debacle, published in 1989, and in a more recent conference paper. Lowe describes the development of the 'Greenprint' in a 1991 article for *Habitat*.

Part 5

Environmental Policy in Australia

Because of the fundamental nature of environmental policy issues, they are always inextricably entangled with other policy domains. Thus tariff policy, industrial relations, gun laws, the relations between business and government, can all have implications for environmental policy. This being so, it is important to understand the context of public policy-making in Australia in order to understand environmental policy.

The analysis of policy is a relatively new activity, not only in Australia, but globally. As a result, the available analytical tools are rapidly changing and improving.

The chapters in this final Part examine, firstly, the processes of policy-making in Australia, bringing to bear some of the most recent ideas and knowledge. Secondly, some of the problems faced by those analysing and evaluating policy are considered.

Until very recent times, there has been a tendency to attempt policy analysis either in terms of the formal political institutions, or by trying to find broad general theories at the national level which can account for all cases. Recent scholarship has come to recognise that the patterns of interrelationship between government and the community vary considerably for different sectors of economy and society; and that they can only be understood at a sectoral level. The emergence of useful 'meso-level' theories which identify relevant patterns of interaction has been an important development.

Policy evaluation has suffered in the past from an over-dependence on a Benthamite, welfare economics model. It has also tended to suffer from atomism: a fragmentation into discrete, project-by-project evaluations. These can readily lose sight of the 'big picture' unless there exists a body of coherent government policy and a will to implement it consistently. A further difficulty is that policy evaluation is often undertaken 'on the run'. Answers are needed urgently, but information is in short supply, knowledge incomplete, and resources for analysis inadequate. Strategies must be developed to cope with these problems.

CHAPTER 16

MAKING POLICY

The perspective of political economy is essentially 'broad-brush'. It very effectively develops an understanding of the major external constraints on state action, as well as the general conditions under which government operates. However, the 'fine grain' of policy processes must be understood by considering the interaction of government with the major forces within the political system. Not only do such forces act as constraints on government policy, but the *pattern* of their interaction with government can act to shape and limit options in significant ways.

Major Forces
Despite their near-monopoly of armed force, governments are not the only sources of power and influence in modern societies. Social values change slowly, and can act as constraints on government action (though they do not always do so: totalistic governments given to murder and rapine do arise). Major religions may have important effects on government policy. Often this is highly variable: the mullahs (Islamic clergy) have very great influence in Iran, but far less in Turkey or India, both of which have large Muslim populations.

263

Chapter 16

Additionally, as has been repeatedly emphasised in the foregoing pages, electoral factors, such as the fear of defeat or the desire to capitalise on potential support, can be very influential. Electoral advantage is a particularly strong stimulus to symbolic politics, because strong popular reactions to symbols often mean that politicians can make big political gains for a very small outlay of resources, whether their own or the public's. Interest groups within the community not only urge their own causes, but also act as points of reference for people who are not members, by articulating widely held points of view. Politicians, seeking election or popular support, can advert to the views of such groups, appearing to support or endorse them, and hence use them to gain power or influence. Some of this influence rubs off on the groups themselves — broadly speaking, the more votes a group can mobilise, the bigger its influence — which means that they too gain in power. Success in placing one issue on the agenda, or in having a proposal accepted, frequently means that the next time round, it will not be so hard.

But by far the most important of the major constraints is what Lindblom has called the 'privileged position of business' in nearly every modern economy. Business is powerful precisely because of the importance of resource mobilisation and 'economic transformation' to government. Private firms, and especially those which control investment, such as banks and major multinationals, therefore have considerable power to endorse or frustrate government policy; how considerable may be appreciated when it is realised that Standard Oil of New Jersey (Exxon-Esso) has an annual turnover larger than the gross national product of all but ten nations.

Many small nations have found to their discomfort that 'bucking' the global financial system is fraught with danger. In Australia, the Ryan Labor Government in Queensland was subjected to an embargo on loan funds by the London market in the 1920s, at the behest of powerful graziers threatened with massive increases in rent. The effect was to drastically modify Labor's development plans for Queensland, forcing the abandonment of a proposed steel industry, and creating widespread unemployment. In modern times, the World Bank and the International Monetary Fund have forced many developing countries to adopt 'economic rationalist' policies — to such effect that the net flow of wealth from such nations to the wealthy West has increased massively.

Command economies such as the old USSR seem at first blush

to be insulated from business influence, but in fact they are not. Internationally, they deal with the same forces as everyone else. Domestically the power of the industrial sector of the economy — their own 'business' — simply became institutionalised within the centralised bureaucratic governmental system. The Soviet Union's military-industrial complex was relatively more powerful than that of the USA, absorbing a greater proportion of GNP, and subject to less scrutiny because of the dictatorial political system. One legacy is the horrendous nuclear waste problem now faced by the nations of the former USSR, a result of casual, uncontrolled dumping.

The institutionalisation of particular viewpoints is quite possible in democratic systems as well. 'Development' in Australian politics is one case in point. Another is the effective takeover of the federal bureaucracy by economic 'rationalists', who

> had captured the promotions system from about 1982 onwards and used it to favour their own kind. From undisturbed positions of power in the central agencies they had — with successive waves of rationalisation and restructuring — demoralised, colonised, and in many cases driven out the real problem-solvers in the public service: the hands-on economists in the industry departments and the other senior policy professionals (engineers, scientists, educationists, medical people, and others) in the program (sic) and social service departments. In this way they obtained an overwhelming domination.[1]

Not only will such a 'captured' bureaucracy have a very direct influence on government policy — in Australia's case, dominating the economic policies of the major parties — but it will also be able to influence and even control the *implementation* of government policy. This permits distortion and reinterpretation of such policy; a phenomenon which often occurs in practice.

Institutions with substantial power can often influence government policy directly, either by threats or by action. But Head points out that the Australian consensus on economic growth has more frequently led to a commonality of interest, in which the language of 'development' is a vehicle for enhanced attention to the interests of industry. Confrontation, as in Queensland in the 1920s, is far less common than cooperation or at least collaboration. But institutional power can also permit government policy to be frustrated, either by outright disobedience or, more commonly, by subversion.

CHAPTER 16

Interactions: Partial Theories

One of the most common forms of subversion is the undermining of government's regulatory capacity. Governments necessarily regulate a vast range of activities. Abattoirs must be inspected to ensure hygienic conditions, corporations regulated so that fly-by-night entrepreneurs do not make off with investors' funds, banks must (or ought to) be prevented from overcharging and sharp practice, builders and contractors must be overseen, safety measures and working conditions in factories must be established and enforced, the sale and use of drugs controlled, schools and their standards supervised; even without environmental regulation, the list is endless. Failure to carry out such duties leads to misery and even death: in May 1993 some 400 women working in a stuffed-toy factory in Bangkok were burnt to death in a locked building without fire escapes. Thai safety and environmental law is notoriously lax, in part as a result of years of predatory military government. In Victoria, the collapse of the Pyramid building society in 1990 deprived many small savers, especially in Geelong, of their life savings. The Pyramid collapse was a direct *consequence* of the abandonment of its regulatory function by government, in line with the prevailing economic orthodoxy.

Even where regulation is attempted, a common means by which it can be undermined is the 'capture' of the regulatory agency. Agencies unavoidably develop close working relationships with the industries which they purport to regulate: forestry departments with the timber 'industry', for example. The 'capture thesis' hypothesises that the regulators increasingly identify with those they regulate, so that they come to see the issues in the same way, and to adopt the same solutions. Foresters, for example, would tend — as indeed they do — to see forests in terms of sustained timber yields, rather than as habitats, recreational facilities, or simply unspoiled wilderness. Where such perceptions are significantly different from the intent of government policy, or the purposes for which the regulations were created, it is clear that both the letter and the intent of the original policy may be undermined.

Some theorists, drawing primarily on US experience, hypothesise that all agencies are captured eventually, and that regulation is consequently a continuing process of tightening up, followed by slow decay. Others conclude that regulation is always doomed to failure, some arguing that therefore it is not worth making the effort in the first place.

But agencies may subvert government policy for more mundane reasons. The New South Wales Forestry Commission found itself short of revenue for managing State forests by the 1970s, due mainly to inadequate returns from sales; it responded by attempting to step up the rate of exploitation, while reducing its investment in reafforestation.

Interdepartmental rivalry can also create serious problems, impeding and even frustrating declared policy. Mant took the dangers of interdepartmental rivalry so seriously that he recommended centralisation of all Victorian water management functions in one 'super-department', to ensure that energies were devoted to carrying out the programme, not fighting over 'territory'. The history of the Victorian Environment Protection Authority (EPA) illustrates how damaging such rivalry can be. Created under the *Environment Protection Act 1970*, it was originally intended to be a semi-independent body with authority over all government and many private agencies. Underfunded from the start, it was never able to carry out more than a fraction of its statutory duties. Creation of a Conservation Ministry two years later led to a tug of war over precedence, and the EPA found its functions and personnel being taken over without notice. Systematically undermined by other departments, denied resources, and rebuffed by the responsible Ministers, the EPA was never more than window-dressing.

Another example of systemic restraint is brokerage politics. First described for Tasmania, brokerage occurs elsewhere as well. It is dominant in Tasmania because of the dependence of the economy and the high level of subsidy received from the Commonwealth under fiscal equalisation. This gives the Tasmanian Government a large pool of money which can be invested or used to support services. Brokerage is a process of 'buying' political support by investment in electorally critical areas, with the aim of building up solid electoral commitment out of gratitude to the political party involved. It cunningly mixes economic payoffs, symbolic rewards, and personal loyalties. In a small state such as Tasmania, such parochial 'pork-barrel' politics is often more important than appeals to national or state-wide interest, let alone ideology. It tends to build extensive networks of mutual support, based on purely pragmatic considerations. Such networks are highly resistant to change, because all concerned — and especially the politicians — tend to see change as a threat to their security.

However, 'development' in Tasmania, as elsewhere in

Chapter 16

Australia, has relied primarily on the exploitation of natural resources. Tasmania's predicament in modern times — in many ways a microcosm of Australia's as a whole — is that as the resources which underpin brokerage become increasingly scarce and prices for them fall, the distributive political economy becomes unsustainable. The change from distribution to regulation is incompatible with continued brokerage politics. This has led to acute tension, not between the established political parties, but between the parties and those groups — such as the Wilderness Society and the Greens — who advocate change.

Such partial views do not account for all the observed policy processes, however.

Policy in a Pluralist State

The truism that important political, social, and economic interests within society have impacts on policy finds its expression in the notion of pluralism. The Australian political system is pluralistic within the normal meaning of the word: a number of interests and pressures continually affect government as it grapples with various problems. They may be organised as pressure groups or interest groups, or they may simply command government attention by their size and power. Rural interests in Australia have long been politically active, through groups such as the National Farmers Federation, the Cattlemens' Union, and so on. Workers have been similarly active in trades unions. Big business in Australia has always had a major influence on government, but it has often been poorly organised or disunited; individual major firms or industry groupings have frequently had more say than 'representative' bodies. These organisations have impacts on politics: both the Labor and the Country Parties have strong historical links with their supporting interest groups. The weaker links between the Liberals and industry pressure groups, despite their obvious connections with big business, are explained both by the historical discontinuity of business organisation, and by the very diverse nature of business demands.

Under such conditions, government's ability to achieve 'economic transformation', as well as to coordinate policy effectively, depends on its relative power. This concept is sometimes expressed as the *capability* of government: i.e., its ability, in the face of more or less determined attempts to make it adopt alternative policies, to get its own way. Implied is a capacity to decide on overall policy, and to plan and implement effectively. Because the state has its own interests and needs — such as Rose's 'defining' functions, and the pressure for 'economic transformation' — it has to be treated as an important independent variable. It cannot be collapsed into one of its components, such as business, labour, a ruling class, or some other social grouping. State strength is itself an important factor.

Strong and Weak States

The Canadian scholars Atkinson and Coleman have suggested that states can be classified into 'strong' and 'weak', on the basis of their capability. They point out that the Anglo-American states, such as Britain, the USA, Canada, Australia and New Zealand, have 'Westminster'-derived political institutions, in which the powers of government are limited. They suggest that such systems have typically been poor at planning, and unable to exercise state power very fully. By contrast, more autocratic systems on the continent of Europe are 'stronger', having fewer controls and restrictions on government.

In weak states, policy-making tends to be reactive, often responding to crisis by seeking a quick, temporary 'fix', without overall policy direction. Stronger states adopt more anticipatory measures, predicting problems and devising policies that will enable them to cope when crisis comes.

It is quite difficult to fit Australian government into the Atkinson and Coleman schema. Much of Australian policy-making seems to fit their 'pressure pluralism' pattern, in which government is weak and business is fragmented and disorganised, each group advocating its own policy agenda. Under these conditions, governments often rely on the competition between such groups to expose policy issues and hence to define their own policy objectives, in a sophisticated kind of 'garbage-canning'. Government, under 'pressure pluralism', is also disorganised, with departments and agencies pursuing their own objectives even when these clash with those of others. Under these conditions, the state is pulled in various directions by the pressures on it. Policy is made incrementally.

'Pressure pluralism' seems to account for many of the weaknesses of public policy in Australia. Planning is skimpy and ineffective; bureaucratic conflict is common; and policy problems are often dealt with on the run, rather than with any overall plan in mind. The policy pattern is reactive rather than anticipatory, and coordination routinely suffers.

Chapter 16

The Australian Paradox

Yet, from the earliest times, Australian government has also behaved like a 'strong' state. It has not been unheard of for government (whether federal or state) to impose specific policies, even against quite strong opposition. Bell instances the imposition of a strong regime of tariff protection in the 1950s and 1960s, under the direction of the then Trade Minister, McEwen, and the more recent switch to a 'free trade' tariff regime under the influence of 'economic rationalists' in the Commonwealth Treasury. Examples from the States include South Australia's industrialisation policy from the 1930s to the 1960s, and the influence of powerful, developmentalist Premiers such as Charles Court in Western Australia — presiding over much of the State's rapid mineral development in the 1970s — and Bjelke-Petersen in Queensland. These are difficult to classify in the Atkinson-Coleman scheme; the long-lived Tasmanian hydro-industrialisation policy is easier, as it is an example of *concertation*, with the powerful and near-independent Hydro-Electric Commission (HEC) organising support from powerful client groups for planned programmes of hydro-electric 'development'.

Yet even these examples are partial, *ad hoc*, and unplanned. In none of the States mentioned was the planning process comprehensive; in general, it affected only a part of each State's economy. South Australia came nearest to comprehensiveness, in that it supported its industrialisation policy with policies of housing and price control designed to keep costs low and hence attract investment. Paradoxically, these policies were implemented by a conservative government. But South Australia was critically dependent on a supportive economic environment, which was out of its control since economic management is a federal government responsibility. In general, the picture is of 'weak' government punctuated by occasional bouts of determination, which frequently were not especially rational. These often reflected the combination of a dominant bureaucrat with a powerful, autocratic Premier; in South Australia, Wainwright with Playford, in Queensland, Hielscher with Bjelke-Petersen.

Bell has suggested that the apparent paradox of simultaneous weakness and strength arises because of a failure to distinguish between the various levels on which the state operates. Central government in Australia is by and large relatively more powerful than in other 'Anglo-American' nations. This reflects both the historical origins of the Australian colonies and the long involvement of government in economic activity. But, at the same time, it has a record of failure in interventionist planning. The poor performance of the Victorian Government in irrigation prior to 1909 is an example, the more striking for the subsequent failure of the New South Wales approach, which took account of the Victorian experience.

Bell concludes that the Australian state often acts in a *weakly* anticipatory manner, but is limited in its capability by a (partly ideological) preference for an arm's-length relationship with industry. This has inhibited development of a capacity for detailed, planned intervention in economic life, for at least two reasons. Firstly, the statism and interventionism of Australian government works largely at the 'macro' level. The state establishes the broad outlines of policy, but it depends critically on private investment to fill in the detail and make it effective. This has the effect of ceding significant power to capital — since investors can always refuse to go along with government policy — as well as leaving to them many of the details of the actual implementation. Secondly, there is considerable variation in state capability from sector to sector of the economy, and of the bureaucratic system. Historically, this reflects past state intervention and participation in economic activity. Thus the state may have a significant capacity to plan and implement in one sector, but lack comparable abilities elsewhere.

This is in part a consequence of the 'two-tier' approach to policy. If government is to make policy at the 'macro' level but implementation is to be by private investment at the 'micro', then the oft-remarked paradox of a 'free market' economy with high levels of government intervention is explained. The 'free market' has 'micro' freedoms, within the overall 'macro' framework established by government. Because of the failure of investment in some areas — transport and communications, for example — government has developed significant capabilities and skills, embodied in developed administrative institutions. In others — for example, detailed control of industry policy — it has few skills and poor resources.

This incapacity directly constrains Australian government capabilities for 'economic transformation', with two major consequences. Firstly, some kinds of policies are not 'macro' in nature, or involve a good deal of 'fine grain'. Environmental policy, especially where it involves ecosystem management, is particularly prone to this kind of difficulty: the necessary actions may be impossible to carry out with the kinds of tools that 'macro' policy has available. In the Murray–Darling, the Victorian Government

CHAPTER 16

found it necessary to establish numerous local bodies for consultation and implementation, precisely because of this problem, while the Tasmanian government's failure to do so was one reason for the stiff local opposition to the Wesley Vale pulp mill proposal. Secondly, if government lacks administrative capability at the 'micro' level, some kinds of policies may be literally impossible to implement, even if fundamentally well conceived. Regulatory policies are especially likely to be affected. The most spectacular example was the recent spate of failures in State banks and building societies; though this was due in part to an ideologically-driven abdication of regulatory responsibility by the relevant Federal Government agency, the Reserve Bank.

COMMUNITIES AND NETWORKS

The policy process is also easier to understand if it is appreciated that particular issues have their own 'public', and that not all issues involve the same people. The simplistic model of a widely attentive audience for the national policy debate, eagerly reading the daily newspapers, is highly inadequate. Rather, individuals and groups have their own preferred issues and problems, which they follow with markedly more attention than others.

Furthermore, the economy in particular, and policy issues in general, can be thought of as being divided into 'sectors', each of which generates different policy problems. The boundaries of these sectors are fuzzy, and often change issue by issue; however, each sector gives rise to one or more **policy communities**.

Policy Communities

Such a community may include government ministers, public servants, and others directly concerned with the making and implementing of policy; members of firms having a direct interest in the policy issues in question — for example, road hauliers, where road transport policy is concerned — journalists involved both in the specialist press and in national reporting; and, possibly, members of the public who may belong to special interest societies or may act as individuals. Its crucial characteristic is that there is a high level of communication among its members. This communication serves both to identify the central issues of relevance to members, and to explore proposed policies and courses of action.

A policy community would therefore typically have a definition of the central problem(s) of concern, some consensus about a range of possible policy measures, and a (generally amicable) ongoing debate about the best policy options. These may be embodied in a series of informal 'rules of the game'.

Most policy communities are organised about government; however, churches or other religious organisations, and in recent years, even international non-governmental organisations (NGOs) such as Greenpeace have played an important part in identifying and defining areas of policy concern. Economic orthodoxy, especially the neoclassicists with their emphasis on 'free trade', makes up another community, one with exceptionally strong influence in politics worldwide.

The existence of such communities means that policy processes can be understood in terms of 'ins' and 'outs': those who are members of the community (and hence 'in') are likely to be attended to, and their ideas considered sympathetically. The policies they propose also have a much greater chance of being adopted. This may occur *even* if one or more 'outs' exist who offer better analyses and better policies.

In fact, policy communities may be very concerned to defend their own orthodoxies, and particularly intolerant of attempts to redefine their central problems. This in turn may become a central criterion for determining who is 'in'. Thus the road lobby might consist mostly of hauliers, truck manufacturers, oil companies, and a froth of sycophantic journalists; outsiders arguing for restrictions on road use to cut accident rates or save fuel might be ignored and excluded as 'crackpots'.

This is important, because pluralistic political systems tend to assume that vested interests are 'legitimate': that is, that they have a claim on government's time and attention directly proportionate to their financial interest in whatever issue is concerned. Governments frequently rely on such interests for information, feedback on policy options, and numerous other functions which constitute privileges of membership in the policy community. If, at the same time, individuals or organisations seeking to raise questions which are either in the public interest or which simply ought to be considered gain no attention, policy biases will inevitably result.

In Australia, this has led to a situation in which environmental groups, such as the Australian Conservation Foundation, The Wilderness Society, Greenpeace, and state and regional conservation councils have a precarious status as members of the policy community. They are fitfully and incompletely consulted on major issues, and they gain some attention through the media. Their full message is rarely heard or comprehended, largely because their construction of the environmental problem does not fit easily into the prevailing growthist, developmentalist mentality. But their exclusion from the policy community means that sympathetic, constructive debate about the environmental problematic, from the point of view of the conservation lobby and the environment movement in general, is stifled. That contributes to an incomplete working-out of the possible policy directions, and an unnecessarily poor analysis of the problems, as a result both of inadequate resources and of lack of contact with the larger policy community. This in turn means that, even when environmentalist proposals are practicable and worthwhile, they are likely to be ignored.

'Epistemic Communities'
Some communities are more closed than others. Typically, it is harder to close a community if it is extensive, and if it has norms, values, or power bases which remove it from the immediate issue. If it is outside the control of any one government, and has internal lines of communication, it is likely to have considerable strength. Haas has drawn attention to the role of **epistemic** (knowledge-based) communities in policy-making. His primary example is the scientific community, in relation to international cooperation.

The scientific community is international, and its work has ramifications for numerous policy issues. It is committed to a set of values which include free scientific communication, objectivity, impartiality, and a critical intellectual stance. While it may at times fall short of all these goals, their existence is an important guarantee of information, evaluation, and opinion emanating from it. Frequently, there is a defined body of scientific opinion on a particular issue, emanating from a subset of the scientific community as a whole, but backed by its prestige. Haas identifies an 'international ecological epistemic community' consisting of scientists in cognate disciplines. In his Med Plan case study, he shows that in those Mediterranean countries which had active scientific communities, there were more likely to be channels, both institutional and informal, for incorporating scientists and their ideas into policy-making. In turn, those countries were more responsive to scientific alarm and concern about oceanic pollution. They paid more attention to scientific advice, and were more willing to engage in international compacts to deal with the problems. Those countries which were least cooperative were the ones which had no or poorly developed scientific communities, and were inclined to deny their responsibility.

In Australia, science is revered and respected, but not always heeded. Its findings are eagerly exploited when they support a desired course of action; they are as often ignored when they do not. The Western Australian Government, throughout the 1950s and 1960s, steadfastly ignored the findings of its own Government Entomologist, who, as long ago as 1945, had discovered that all known major pests of cotton could be found in the area of the proposed Ord River irrigation scheme. Yet cotton was chosen as the preferred crop when the scheme opened in 1963; cultivation collapsed in 1974. The proponents of the controversial Wesley Vale pulp mill proposal in Tasmania were highly critical of scientific evidence purporting to show that unacceptable contamination of Bass Strait and its fisheries would occur. Kellow concluded that technical decisions in the electricity industry in Australia and New Zealand were rarely, if ever, based on technological factors; rather, arguments from technological necessity were used to justify decisions reached on other grounds.

In a more recent case the Government of New South Wales, through its Minister for Agriculture, the State Pollution Control Commission, and the Sydney Water Board, attempted to prevent the publication, not of raw scientific information, but of discussion and analysis which would make it intelligible to laypersons. Fish caught in the vicinity of Sydney's notorious sewer outfalls were proving to have very high levels of pesticide and heavy metal contamination, the pesticides being on average 122 times the National Health and Medical Research Council (NH&MRC) limits. The State Government opposed publication of the results on the ground that they might unduly alarm the public!

By contrast, the Victorian Government developed a forest policy in the period 1982–91 which depended heavily on a fine-grained consultative process involving both local community and scientific input. This promised a sensitive management régime capable of taking ecological factors as well as local community needs into account. Regrettably, it did not long persist: expedient changes made by the Kirner and later Kennett governments, primarily in order to increase State revenue from timber-cutting, drew sharp protest from scientists and conservationists. By 1993, the combination of clerical error and pitifully inadequate reserves led to loggers destroying an important habitat for the endangered Leadbeater's possum.[2]

This Victorian example is one in which the views of the epistemic community were first sought and then ignored for political reasons. In New South Wales, outright suppression was attempted; in the Ord case, the Western Australian Government simply

Chapter 16

ignored data gathered by one of its own instrumentalities.

The Terania Creek Inquiry of 1979–81 dramatised yet another kind of barrier to the communication of scientific knowledge; that posed by the adversarial methods of the legal system. This Inquiry, which had a significant impact on the future of the northern New South Wales rainforests, was conducted by a retired judge of the Industrial Court, who not only oversimplified the issues by an insistence on dividing all witnesses into 'pro' and 'anti' logging groups, but also assumed that the State Forestry Commission's witnesses were impartial. His handling of respected senior scientists in the witness box was so insensitive as to spark an indignant public protest from those affected. Taplin concludes that the conduct of the whole Inquiry impeded the rational assessment of the problem; certainly the final report showed no sensitivity to the scientific issues or the environmental values involved.

Even the institutionalisation of scientific access to government does not guarantee the best advice. Informally, and even in formal advisory institutions, there is a risk that 'old-boy' networks will emerge, in which retiring members tend to nominate their own protégés. This will naturally reflect the past pattern of scientific research investment, since new members will typically be senior scientists, retired from active research. It may also self-select for politically active scientists. Scientific establishments are often very conservative; scientists, themselves highly specialised, are often subject to tunnel vision, with its accompanying inability to see the 'big picture'.

Furthermore, the epistemic community itself may not agree on an issue. Wildriver, discussing the Mount Etna caves issue in central Queensland, assigns considerable significance to disagreement. Where scientists can be hired by either side, and scientific information is incomplete or its interpretation contested, the resulting political uncertainty is readily exploited by vested interests on one side or another of a dispute. Uncertainty in the international epistemic community — for example, over the effects of climatic modification — is often seized on by political forces seeking to push particular lines. Frequently, such exploitation goes beyond the scope of legitimate scientific disagreement, casting doubt on matters which all informed parties agree on.

It is not clear whether economists form an 'epistemic community' or are simply a very powerful international pressure group or policy community. To the extent that their theories are based on myth, as many of their critics claim, they clearly cannot claim true epistemic status. But their penetration of Australian national politics has been far more successful than that of the scientists. Pusey's finding that half the senior federal public servants in the most powerful departments were university-educated economists owing allegiance to neoclassical economic orthodoxy helps to explain their dominance in both major parties' policies of deregulation, 'free market', and anti-welfare state ideas: so-called 'economic rationalism'. While their greatest concentration is in Treasury, the spread of their influence throughout the 'key' — most powerful — departments gave this group control over virtually all government policy. Most crucially for this discussion, the bureaucrats — themselves part of a powerful policy network — derive their support from a community of international scope, embracing not only a large sector of the international academic economics community, but also their supporters and sycophants in the Press and media. Pusey's brilliant analysis shows how this intellectual orthodoxy is both irrelevant and damaging to the interests of ordinary Australians, and how it also works against the national interest in overseas trade.

Epistemic communities and their beliefs, where these are communicated to government, are therefore of exceptional importance in understanding the *direction* and *content* of government policy. Even study of the distortions in the transmission process is revealing.

Policy Networks
Policy and epistemic communities influence policy precisely because they play a significant part in determining both how problems are perceived and what policy measures are considered. When members of a community are part of the policy-making process they are said to be part of the **policy network**. The conception of a network differs from that of a policy community in that members of the network all have a direct role in formulating and implementing policy. This does not mean that they are necessarily members of the government, or even of extended state instrumentalities such as statutory corporations or advisory boards. In road haulage, major trucking firms or unions may have a very direct input into policy, especially if government consults them directly, or if 'garbage can' policy-making is in train. Advisers to political parties, in or out of government, may also be network members.

A policy network may embrace one or more policy communities; it may exclude some members of a particular community,

CHAPTER 16

but include others. It may focus on a specific policy problem, or a cluster of related problems; or it may form about a particular policy process 'such as budgeting, auditing or planning'.[3] A policy network may be defined by describing the interactions among its members, as Pusey does with the 'economic rationalists' in Canberra.

Once again, 'ins' and 'outs' matter. The policy network which determines environmental policy in the Australian federal government, for example, consists of ministers, bureaucrats, and representatives of the affected sectors; it is quite fluid, changing its membership depending on which issues are involved. The forest 'industry', for example, may help determine questions concerning native forests, but it is the tourist 'industry' that becomes involved when coastal development issues are concerned.

What is quite noticeable about this policy network, however, is that it does not include the major environmental groups. While they enjoy some status in the policy community, they tend to be treated by government as 'single-issue' groups to be given some sops, but not an integral part of the 'important' business of economic growth. This is the more extraordinary when it is appreciated that, for example the Australian Conservation Foundation has many more members than the Australian Labor Party.

Exclusion was underlined very dramatically when the Federal Government's Working Parties on Ecologically Sustainable Development were deliberating during 1992. The Working Parties, of between 12 and 18 members, consisted of representatives from the Australian Council of Trade Unions (ACTU), industry, relevant *governmental* authorities (state and federal), the Australian and New Zealand Environment and Conservation Council (ANZEC, an intergovernmental body), Commonwealth Government Departments, the CSIRO, and, in one case, two independent scientists. The Australian Conservation Foundation had two or three members on each committee and Greenpeace was represented at the initial deliberations. However, Greenpeace withdrew at an early stage, complaining that *ecological* sustainability was not being seriously considered, the predominance of industry and government resulting in an emphasis on sustainable *economic* development. No attempt was made to bring Greenpeace back into the process, nor were its criticisms addressed. Both environmentalist and independent scientific opinion remained significantly underrepresented. Criticism of the resulting reports has repeatedly drawn attention to this fact.

Communities, Networks, and Closure

Dror's critique of incrementalism was that it tended to be reasonably satisfactory (at least to the major client groups) as long as the situation did not call for major change. Schaffer's criticism of 'mainstream' models of the policy process, which includes all 'decision strategy' approaches, is similar. In fact, he goes further, arguing that, because specific decisions do not simply occur at particular moments in time, it becomes very easy for bureaucrats to evade responsibility by pushing it off onto another, mythical, superior, or by taking refuge in technique.

When the situation is undergoing significant change, the most important policy imperative is to find alternatives: new ways of constructing problems, innovative ways of addressing them, novel solutions and ways to implement them. Schaffer calls this 'room for manoeuvre': an expanded 'policy space' made necessary by the need for innovation. His critique of decision strategies is that they aid and abet the formalisation of decision-making into a bureaucratised routine which may ignore important ways of looking at the problem, block novel options, and frustrate radical initiatives. Governments, fearful of upsetting stable patterns of support by antagonising powerful groups, may actively seek to avoid alternatives. In short, governments and their bureaucracies practise agenda control, hiding it behind the technicalities of 'mainstream' decision analysis, or justifying it as technological necessity.

Policy communities and networks, to the extent that they are 'closed', necessarily exercise agenda control in a similar way. The exclusion of some groups or individuals, as already noted, means that certain constructions of problems will be ignored, some issues will not be recognised as problems at all, and some viable solutions will be ignored.

OVERVIEW: MAINSTREAM AND MARGIN

Pluralism lends legitimacy to pressure and interest groups, treating their concerns as proportionate to their financial commitment. Major economic players, such as the various industries, therefore gain considerable attention, and their views are accepted by government as knowledgeable and authoritative. This attitude fits in well with developmentalism. It permits discreet 'garbage canning', in which the government allows the interplay of interests to define policy dimensions for it: in short, 'pressure pluralism'.

Community groups, on the other hand, are generally treated as marginal. They may get what they want if major conflict does not

Chapter 16

arise, or if the issue arouses sufficient public concern. But frequently, they are either placated or ignored, depending upon the political climate.

The pattern of policy in Australia has reflected the paradoxical, sectorally variable nature of state capability, reflecting both the broad overall power of the state in Australia, and its lack of capacity for detailed intervention in many, if not most, areas.

Australian governments have developed no special capability in environmental management. The implication is that policy in this area will tend to be developed in an *ad hoc*, reactive fashion, with poor intersectoral coordination, and a tendency to 'garbage can' solutions.

The policy communities with most relevance to environmental issues either exclude many voices concerned with ecological rationality, or accord them only peripheral status. The science-based ecological epistemic community is at best fitfully influential. And the policy networks which primarily decide environment-related issues are dominated by bureaucracy and industry.

This situation has tended to disadvantage groups concerned with environmental or ecological issues. They tend to be treated as single-interest groups, to be recognised along with numerous other 'hobby' groups. Predictably, therefore, environmental 'wins' have tended to be on *single issues*. Large areas of forest have been gazetted as natural parks or placed on the World Heritage register, but neither the Commonwealth nor any State has a comprehensive forests policy which addresses the steady loss of 'old-growth' forests and the attendant destruction of habitat. Tasmania, prevented by a High Court decision from constructing the Gordon-below-Franklin dam, did not reconsider its hydro-industrialisation policy; it simply moved on to the next project. The larger question of incorporating an ecologically rational perspective into government decision-making has been neglected, in direct conflict with the need to treat ecological rationality as fundamental. Environmental issues are treated as peripheral and not even seriously evaluated.

Further Exploration

K. Davidson's chapter 'The Failures of Financial Deregulation in Australia' is a succinct exposition of the need for regulation in the financial sector; Falk pursues the same theme in his chapter 'Economic Rationalism and Environmental Regulation'.

Wright's paper 'Policy Community, Policy Network and

Comparative Industrial Strategies' is a useful short overview of the relatively new concepts of policy community and policy network. Haas introduces his concept of epistemic communities in his work on Mediterranean pollution control, usefully extending it to the international regime on air pollution in more recent work. The distinction between strong and weak states, though widely applied, is systematically taken up by Atkinson and Coleman, who marry it with the policy network concept. In Australia, Bell has been critically exploring the application of these ideas, most recently in *State, Economy and Public Policy in Australia*, co-authored with Head. Schaffer's essay 'Towards Responsibility: Public Policy in Concept and Practice' appears as Chapter 9 in *Room for Manoeuvre*, jointly edited with Clay.

The various factual examples supporting the theoretical analysis have been drawn from a number of sources. The Ord river example comes from papers by this author; electricity and forests are treated by Kellow; the New South Wales Government and its attitude to scientific studies of fish are documented by Beder; Terania Creek is the subject of a PhD thesis and several papers by Taplin; and McEachern and Economou offer contrasting views on Wesley Vale which are nonetheless not entirely in conflict. The sorry history of the Victorian EPA is detailed in Russ and Tanner, *The Politics of Pollution*. Ziman's treatment of conservatism in the UK scientific establishment, *Knowing Everything About Nothing*, has relevance for the Australian case.

CHAPTER 17

EVALUATING ENVIRONMENTAL POLICY

There is little point in understanding *how* policy is made if no attempt is made to use that knowledge to evaluate proposed actions. Policy evaluation is logically distinct from policy analysis, but it faces many of the same problems. A major difficulty is that policy evaluation operates in an environment of acute uncertainty, in which not only is available knowledge poor, but the consequences of actions may be counter-intuitive and difficult to foresee. Unfortunately, many of the methods applied to environmental policy evaluation have failed to take account of these difficulties, and consequently have severe practical disadvantages.

COMPREHENSIVE POLICY EVALUATION

In response to perceptions of 'environmental crisis' in the late 1960s and early 1970s, many writers called for fundamental political and social reform. Others responded by attempting to find techniques for dealing with environmental problems within the existing political framework.

This latter approach found quick acceptance, promising as it did minimum disruption to existing social systems, and avoidance of the painful confrontations anticipated by the radicals. By far the most important development in this process was the adoption by the US Government, as part of the *Environmental Protection Act 1972* (EPA), of **'Environmental Impact Assessment'** (EIA). This required the production of an Environmental Impact Statement (EIS) for every proposed activity which came under the ambit of the Act. A meter-reader's dream, it stimulated the rapid proliferation of environmental consultancy firms and other 'expert' activities. It was not long before EIA practices were labelled a 'boondoggle', and criticised for lack of scientific integrity, insensitivity to major advances in relevant fields of study, and, because of their limited circulation, the lack of informed critique of their findings.[1]

In Australia, the Whitlam Labor Government's *Environmental Protection (Impact of Proposals) Act 1974–75* was followed by similar legislation in each State. Its provisions are similar to those of the EPA, especially in requiring those proposing a project to undertake an assessment.

The model for EIA was **Cost-Benefit Analysis** (CBA), a technique from economics which had been brought to a reasonable level of serviceability by the 1960s. CBA involves adding up the economic costs and benefits of a particular proposal, and (ideally) comparing it with those of other similar proposals — including doing nothing — to see firstly whether any net gain results, and secondly, which of a number of options is best. EIA attempts to do much the same thing, with reference to ecological and human environment factors.

Both techniques are essentially variants of the 'rational-comprehensive' decision strategy. They suffer from misplaced comprehensiveness, a fragmentary approach to the problem, and serious risks of bias and misrepresentation. Like the 'rational-comprehensive' decision strategy, they are essentially Benthamite: attempts at a decision calculus based on an attempt to 'sum up' everybody's utility. There are serious flaws with this approach. Curiously, however, their effect is piecemeal and incrementalist, primarily because of the context in which they are usually applied, and because of the failure of comprehensiveness in practice.

The Pursuit of Happiness

Jeremy Bentham was a contemporary of a number of important early nineteenth century figures, such as James Mill — a leading light in the East India company and father of John Stuart, the greatest exponent of Liberalism — and Ricardo, second only to Adam Smith in the development of classical economics. Like them, Bentham had a considerable impact on the politics and

CHAPTER 17

thought of nineteenth-century England. His ideas on administrative and legal reform had an enduring impact in the British Empire, and his doctrine of **Utilitarianism**, while not uniquely his invention, remains the most influential strand of English ethical thought.

Bentham's ethical system was based on the idea that humans sought pleasure and avoided pain — the 'two sovereign masters' — and that an ethical system for society should seek the 'greatest happiness of the greatest number'. Bentham proposed a **felicific calculus** — a formula to calculate happiness — which could sum up all the goods and bads, finally coming to a conclusion about the best possible course of action.

But counting heads in this way raises numerous highly intractable paradoxes; some were touched on in Chapter 12. Furthermore, it requires considerable *commensurability* between individual needs, wants, and tastes: it must be possible in principle to reduce all to the same basis, and ultimately the same units of measurement. Otherwise, attempting to aggregate pleasures and pains would be like lumping apples in with oranges. Bentham rejected all qualitative distinctions, insisting that 'pushpin [a board game of the period] is as good as poetry'. But except for the most basic of human needs, it is now generally accepted that the *amounts* of pleasure and pain derived from various activities are highly variable, and that the marginal gain from an extra unit (be it monetary, or a unit of effort) may vary widely depending on the condition of the individual. An extra dollar may mean the difference between starvation and survival to a Somali, but have no detectable effect on a millionaire. In consequence, it is very difficult to make interpersonal comparisons of utility.

But Utilitarianism has other faults. Maximising the good for the greatest number can result in a 'tyranny of the majority'. If 99 people would be better off if the 100th were killed, Utilitarianism says that person should be killed; it has no doctrine of individual rights. Bentham thought rights were 'nonsense on sticks!' Minorities, under this rule, could expect no mercy, however just their claims.

And Utilitarianism is a *consequentialist* doctrine: it assumes that the consequences of each action can be predicted, and that as a result, the action can be weighed against alternatives, the consequences of which are also reliably known. This assumption does not allow for uncertainty, nor for unintended and unforeseen consequences. Unless uncertainty can at least be limited, no consequentialist ethical position is tenable. The uncertainty, incomplete data, complexity and unpredictability so characteristic of ecological systems and of environmental problems make techniques based on Utilitarian assumptions highly vulnerable.

Though there are good reasons for adopting a modified Utilitarian position on many issues of environmental policy, a simple 'counting heads' approach is far from adequate.

Welfare Economics
Welfare economics is very attractive to those engaged in evaluating environmental policy because it grew out of a concern with market failure, the very phenomenon that cripples all market-reliant approaches to environmental policy problems.

By the late nineteenth century, economists were responding to the problem of 'market failure' by suggesting that perhaps economics could identify a *social welfare function*: that is, a sum of the economic welfare of everyone. It would then be possible to compare potential policies, based on the expected amounts of economic welfare generated. This promised to eliminate the random, reactive approach of the market, by permitting elements of anticipation and coordination to be built in.

Though its inspiration was Benthamite, the reduction of welfare to economic terms is highly impoverished by comparison with a felicific calculus embracing all human values. Ethics is eliminated, as are other non-economic values such as æsthetics or affection. Essentially, welfare economics attempts to solve the problem of commensurability by reducing all values to economic ones. But it suffers from the same technical problems, and remains open to the same objections, as Utilitarianism.

The crippling objection is the problem identified by Arrow: that there is no way of devising a procedure which would reliably translate unrestricted individual preference orderings into a complete and transitive collective ordering, while remaining responsive to individual preferences.

But, long before the implications of Arrow's work were generally appreciated, welfare economists had started to develop techniques for applying their approach. The most important of these was cost-benefit analysis, described above. CBA is a *comprehensive* technique. It seeks to assess all the benefits, and all the costs, of a policy, proposal, or project, and assess definitively whether it is worthwhile. If the benefits exceed the costs, then the project should proceed. There are several difficulties, however. One is that CBA can only measure non-economic costs and benefits by assigning them monetary values — using techniques such as 'shadow pricing' — and the technique used to assign the values is obviously critical. In practice there is considerable disagreement about their validity. Another is the continuing controversy among economists about 'flow-ons', such as 'secondary' benefits, defined as stemming from the primary activity. For example, a shopkeeper who enjoys increased profits as a result of an influx of workers to a specific project is said to be gaining a secondary benefit; the primary benefit being the employment of the workers. Much of the controversy is about what can be counted as secondary benefits; some of it about the proper multiplier to use in determining their probable extent. Very often, highly optimistic multipliers are used, attributing, for example, four new jobs to every one directly created by a particular project; figures of one or even less are far more realistic.

But the critical weakness of CBA is that unless *all* costs are considered, or all benefits assessed, then the analysis will be faulty. There are many such cases: an amusing example was one of the first large-scale CBAs in Australia, the 1957 study by US consultants Ford, Bacon, and Davis, on the proposed upgrading of the Mount Isa railway. One of their predictions was that the refurbished line would make profits of some £1,250,000 per year. In the event, it did, but ore traffic fell short of expectations even before the bitterly divisive strike of 1964–65, which lasted 8 months. The profits stemmed instead from graziers in the Julia Creek area, who transferred stock to and from agistment and the meatworks in Townsville via the upgraded line, which had been unusable due to unreliable, excessively slow transits. The consultants were right, but for the wrong reasons; an element omitted from their analysis had confounded their predictions.

Davis, Wanna, Warhurst and Weller see this as a central and unavoidable problem with rational-comprehensive decision strategies of all types, arguing that a flawed comprehensive analysis may be *worse* than a 'quick and dirty' analysis that just happens to get the important variables right. Indeed, so long as formalised tools remain unreliable, a more intuitive approach may be the best option.

CBA generally finds it easier to quantify the *benefits* than the *costs*. This is particularly true where the benefits are financial, and accrue to private pockets, while costs are non-economic, and are borne by the community at large. Critics claim that CBA consistently underestimates costs, especially where they are non-private, and that some means — such as using a benefit-to-cost ratio of greater than 1 — must be found to compensate. And Pearce has further shown that, because of the difficulties of valuing ecological inputs and environmental impacts, CBA is consistently unable to cope with 'stock' resources and 'stock' pollutants. Application of the technique usually results in greater-than-optimal rates of depletion and of pollution.

THE TRAP OF TECHNIQUE

Basing techniques on Utilitarian or Welfare Economics principles is therefore fraught with pitfalls, many of them potentially fatal. Yet the widespread adoption of CBA and EIA has led to repeated failures in policy evaluation, many of them for the very reasons predicted above. The problems include:

- failure to recognise and account for the pitfalls of social choice;
- failures of comprehensiveness;
- a piecemeal approach, leading to an incrementalist approach by default;
- failure to appreciate effects of inadequate data.

The Pitfalls of Social Choice

These are sometimes called 'Arrow problems'. Simply stated, as the number of alternatives increases and/or as the number of participants in a social choice process increases, the more likely it is that any attempt to aggregate their preferences — for example by voting or through 'revealed preference' in a market — will result in an intransitivity. The problem is trivial as long as single issues or clusters of issues are involved, but it becomes very serious as soon as coordination is needed, especially across issues. Environmental problems — because of interpenetration — almost always require coordination.

Therefore it is not possible to 'conduct a survey' and simply apply the results to policy development, and any attempt at EIA or CBA which does not carefully take into account (at the very least) the side-effects of the problem being addressed, is not only radically incomplete, but risks clashing with other policies reached by similar processes, also in good faith.

Failures of Comprehensiveness

This is why a comprehensive decision-making strategy must always be completely comprehensive, or it will fail. In fact one of the commonest criticisms, both of CBAs and of EIAs, is simply that they have overlooked important data. Yet the standard criticisms of comprehensive decision strategies stress that the data are frequently too complex for analysis; that necessary data are frequently unavailable; that suitable theoretical interpretations for handling the data are also absent.

To adopt a comprehensive strategy, therefore, is to court failure in the event either that important data are not incorporated, or that the analysis of the data is inadequate. There is no fall-back strategy in these techniques: they are 'all or nothing'. Thus their comprehensiveness is *misplaced*: given that, in the real world of highly fallible humans, errors are made as often as not, comprehensiveness is not a very appropriate strategy.

This leads in practice to incrementalism. Because the comprehensive approach is impracticable, its practitioners become atheoretical incrementalists, tied to unsuitable analytical techniques the weaknesses of which they will not recognise. This in turn leads inevitably to a piecemeal, reactive approach to policy evaluation, responding to proponents of environment-modifying projects in an essentially *ad hoc* fashion.

Risks of a Piecemeal Approach

But the piecemeal approach is just as crippling, particularly with environmental impact assessment, though the use of CBA for isolated projects will result in similar problems. Precisely because ecological problems are often both extensive and far-reaching, attention to a single problem, without consideration of its ramifications, ignores Commoner's two basic rules that 'Everything is Connected to Everything Else', and 'Everything Must Go Somewhere'. For example, at one time there were more pending applications to create tourist resorts in bays on the east coast of Queensland than there were bays to accommodate them. Obviously, in such a case, there must be an overall policy, both to control the allocation of opportunities to developers, and to consider the overall impact: in particular, whether some bays should remain unspoiled. An EIA for each in isolation would not be able to take such questions into account. Similarly, a decision (say) to pipe sewage out to sea would be mere displacement, *unless* the proponents knew for certain that no adverse effects would follow. Sydney, faced recently with the problem of sewage washing back onto famous inner-suburban beaches, simply built a longer outfall sewer. As with taller smokestacks in Britain, this probably ensures that the bad effects will occur further away, but it does not guarantee that they will be eliminated. A better solution might have been to explore the possibility of keeping the nutrients on land and turning them into much-needed fertiliser.

Many such decisions, each in isolation, lead to a phenomenon known in planning and environmental circles as the 'tyranny of small decisions': no *overall* decision is ever reached on an issue, but the accumulated weight of a myriad trivial decisions preempts the issue. Readers familiar with South-East Queensland will know how the failure of town planning has led to steady erosion of amenities; in particular, the lack of 'green belts' north and south of Brisbane is leading to steady urbanisation of a strip of the coastal zone more than 160 kilometres long. No serious evaluation of the social, environmental and aesthetic costs has ever been made. Instead, residents suffer a steady loss of amenity punctuated by occasional 'scares', such as those over toxic waste disposal and loss of koala habitat.

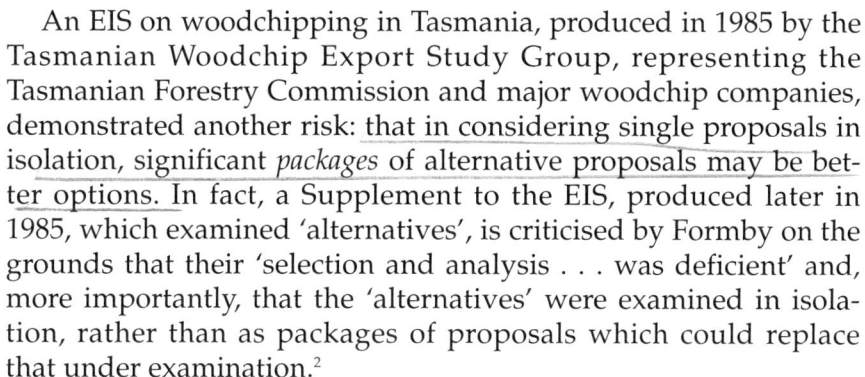

An EIS on woodchipping in Tasmania, produced in 1985 by the Tasmanian Woodchip Export Study Group, representing the Tasmanian Forestry Commission and major woodchip companies, demonstrated another risk: that in considering single proposals in isolation, significant *packages* of alternative proposals may be better options. In fact, a Supplement to the EIS, produced later in 1985, which examined 'alternatives', is criticised by Formby on the grounds that their 'selection and analysis . . . was deficient' and, more importantly, that the 'alternatives' were examined in isolation, rather than as packages of proposals which could replace that under examination.[2]

Ignorance is not Bliss

Inadequate data lead to unknown risks. Especially if the ecological effects are not understood, or the connections between them and human activity are unforeseen, the risks can be serious. One of the most publicised recently is the damage to the ozone layer caused by chlorinated fluorocarbons (CFCs). CFCs were developed as an alternative to refrigerant fluids which were inflammable and consequently dangerous. They were tailored to be as inert as possible, while retaining the property of absorbing and giving up latent heat at temperatures suitable for refrigeration engineering. Their inert properties later recommended them for other uses, such as spray cans and the foaming of plastics for use as insulation and packaging. Not until some 10 or more years *after* their introduction — to general approval — was it realised that CFCs, once released into the atmosphere, rose into the stratosphere and there reacted with ozone under the catalytic influence of solar radiation.

Chapter 17

This in turn damaged the ozone layer, exposing the earth's surface to greater than normal levels of ultraviolet radiation. The blind sheep of Tierra del Fuego are victims of well-intended technological innovation.

This underlines the fundamental importance of knowledge in environmental policy, and it also throws into striking relief the inadequacy of present-day knowledge. Given the incomplete knowledge of ecology, climatology, and many other relevant sciences, further surprises, some unpleasant, can confidently be expected. Policy evaluation techniques must allow for imperfect information, by adopting strategies that will allow for failures of comprehensiveness. Issues such as the *pace* of innovation must also be considered.

The Politics of Policy Evaluation

If cost-benefit analysis and environmental impact analysis have so many deficiencies, why have they been so widely adopted? The answer is that they create work for 'experts', and, even more important, they are very convenient. 'Off-loading' decisions to technical experts is politically highly rational. Any opprobrium then attaches to the expert, not the politician (who is merely 'acting on advice'), and very often the decision can be presented as unavoidable. But, as Kellow and Winner have shown, apparently technical criteria are often stalking-horses for political motives.

It is also politically rational to require the proponent to carry out the EIA. The advantage for government is that the cost of analysis is shifted to the proponent, saving the expense of setting up its own monitoring apparatus and developing the necessary expertise. But making proponents undertake assessments themselves opens EIA up to great stresses and distortions. Furthermore, if government is unable to evaluate the resulting EISs, the whole process is wasted. The consequent need for an evaluative capability may mean that savings on developing monitoring are minimal.

The Sponsor as Judge

Permitting proponents to evaluate their own projects introduces risks of bias and of lack of diligence. The use of EIA or CBA techniques, even if the strictly technical pitfalls are avoided or their impact minimised, exposes sponsors of particular projects to the temptation to distort and to omit, in their own interests.

Distortion will occur because of a natural tendency for the

interpretation of data to be biased in favour of the proposal. Omission may occur either through actual suppression of information inimical to the proposal, or through failure to seek out information likely to be so. Deliberate suppression is probably quite uncommon, since it requires conscious public deceit; some unscrupulous operators may practise it, but developers of integrity are not likely to, nor are consultants or other reputable specialists. Far more probable is failure to seek out information. For example, many impact assessments contain statements such as 'no endangered species will be affected'. This can mean what it says *only* if very extensive research has been carried out and the species composition and ecological relationships of *all affected habitats* are known in detail. This is quite unlikely; the statement will more probably mean that nobody knows of any such species likely to be affected — not quite the same thing! Couple it with inadequate estimates of habitat size for known species, underestimation of the impact of domestic animals, human disturbance, and so on, and there is a recipe for bias. Omission may also occur if necessary research is neglected.

The normal practice for EIAs is that the sponsor hires a firm of consultants, who then prepare a study which is submitted to government in fulfilment of the legal requirement. Consultants make their living by pleasing clients; otherwise there will be no repeat business, and they will soon be facing the dole. Consequently, it is not in the consultant's interest to uncover information which will result in the project's rejection. It is a striking fact that very few EIAs or CBAs find that the proposal they are examining is not viable. The consultant therefore treads a very fine line between professional ethics and 'giving the clients what they want'. 'Inadvertent' omission of data is bound to occur, and failure to identify, let alone address, major research needs is normal. This is exactly what occurred when proponents of the Wesley Vale kraft pulp mill produced a favourable EIS in 1987. Opponents criticised it for neglecting risks of pollution associated with processing; and the Tasmanian Department of Deep Sea Fisheries was extremely critical, alleging that the EIS failed to take into account tidal effects on effluent flows in Bass Strait.

Thus the practice of making the proponent responsible for assessment of proposals is fraught with risks of bias and distortion. While the objective of those who adopted it may have been laudable — for example, it can be claimed that the assessment exercise is a valuable learning experience for the proponent — what EIA cannot do is offer a dispassionate assessment of a proposal.

Chapter 17

The Analytical Capability

EIAs and CBAs, once completed, must be evaluated. In practice, this means that the relevant government department, typically of Environment, must scrutinise each report. This means that to discharge its function effectively, it must have evaluative capability. This requires, at the minimum,

independent sources of knowledge; and

an independent capability for analysis.

To do this, it must therefore employ professional staff with similar expertise to those who write the EIAs and CBAs in the first place; and it must have ongoing and effective monitoring and research programmes, so that it can critically evaluate claims made in sponsor studies. In practice, very few government environment departments, in Australia and elsewhere, have the capacity to carry out such functions thoroughly and conscientiously. Monitoring is neglected, research is inadequate, and staff are frequently overloaded. Not only does this mean that projects cannot be thoroughly evaluated; it also poses the risk of atomism. Overstretched and under-resourced, the department risks losing sight of the big picture altogether, becoming mired in detail instead. The central issue is government capability: policy cannot be made and implemented effectively unless some knowledge, skill, and detachment exists.

Economou, however, points out that in cases such as Wesley Vale, where there is 'a strong sense of Government–developer collusion', government may choose not to be critical of an EIS and even to neglect obvious flaws. He suggests that the model of an impartial government conscientiously evaluating EISs is 'rather naïve', given the strong mutual interests of government and business in development.[3] In fact, so close was collusion between the Tasmanian Government and the Wesley Vale joint venturers that a press release announcing that Parliament would be recalled to pass legislation ensuring that the project could proceed was issued on the consortium's notepaper, not the Government's! McEachern points out that the Tasmanian Government eventually caved in to all the joint venturers' demands. It was opposition to the proposal from powerful industry groups such as farmers and fishermen that appears to have stalled the project, not government insistence on environmental standards; though McEachern suggests that marginal profitability may also have been a factor in the final shelving of the proposal.

This in turn bears directly on the need for regulation.

The Role of Regulation
Many environmental problems, as we have seen, involve a regulatory issue. Toxic waste, soil degradation, food and drugs, all require oversight. Governments are often the only bodies capable of regulating, because only they have enough power. And to regulate effectively, they must have a capability to monitor, analyse, and enforce. This involves scientists, laboratories, legislation, and enforcement personnel. There are implications in terms of political will, budgets, and individual skills.

Failure in these areas may occur even when progress is being made in other ways. The Goss government in Queensland has nearly doubled the area of national parks (from a mingy 2% to an exiguous 4% of the State's area!) since its election in 1989; but water quality monitoring, pursued at best halfheartedly by its predecessors, has been virtually abandoned. Examination of policy trends shows that popular, 'cuddly' issues have been pursued at the expense of the tougher, more mundane regulatory ones. That the latter are less popular with industry can scarcely be expected to enter into the calculations of a 'development'-oriented government.

But in the long term government cannot avoid regulatory responsibility: ecological stability requires it, and political rationality may also dictate it, especially if public concern over environmental issues runs high. In cases where the EIA process has been controversial, such as Wesley Vale, it is frequently the tension between government desires for 'development', and its unwanted rôle as protector of the environment, that is the source of the conflict.

INCREMENTALISM

The failure of comprehensiveness typically leads policy makers to fall back on incrementalism, even though it courts serious coordination problems due to its arbitrary and piecemeal approach. Further, Goodin has argued that the advocates of incrementalism as a technique advance three inadequate arguments for it. The attendant strategies, he claims, are consequently defective.

The Atheoretical Approach
Atheoretical incrementalism does not attempt to understand the processes with which it deals. Instead, it seeks to reinforce good trends by repeating the behaviour that led to them, and to eliminate bad by not repeating. It treats the actual process linking

behaviour and outcomes as a 'black box', the processes within which cannot be known. This, say proponents, means that theory — even in the simplest sense of an understanding of the processes in human activity — can be dispensed with. Instead policy changes are made in small increments, with the idea that the outcomes will also be small, though detectable. The resulting strategy, at its simplest, is to:

1. Arbitrarily select an incremental intervention.
2. If the results are positive, repeat it on a slightly larger scale.
3. If the results are negative, switch to some other arbitrarily selected intervention that thus far has not produced any significant negative results overall.[4]

But deprived of any model of the process, the atheoretical incrementalist does not know when or where to look for the results. The only solution to the 'when' problem is to impose an arbitrarily selected time delay, which may allow too much or too little time; the identity problem is even more serious, since any change in the area of interest must be assumed to be a result of the policy increment. Goodin argues that in fact the atheoretical practitioner does not even know what is a properly incremental step: if there is no model of the effects of the action, however intuitive, the decision-maker cannot gauge what is adequate and what is excessive.

Given the vulnerability of incrementalism to threshold and 'sleeper' effects, this is especially telling. If incrementalism cannot detect thresholds, it will fail to allow for their effects. It may therefore mistakenly identify a change as caused by a policy, as well as evaluating the policy positively or negatively. It may also fail to appreciate a build-up of adverse impacts — as in the case of 'stock' pollutants — and consequently make a series of 'wrong' and progressively 'wronger' decisions before the ill-effects show up. This is just what happened with CFCs.

Decision-makers recognise this risk in practice: Goodin cites the case of the NASA moon programme, which needed some years' investment and considerable experimental work before any results could be shown. Similarly, governments have often invested large sums of money in major dam projects which would not be completed for many years. The Burdekin in Queensland was so funded, from State sources, in dribs and drabs over a period of some 40 years.

In short, even the most pragmatic policy maker needs a model of the process which is being addressed. It may be very crude, but it is quite indispensable. 'The first — and, indeed, the only — responsible course for a decision-maker presented with a black box is to pry open a corner of it.'[5]

Society as Laboratory

A second variant of incrementalism recognises the need for a model, but mistrusts theory, and particularly its predictive abilities. It argues for incrementalism as experiment, using policy as a means of verifying the predictions of theory. In this model, small policy steps are justified as a means of controlling variables — making changes to only one, while as far as possible holding others constant. This is, of course, difficult to do, and the chances of exactly replicating such a policy 'experiment' are very small, casting some doubt on the utility of the method. This is essentially 'piecemeal social engineering', on Popper's model. It suggests the following strategy:

1. Start with a theoretically informed hypothesis about the system.
2. On that basis, nonarbitrarily select an incremental intervention chosen for its practical and epistemic utility.
3. Observe the results of the intervention.
4. Revise the hypothesis, or change it if necessary.
5. Repeat the experimental procedure.[6]

This approach is vulnerable both to variation in the observed data, so that verification is uncertain, and to changes in the underlying relationships in the policy field. Thus a policy may be judged 'effective' in producing a change that can be accounted for by normal statistical variation. Or, in a rapidly changing system, the hypothesis, by the time it has been verified, is obsolete and cannot serve as a guide to future policy.

Incrementalism as Adaptation

A third argument for incrementalism is adaptive, claiming that incrementalism permits policy makers to learn by trial and error. Small policy steps are advocated so as to avoid making irreversible mistakes. The resulting strategy is summed up by Goodin thus:

1. Start with a theoretically informed hypothesis about the system.

2. On that basis, nonarbitrarily select the incremental intervention that is expected to maximise utility subject to the constraint that whichever course of action is pursued must be reversible.

3. Observe the results of the interventions to obtain data regarding the comparative advantages of alternative courses of action.

4. Revise the hypothesis, or change it if necessary.

5. Repeat the procedure, backtracking and pursuing an alternative course of action if the revised theory so indicates.[7]

This strategy assumes that the smaller the intervention the more reversible it is; and that it is always desirable to avoid irreversibility. Both assumptions are open to question. Some very small interventions, especially when close to an important threshold, can cause very large consequences. And their irreversibility sometimes makes policies desirable. A very small increment of nutrients might trigger eutrophication or an algal bloom; and, as Goodin points out, the irreversible eradication of a serious disease such as smallpox might be eminently desirable.

The Failings of Incrementalism
Goodin's critique of incrementalism shows that there is always a need for some theoretical understanding of the policy field — i.e., the subject matter of the policy process — and that there is no optimal size of increment in a policy intervention. Like Schaffer, he warns against failure to think seriously about policy at all.

He also challenges the libertarian assumption that individuals always know what is best for them. In real life, a multitude of pressures make for hasty, ill-informed, ill-considered or simply emotional decisions. Frequently these respond only to immediate pressures and not long-term well-being. And individuals are prone to lapses, either under peer pressure or simply because of transitory weakness, that will lead them to do things they may later regret. For these reasons, it may be necessary to legislate to 'save people from their former selves'; in addition, of course, government is needed to establish the conditions for cooperation and coordination in the large number of cases where individual effort cannot supply the good.

Goodin suggests that the task of collective policy-making is greatly eased if it is recognised that a modified Utilitarian position does make possible some reasonable judgements, which can be in the interests of all. For example, while many pleasures are incommensurable, there are pains which must universally be avoided; shelling defenceless civilian populations, for example, is a war crime however good the cause in which it is committed, and there should be little difficulty in deciding what to do about it. Similarly, it is often possible to anticipate the outcomes of particular kinds of actions, since 'common sense' — the knowledge nearly everyone has of the way things work — will afford sufficient information. In that case, neither atheoretical nor 'trial-and-error' approaches are necessary. Goodin therefore suggests that environmental decisions ought to seek:

1. to bias decisions against *irreversible* choices (which may sometimes be permissible, but only after much more careful scrutiny than they receive in the ordinary expected-utility calculus);

2. to bias decisions in favour of offering special protection to those who are especially *vulnerable* to our actions and choices;

3. to bias decisions in favour of *sustainable* rather than one-off benefits; and

4. to bias decisions against *causing harm*, as distinct from merely foregoing benefits.[8]

Not only does this direct attention to the overarching issues, but it also helps to avoid — for example — the 'tyranny of small decisions' by which important values suffer a 'death of a thousand cuts' from incremental attrition.

METHODOLOGY

The weaknesses of policy evaluation reflect both the history of the social sciences and the specific socio-intellectual environment in which the techniques first emerged. A striking feature of most sociological theories, as well as of classical and neoclassical economics, is their general irrelevance to ecology and environment.

Weaknesses of Theory
This weakness cannot be remedied by adopting the currently fashionable argument for a 'biocentric' perspective on environmental questions, with its injunctions to 'think like a mountain'. Neither mountains, plants, nor animals think in ways accessible to

Chapter 17

humans. Only humans (so far as is known) think analytically, so environmental predicaments have to be analysed by humans. While it may choose to be *humane* and consequently highly considerate of other species, human understanding is unavoidably anthropocentric. Attempts to extend 'rights' to non-human species constantly run into the objection that such species cannot assert or understand a right in the sense that humans do. Since most environmental problems are perceived diseconomies of human activity, brought to notice by some unpleasant feedback from the affected natural environment, this is not inappropriate. Humans must create the solutions to the problems they themselves have caused. The resulting perspective is bound to be human in its mode of understanding; it need not necessarily serve only human interests.

The anthropocentrism of most 'mainstream' social science, particularly sociology, is far more thorough-going; it is either pre-ecological or explicitly denies ecology's importance. Indeed, the choice of major analytical dimensions in historically important social science theories often seems remarkably capricious or arbitrary. Examples include Hegel's notion that the state incarnates rationality; Marx's views on the class struggle as the 'motor of history'; Auguste Comte's 'natural' stages of development, Pareto's social 'optimalities'; and so on. Most eighteenth and nineteenth century thinkers shared the untested assumption that 'development' or 'progress' was both automatic and beneficent; a powerful superstition with vast influence on modern public policy. It has obscured consideration both of population growth and of the related question of living standards in a finite environment. And social theory long ignored the size of institutions and systems, assuming that scale had no effect on complexity or structure.

As long as social science primarily addresses questions derived from broad general hypotheses about the nature of social systems, their structure, or their mechanisms of social causation, its methods will remain too general for application to specific problems and hence unsuitable for the evaluation of policy. Yet the policy dimension of environment and human resource studies is often critical. Governments and politicians demand immediate advice on intrinsically urgent issues.

A Pragmatic Approach

The political economy approach adopted throughout this book avoids these pitfalls of irrelevance and atheoretical technique. By recognising that human productive systems are always embedded

Evaluating Environmental Policy

in physical and biological systems, on which they depend, it grants standing to relevant environmental factors in understanding human socio-economic systems.

By underlining the stresses and pressures on nation-states in competitive multi-state systems, it establishes the 'macro' constraints on policy, and relates them to the complexities of individual issues. It avoids the sterility and formalisation of decision strategies and evaluation techniques by recognising the complexity of political systems, the great flexibility of their responses to policy dilemmas, and the implications of the distribution of power within them.

This enables it to support a policy-oriented approach to environmental problems, which must necessarily be concerned with pragmatic, nitty-gritty issues, but cannot lose sight of the need for theory. While promoting short, medium and long term research which addresses the substantive issues, and on which longer term policy settings can draw, it must also urgently generate tentative but useful answers to policy questions. At the same time, it is compatible with the principles put forward by Goodin, which must be satisfied if policy is to be ethically sound as well as ecologically rational.

The question which remains, and to which there is no universally accepted answer, is whether, given the known deficiencies of all evaluation techniques, there is any way in which policy evaluation can be fruitfully approached. The answer is yes, by attempting to combine the perspectives of the long view with a 'quick and dirty' approach to answering the urgent questions. But to do so requires some guidelines.

Specification for a Methodology

The lack of a universal methodology which can cope with all contingencies may never be remedied, given the complexity and variety of issues that it would have to tackle. The alternative of attempting to 'walk on two legs' by combining short and long term considerations in a single approach can itself only be done if the ultimate aim of the operation is kept firmly in mind, and rules specified which evaluation processes must satisfy. These would at least permit the weeding out of unsatisfactory or incomplete approaches. As a minimum, a suitable methodology would be able to:

1. Employ a range of methods, from sophisticated to rule-of-thumb.

Chapter 17

2. Cope with poor information.
3. Identify gaps in existing knowledge and suggest suitable research.
4. Comprehensively link immediate problems to their wider context.
5. Identify and channel feedback from the natural and human environments.
6. Repeat the process when needed.

The last item, the capacity for iteration, is particularly important, given that scientific uncertainty and the failure of comprehensiveness mean that 'once and for all' decisions are nearly impossible to achieve.

In identifying suitable methodologies, four basic principles should be applied. These can be simply stated:

1. *Pragmatism*. Methods for research and analysis should be judged and adopted on the basis of their utility and effectiveness for the tasks in hand. Unsatisfactory approaches should be freely discarded or modified. For example, conventional sample survey/interview methods are of relatively low utility where the sampled population does not understand their purpose, is illiterate, etcetera. Bias may enter results due to agreement tendency, suspicion of 'officials', and so on. Alternative methods are more satisfactory in such cases.

2. *Eclecticism*. It is important to draw methods and insights from all relevant disciplines, while adopting the particular perspective of none. This helps to maintain a more comprehensive overview of the problem, yet permits the effective use of various techniques — the more effectively, the better their limitations are understood — without the narrow perspective of the parent discipline.

3. *Versatility*. Especially in data analysis, the criterion for the use of a particular technique is suitability; both qualitative and quantitative methods, or any appropriate mix, should be used.

4. *Simplicity*. Without doing violence to the complex web of natural and human relationships, the analyst should not lose sight of the need for simplicity. An elegant, cost-effective analysis does not overextend itself; data are not collected for

their own sake, nor is analysis pursued to impossible lengths. Very often a simple analysis using well-chosen parameters and selected data will be more revealing than a complex one with ill-chosen parameters and a mountain of data.[9]

These principles place a good deal of stress on the choice of appropriate and relevant variables. The question of relevance has caused a good deal of worry, for there is no universal, simple and certain method by which to identify suitable variables. Fortunately, virtually all research into environmental issues is problem-oriented. Because of this, to an extent which might not be possible with fundamental research, the problem itself can be allowed to guide the selection of parameters. While it might be argued that this merely shifts the relevance problem back one stage, it can be countered that in general it is easier to define a real problem than to select analytical parameters in a vacuum. And while a problem may be defined totally wrongly — as 'economic rationalism' so strikingly demonstrates — this, too, is easier to detect.

OVERVIEW: THE ULTIMATE TEST

The most commonly accepted project evaluation techniques, CBA and EIA, suffer from deep flaws in their basic principles and in their implementation. These stem from a range of problems, including the Arrow impossibility result, the failures of rational-comprehensive decision models, the built-in biases consequent on policy evaluation by proponents, and a piecemeal, fragmented, *ad hoc* approach to decision-making. Because of the failure of comprehensiveness, the technicians of CBA and EIA are forced back into incrementalism, which also fails to solve their problems of completeness, and coordination.

A political economy approach which recognises the fundamental dependence of human productive systems on their natural environments is a basic requirement for a workable methodology for environmental policy evaluation. While no methodologies have been devised which can operate within this framework, it is possible to specify some of the characteristics of a suitable approach.

Policy evaluation is a sternly practical activity. The consequences of poor evaluative techniques are often not experienced by those who apply them; but someone bears the cost ultimately. Since one of the broad aims of environmental policy is to reduce the social cost of political decision-making, it is desirable to avoid

Chapter 17

bad techniques and methodologies for policy evaluation. And that is the final litmus test. Policy evaluation must always address the problem in hand, if it is to have any utility.

Further Exploration

The weaknesses of EIA and CBA are beginning to spawn an extensive critical literature. In Australia, Formby and Economou have both explored the weaknesses of EIA; there is an extensive technical literature of surpassing dullness.

The critique of rational-comprehensive techniques is also extensive. The most complete is still to be found in Braybrooke and Lindblom; Mackenzie offers a very useful potted overview in his *Politics and Social Science*; despite the book's age, its conclusions remain surprisingly apt. There is also a short critique in Davis, Wanna, Warhurst and Weller.

The literature on cost-benefit analysis is by now extensive, with such prominent figures as Mishan and Pearce making valuable, if incompatible, contributions. Pearce's *Environmental Economics* has a comprehensive overview of economic approaches to policy evaluation, including CBA, in a theoretical context similar to that canvassed here. His more recent collaborations, *Blueprint for a Green Economy* and *Blueprint 2* attempt, not without controversy, to apply economic principles to the valuation of economic costs and benefits in the British and world economies respectively.

The story of the Mount Isa railway reconstruction is in the author's MA thesis, *Development Politics*; a shorter version appears in a paper by Knight. A short account of the strike is in Fitzgerald's *A History of Queensland, from 1915 to the 1980s*, at pp. 229–236.

The criteria for judging methodologies were first put forward by this author in 1986, in a paper originally written for the benefit of Indonesian practitioners with little theoretical background.

Chapter 18

The Challenge of Environmental Policy

This book has attempted to show how both day-to-day policy and the ongoing pressures of statecraft frequently conflict with ecological rationality.

In moving to a broader canvas, this final chapter will try primarily to clarify the analysis the book offers. In the process, some judgements about the outcomes of policy processes are unavoidable. However, far too much of the environment literature has succumbed to the temptation to offer prescriptions, some apocalyptic, others highly reformist. This chapter's themes draw on the analysis already offered, avoiding *ex cathedra* pronouncements on issues which have not been touched on. In short, while any opinions expressed will not evade controversy, they will be soundly based.

But the point of this book is not to offer prescriptions for specific problems, nor to make a final judgement about the capacity of any particular political system to cope with environmental problems. Instead, it offers an analytical approach to thinking about environmental policy.

Is Environmental Policy Different?

Some have argued that environmental issues are similar in most basic respects to other policy issues, and that they can therefore be

Chapter 18

fitted into the framework of 'business as usual'. But this is not really so. In the first place, the underlying questions of ecology are more fundamental than most issues of day-to-day politics. In the second place, the organisation of state societies, their modes of production (including technology) and their social belief systems have been geared predominantly to environmental exploitation, not to its rational long-term management. The results can now be seen almost everywhere.

This is why environmental policy problems have such radical implications. Nation-states, organised around 'defining functions' which emphasise their competitive relationship with other nations, have tended either to take environment on board as a 'special interest', single-issue problem, or to ignore it as irrelevant to more 'fundamental' issues such as national defence or economic growth. They now find themselves faced with a need for institututional re-learning on a scale which is literally unprecedented. Insulated by technology and by the attitudes engendered by the seeming abundance of the Industrial Revolution, they have failed to confront the issues of overpopulation, resource depletion, loss of biodiversity, and pollution. By ignoring the most fundamental questions of all — those which make up the ecological problem — they create a tragic paradox. The very conditions which are essential for them to pursue their defining functions no longer exist. The biosphere can no longer be exploited without thought for the consequences; failure to address the problem (and it may already be too late, given the problem of overshoot) will result in modern civilisation suffering the same decline that has affected many other elaborate biosocial systems.

Environmental policy, therefore, is different, in the sense that it is more fundamental, and frequently raises issues about which there is great uncertainty and acrimonious dispute.

Free Speech, Free Thought, and 'Good Science'

In modern nations, the dominance of policy-making by restricted networks of notables seriously impairs their capacity to deal with many of the central issues thrown up by environmental degradation and looming ecological crises. Politicians, by and large, gear their activity to exploiting a system of periodic elections. Bureaucracies — in private and multinational firms as well as government — are preoccupied with empire-building and 'territory'. There is little provision for serious and rational discussion of the greatest problems of the day. Nor do the media of communication promote open discussion. Mass media, owned and operated as commercial enterprises, trivialise and sensationalise issues. The exact extent to which this impedes rational intellectual discourse is the subject of learned debate, but it is quite evident that at the very minimum large amounts of irrelevant 'noise' clutter the system, and that at worst large amounts of relevant data and interpretation are simply not aired.

Only the scientific community — via its conventions of publication and communication — has channels which are reasonably open. But much scientific communication is technical, concerned more with the minutiæ of theory building and interpretation of data than with discussion of their implications for policy. Few scientists are engaged with public affairs.

The importance of these constraints is very great. The standard argument for freedom of speech from utility, due to John Stuart Mill, is that unless there is a sort of 'free market' in ideas, in which all are critically evaluated and tested, then the best will not emerge, and society will be the loser. This idea, sometimes called the practice of 'critique', is fundamental to free academic discourse in institutions such as universities. Science is based on an almost identical reasoning, which is why it insists that all conclusions be testable, both in principle and in practice, and that all experiments be capable of replication.

The principle can be extended to argue that ways and means for coping with environmental problems in general, as well as with local, regional, and national problems at every level, must be freely submitted to critical scrutiny and thereby tested. This ability to conduct 'mental experiments', it should be remembered, is an essential part of human adaptive capability, and a consequence of the capacity to abstract, conferred by language.

In turn, this suggests the desirability of human political arrangements which are significantly more conducive to free, full, and pointful discussion than any that exist today. The debate over this issue is outside the scope of this book, but it is the driving force for numerous discussions of 'green democracy' and the like.

GLOBAL PROBLEMS

The world became more, not less, unstable after 1990. The collapse of the Soviet Union in 1991 ushered in, not George Bush's 'new world order' of unprecedented prosperity, but increased turmoil, including territorial conflict in the former USSR; the intertribal strife and consequent famine of Ethiopia and Somalia; continuing military conflict and environmental damage in Burma, Thailand,

Chapter 18

and Cambodia; and the ghastly war crimes in Bosnia. Europe in particular faces the spectre of the resurgence of a fascism which many had thought finally vanquished in 1945. The problem for many countries is to maintain defining functions in the face of sharpening conflict over land and resources, while the need to retrieve the ecological underpinnings grows ever more urgent. It seems highly likely that the twenty-first century will be even more bloody than the twentieth, already the bloodiest in human history.

If cooperative behaviour among nations can reduce the destructive competitiveness of interstate rivalry, an understanding of game theory and an appreciation of the virtues of cooperation and coordination become correspondingly important. But, as Haas demonstrates, cooperation requires hard work, suitable intellectual and political preconditions, and considerable time. Furthermore, the goals are very important. Barbara Ward, writing in the 1960s, drew attention to the possibilities of increased global trade, and the links of mutual self-interest it created, for achieving greater international cooperation and improving environmental management.

Unfortunately, the expansion of trade does not necessarily result in improved environmental management. The General Agreement on Trade and Tariffs (GATT) Uruguay Round of 1991–93, billed as the greatest single move towards free trade since 1945, has come under increasing criticism for *reducing* environmental standards, and making it very difficult for national governments to legislate for improvement. This is achieved by binding all participating nations to a 'lowest common denominator' set of environmental standards for traded goods, and then providing that any national legislation beyond the treaty provisions can be legally challenged as 'in restraint of trade'. For example, US laws banning the import of rainforest timber from Thailand, Burma, or Malaysia could be overturned by invoking the GATT provisions. The effect is to make free trade more important than ecological rationality: a clear victory for the 'wrong', unsustainable, rationality.[1] The Uruguay Round concluded on December 1993, its central proposals significantly watered down, but still without sufficient recognition of its potential for environmental damage.

Trade, and the attendant transport systems, pose other dangers, too, most of them systematically ignored by its enthusiasts. In particular, they threaten the integrity of ecologies and facilitate the spread of pests and diseases. One of the most striking changes in modern trade is the way in which increasing quantities of *fresh* foods are traded. While European nations in particular have long

The Challenge of Environmental Policy

been reliant on imports of primary products from the colonised world, until very recent times they were mostly durables. Even these were not without danger: the introduction of rats and other pests into many island ecologies — with devastating effects — was an unintended side effect of visits by European shipping. Insect pests have come to Australia via infected timber in softwood packing cases, as well as in grains and seeds. Increased trade in fresh foods makes quarantine more difficult, both because of the increased volume and because of spoilage; and, for example, hopes that Australia could exploit its reputation as a 'clean' supplier may founder if exotic pests proliferate rapidly. Similarly, concern has frequently been expressed at the risk of swift spread of disease created by air transport and the easing of travel restrictions. The rapid spread of the AIDS virus is a contemporary example, aggravated by such charming colonial behaviour as 'sex tours' for European males to countries such as Thailand and the Philippines.

Free trade supports continuing over-consumption by rich countries of the resources of the poor, a hangover of the colonial era. Accelerated ecological damage results. One of the best known examples of neocolonial exploitation of overseas resources is the clearing of Amazon rainforest by cattle ranchers in Brazil, primarily to supply the US market for hamburger meat. But there are many other cases. Furthermore, the economic pressure leads to habitat destruction, species loss, and other forms of environmental degradation. The Amazon is well publicised, but Burma and Thailand are at least as seriously threatened, and in Kalimantan (Borneo) the responsiveness of timber-cutting to foreign prices is well documented. Globally, the impact of foreign markets on deforestation has been dramatic.

The consequent need to insulate local ecologies against undesirable impacts pits ecological against economic rationality. If 'free' trade, without significant barriers, will impose serious ecological costs, then deterring it will create economic costs, either through barriers such as quarantine and restrictions on the movement of particular goods, or through the need for rigorous testing and monitoring. These all add costs which economic irrationalists would consider unwarranted.

To achieve *ecological* sustainability would therefore require that all international agreements incorporate measures designed to promote, rather than defeat, ecological rationality. It would also require rearrangement of the present terms of trade, which discriminate both against the less industrialised nations, and those which have more environmentally responsible policies. Nations with poor or no environmental regulation can produce goods more cheaply than those which impose stringent pollution controls, which confers a trade advantage. If means are not found to 'level up' the terms under which they compete, nations adopting responsible pollution control measures are likely to be penalised, the more so the more 'free' the international trade regime.

A further serious problem is international competition for resources. Few appreciate how quickly the major Western societies, especially the USA, have moved from being self-sufficient in resources to being highly dependent on imports. This dependency is the primary reason for the 'neocolonialism' of modern times, in which nominally independent ex-colonial nations acquiesce in the plunder of their natural resources. These, Catton's 'ghost' resources, have been the occasion of numerous conflicts, most strikingly in the Gulf War of 1990. A former CIA protégé, the Iraqi dictator Saddam Hussein, chose to invade Kuwait, a tiny oil principality at the head of the Arabian Gulf. Kuwait was a creation of Western imperialism, hived off from Iraq in the 1920s to ensure British control of its oil. It is ruled by a single family, without pretence of democracy or human rights. Before the Gulf War, nearly 50% of its population was immigrant, mostly Palestinian, without any rights whatever. US response to Iraqi aggression was swift: within weeks a campaign had been launched, with the assistance of the Western European nations, to 'liberate' Kuwait and push back the Iraqis. Significantly, no attempt was made to oust Hussein, who was left in place to continue his policies of genocide against the Shi'ite and Kurdish minorities. And the biggest losers were Kuwait's Palestinians, who were accused of siding with the Iraqis and persecuted in the aftermath of the war. The ecological effects of the Gulf War were serious; in particular, major oil spills resulted from Iraqi attempts to destroy Kuwait's oil wells. The most interesting fact, however, was that *both* parties to the war were sometime US protégés. The US and Europe underlined their priorities unmistakably when, only months later, they failed entirely to intervene in Bosnia to prevent the worst genocide Europe has seen since Hitler.

These problems raise the fundamental issue of environmental choice: whether to wait for war, disease, and famine to exercise 'positive' checks on human populations, or to limit population, and hence misery, by voluntary action? The very question challenges the *laissez faire* assumptions of classical and neoclassical economics: if the future has to be the subject of hard choices, then

The Challenge of Environmental Policy

drift and tinkering are not viable options. Given the paradoxes of social choice, a good society *has* to be something more than one in which people are 'free to choose' among random fads.

And good environmental management at the international level will clearly require much *more* cooperation than currently exists.

General Domestic Problems for Nations

The tension between economic and ecological rationality has a number of implications for nations. One has already been mentioned: the need to insulate regional ecologies against the impacts of over-free trade. Another is the problem of stability. Economic 'rationality' demands highly open markets, and an acceptance of their fluctuations as natural phenomena with which no-one should tamper. But ordinary human beings need a certain amount of stability in their lives. They cannot plan effectively, let alone discharge responsibilities to their families and associates, if their economic environment is constantly fluctuating. The state of total turmoil so beloved of economic 'rationalist' policy makers is actually inimical to effective productive work by individuals, since they have to spend so much time reorganising themselves that they cannot produce.

This is relevant to environmental management in at least two ways. The first is that effective management of land, habitat, vegetation, and numerous other aspects of ecosystems requires stability and continuity; as Paul Ehrlich has observed, animals and plants haven't heard about economics and are disinclined to obey its 'laws'. The second is that if policy makers, and even ordinary citizens, are continually coping with the short-term buffeting of extreme fluctuations in their immediate environment, they will have little time to consider longer-term issues, and come to considered assessments of them. The larger policy issues will suffer, as policy is made on the run. Indeed, some have suggested that the continual turmoil of 'micro-economic reform' and rampant marketeering is a radical political stratagem designed precisely to emasculate future-oriented thinking.

A second, central, unresolved tension is between the 'defining' activities of the state, and its unavoidable rôle of environmental stewardship. The paradox here lies in the fact that the short-term rationality of political and military survival places at ever greater risk the ecological underpinning of the social system. The most dramatic example of this recently has been the civil war in Somalia, where the military rivalry of the warlords has prevented the peasant population from carrying out their day-to-day tasks; the result has been widespread crop failure and devastating famine. Similar problems have arisen in Angola and Mozambique, fuelled by external aid to competing military movements. The tension is less stark for 'civilised' countries of the West, but as the extent of radioactive contamination from military nuclear activity, and of massive resource depletion for purposes of 'economic transformation' are revealed, the costs become increasingly apparent.

A further unresolved difficulty revolves about the issue of minorities and their status. As noted in Chapter 15, modern states have handled minorities badly, placing strains both on social welfare and on ecologies. The effect of excluding minorities and their concerns has frequently been to locate policy decisions in remote bureaux, which is maladaptive in the medium to long term, and sometimes even in the short. Conflict uses up energy and resources that would otherwise be available to develop cooperative strategies for environmental management. In effect, it dissipates useful energy in wasteful conflict.

This is a suicidal strategy in times of rapid change, especially when ecological stresses make rational environmental management urgent. Minorities are important both because of the need for a 'free market in ideas' in science and public affairs, and because their cooperation is essential for effective environmental management. The nation-state's failure to resolve the problem of minorities is therefore inimical to its adaptive flexibility, precisely because it diverts attention from important and fundamental issues to 'artifical' questions of politics.

Australia

Australia's problems resemble those of other nations, but with some highly specific twists, some a consequence of geography, others of ecology.

Geopolitically, Australia could be vulnerable to attempts to annex land or resources, especially by Asian nations suffering overpopulation and resource shortages. Economically, it is as heavily colonialised as any Third World nation, due to supine government policy and foreign control of the mass media and major extractive industries. In a world in which natural resources were scarce and high-priced, Australia's trading position would greatly improve, provided it were not invaded first. But at the present time, the continuing long-term decline in resource prices, plus dependency on exports of primary products, has resulted in a corresponding decline in Australia's trading position and living standards. As Australia's population rises, the exportable surplus will decrease, until all production is for domestic use; this will make a complete nonsense of the primary export economy.

Government policies based on 'more of the same' have attempted to make up the shortfall in export earnings by exporting greater volumes, whether of minerals or produce. But this policy is economically irrational because it requires the country to run ever harder simply to stay where it is; it offers no hope for future well-being, let alone a chance to regain Australia's position of 1900, when it had the world's highest *per capita* standard of living. It is also ecologically irrational. Extensive damage to rangelands due to chronic overstocking continues unabated, with the probable long-term consequence of extensive desertification. Nutrient loss, even from the richest farming areas, remains an unrecognised 'sleeper' problem. And habitat destruction, coupled with the baneful effects of introduced species, continues to fuel the extinction of native flora and fauna.

Australia confronts the ecological paradox especially sharply because of its unique and very precious ecology, with its rich store of ecosystems and rare species. This places bigger demands, both of adaptation and of forbearance, on the politico-economic system. In particular, once the aridity of the Australian climate is appreciated, the limited biological surplus available for human exploitation comes sharply into focus: Australian ecosystems simply cannot sustain high rates of extraction. The challenge is therefore greater than in ecologically 'rich' areas such as Europe, and it cannot be met by 'cultural cringe', borrowed ideas and philosophies, or economic *laissez faire*.

The continuing dominance of statist developmentalism has seen the embracing of economic 'rationalism' in an attempt to restore the country's fortunes. But this, because it ignores ecological considerations, places excessive emphasis on trade, and treats all environmental goods as infinite, costless and substitutable, has actually increased environmental stress. This has aggravated economic problems directly due to ecological factors, such as land degradation and nutrient loss, which have led to the failure of farms and businesses.

There would seem to be as great a threat to rational environmental management from political 'rationality' as there is from economic. This is not to minimise the deficiencies of economics when faced with environmental problems, especially those involving collective goods, threshold and 'sleeper' effects, or discounting over time. But, as noted in the foregoing chapters, economically rational management practices are frequently closer to ecological rationality than existing, politically motivated, arrangements. Particularly noticeable is the political power of some special interest groups, and their ability to translate it into concessions and subsidies to themselves.

Australia has not so far been very successful in meeting the challenge in terms of government policies or structural adjustment. The continued subsidisation of energy wastage, land clearance, overstocking, and numerous other well-documented ecologically irrational practices inhibits creative change, making it expensive and often futile. Despite government inflexibility, there is an amazing amount of activity in research, development, and small-scale innovation, whether in renewable energy, low-impact technologies or novel land use systems. But its potential is not being realised.

FACING THE CHALLENGE: RESPONSIBILITY IN PUBLIC POLICY

B.B. Schaffer has argued that public policy in nation-states frequently becomes ossified and overbureaucratised. Its agenda, he argues, is increasingly limited by the perspectives and preoccupations of narrow, highly institutionalised policy communities. This leads to irresponsible policy-making, which ignores viable alternative policies. Stagnation results, and, most importantly, failure to solve urgent problems. In Australia's case, Pusey's critique of the policy dominance in Canberra of economic 'rationalists' suggests that this is precisely what has happened.

As noted in Chapter 12, Schaffer accuses policy makers of hiding behind the formalisms of 'cut-and-dried' analytical techniques in order to evade responsibility for their own failures. He argues that responsible policy makers have a duty to seek out viable alternative policies and evaluate them fairly. He claims that alternatives almost always do exist, and frequently it is simply inertia which prevents their being adopted. He therefore urges a highly self-critical stance on policy makers. His critique has particular relevance for environmental policy-making, granted the very substantial attitudinal and structural changes that ecological rationality demands.

PROSPECT: ALL IN THE MIND?

The challenge for Australian environmental policy, as elsewhere, is undoubtedly to establish sustainable, stable systems of environmental exploitation, capable of resisting undesirable external pressures and maintaining good living standards for all. To do so will require the development of novel forms of husbandry, as well as a dedication to more economical use of energy and minerals.

The central problem at the time of writing remains the gulf between the centres of policy and the scientific ecological epistemic community. Important changes are under way in the latter; but government and its policy networks at present take little account of these.

The changes in the ecological epistemic community are nothing short of exciting. Considerable attention is being given to innovative farming techniques, designed both to make better and more sustainable use of existing Australian resources, and to develop new ones. One trend is to exploit novel or unusual crops, longer or off-beat growing times, and diversifying markets in nearby countries, such as in South-East Asia. Associated with this is a move away from 'fair average quality' (FAQ) exports selling at low prices, to novelty exports selling at premium prices: limited production for a higher return per unit, rather than extensive production for low unit returns.

Research into novel systems of cropping has followed widespread changes of attitude among farmers and others connected with rural industry. Innovative crops, many offering attractive returns, are now being explored. The possibility of exploiting native species is being taken more seriously than ever before. The 'farming' of kangaroos and emus for meat is the best-known example, but investigations into indigenous plant species for crops and pasture are probably more significant. These developments do not all arise from a willingness to work within the ecological constraints of the continent; some are simply responses to the near-collapse of traditional rural activities under pressure of drought and poor markets. However, farming techniques which are environmentally benign can have desirable effects whatever the motivation.

The significance of these changes lies primarily in the changes in the conceptual framework which they represent. The idea of sustainability by working with the ecosystem is catching on among farmers and scientists. The long-term implications for sensitive land management are very exciting.

Similarly, appreciation of the need to change the pattern of energy use is growing steadily. The most obvious, of course, is the response to the greenhouse effect, which has caused growing concern about the use of fossil fuels. Australia has always used proportionately large amounts of fuel *per capita*, primarily because of the importance of transport to the economy; close to 40% of all energy use. Motor vehicles alone cause 50% of all Australian air pollution. While liquid fuels may prove difficult to replace, shifting freight to rail rather than road, easily achieved by small changes in the tax régime, could effect very significant savings. More progress is being made in electricity, where economy measures such as insulation, solar hot water and heating have already made an impact and can make far more. Most important is the realisation that energy generation, whether from fossil fuels, nuclear energy, or any other source, will release more heat into the biosphere, thus worsening greenhouse warming. Slowly, the emphasis is switching to energy *capture*, exploiting the energy which is available from the sun, either directly or indirectly. These energy forms also lend themselves to substantial decentralisation, which offers the possibility of significant change in the electricity industry, including major savings in investment.

Similar changes are taking place in industry. Once again, the move is from minimally processed bulk commodities to 'elaborately transformed' products embodying high levels of knowledge. Many recent successes, especially in the export market, have been in this area. In particular, Australia has a lead in medical technology and some other related science-based industries. Ecologically, such developments are important, because the environmental impact of modern, 'knowledge-heavy' industry is often markedly lower than that of traditional heavy industries based in metallurgy and petrochemicals. On environmental grounds, the encouragement of such relatively benign industry is clearly desirable.

A Last Look

Environmental policy presents a fundamental challenge to the modern nation-state. Geared to survival in a world of competing states, nations are organised firstly about the maintenance of the 'defining' functions, and secondly about resource mobilisation, which in modern times is essential to the first. Ecological balance, always a potential problem for statecraft, presents a major difficulty because it is prior to either political survival or resource mobilisation. It therefore clashes with military and economic 'imperatives'.

No modern state has successfully adapted to this challenge. Dependence on 'ghost' acreage and resources has increased the potential for resource conflict as well as exacerbating ecological imbalance in both exploited and exploiting nations. Freer world

The Challenge of Environmental Policy

trade by itself can do nothing to relieve this problem, and may worsen it.

Australia, settled as part of the second great wave of European colonial expansion, was woven very quickly into the imperial system. Its pattern of environmental exploitation, derived from that of Britain, was even more unsuited than in other places, due primarily to the unique environment. The pattern of 'statist developmentalism' which evolved during the nineteenth century led to deep involvement by the state in economic development and exploitation of natural resources.

Australian public policy has tended to follow a 'pluralist' model, but one in which the state has often been able to display considerable strength vis-à-vis domestic forces such as business, while in general not posing a significant challenge to foreign power, especially in its corporate form. The cosy relationships between government and the more powerful interest groups have meant that much policy-making has been consensual, seen through the lens of 'development' and the mutual self-interest of government and business. Groups not included in this concensus — including environmentalists — have tended to find access to power difficult, if not impossible.

Both in Australia and elsewhere, the exercise of state power has been in the hands of a relatively small network of individuals with privileged access. This has tended to reinforce conservative attitudes and accepted constructions of the policy problems faced by the nation. It has also excluded minorities and their views. The resulting inflexibility is one of the striking characteristics of governments, especially in the environmental policy field.

The result, especially in recent times, has been that the international ecological epistemic community and its various national offshoots have tended to be far more aware of the extent and seriousness of environmental problems than the politicans. Consequently, technology has outstripped policy change, both because technological questions are easier and less complex, and because the structure of the policy system is so resistant to change.

Further Exploration

John Stuart Mill, the great nineteenth century philosopher of Liberalism, wrote very extensively, and editions of his collected works are numerous. Most include his essay 'On Liberty', perhaps the most widely read of all his works. The issue of 'truth' in science is hotly contested — see, for example, Chalmers' *What Is This*

Chapter 18

Thing Called Science? — but the importance of communication is generally agreed; Ziman's *An Introduction to Science Studies*, especially pages 9–11 and Chapters 4 and 16, is a useful scholarly exploration of science as 'public knowledge'.

Barbara Ward, in Chapter 22 of *Progress for a Small Planet*, canvasses the integrative properties of global trade; recent articles in *New Scientist* and *The Ecologist* have stressed the potential for ecological damage implicit in the provisions of GATT's Uruguay Round. Part Four of Boyden, Dovers and Shirlow, *Our Biosphere Under Threat*, has some useful figures on Australian energy usage and its ecological significance. Cocks' *Use With Care* includes data on the energy use implications of Australian transport patterns. Kartawinata and Vayda showed, in their studies of forest conversion in Kalimantan (Borneo) that timber-getters were more sensitive to prices on the Tokyo market than to factors traditionally emphasised by anthropologists.

NOTES

Introduction
1 O'Neill, G., 'Land for Food is the Problem Ahead', *The Age* (Melbourne), 10 June 1992, p. 6, quoting Dr R. Swaminathan, former Secretary of Agriculture to the Government of India.
2 Suzuki, D., *Inventing the Future*, p.149; the World Wide Fund for Nature (WWF) was quoting a figure of 50 per day in mid-1992. Numbers like this are always very rubbery, not least because many species have never been described by scientists, and as a result, it is impossible to know reliably how many are being lost. What is quite certain is that the quoted numbers are underestimates: the actual situation is probably far worse.
3 Gould, S.J., *The Flamingo's Smile*, Chs. 28–30.
4 Gould, S.J., *Hen's Teeth and Horse's Toes*, Chs. 25 & 27.
5 Walker. K.J., 'Methodologies for Social Aspects of Environmental Research,' Social Science Information, Vol. 26, No. 4, December 1987, pp. 759–82.

Chapter 1: Toxic Waste Pollution
1 Envirotest, Draft Report: 'Environmental Audit of Kingston Properties', Mt. Taylor, p. 2.
2 Tamvakis, M., 'The Kingston Toxic Waste Problem: Government Control of Pollution and Land Development', p. 49.
3 Mokihiber, R., & Shen, L., 'Love Canal', Ch. 8 in Nader, R., Brownstein, R., & Richard, J., *Who's Poisoning America?* p. 279.
4 Ashworth, W., *The Late, Great Lakes*, pp. 168–70.

Notes

Chapter 2 Human impact
1. Weiskel, T.C., 'The Ecological Lessons of the Past: An Anthropology of Environmental Decline', *The Ecologist*, Vol. 19, No. 3, 1989.
2. Commoner, B., *The Closing Circle*, p. 39.
3. *Ibid.*, p. 163.
4. *Ibid.*, p. 259

Chapter 4 Salinity in the Murray–Darling
1. Murray–Darling Basin Ministerial Council, Murray–Darling Basin Environmental Resources Study, p.5.
2. Burton, J.R., 'Community Involvement in the Management of the Murray–Darling Basin,' Canberra Bulletin of Public Administration, No. 62, October 1990, p. 76.
3. Williams, B.G., 'Salinity and Waterlogging in the Murray–Darling Basin', in ASTEC, Environmental Research in Australia, pp. 87–120; esp. pp. 92–3.
4. Boyden, S., Dovers, S., & Shirlow, M., Our Biosphere Under Threat: Ecological Realities and Australia's Opportunities, p. 118.
5. Pearce, F., 'British Aid: a Hindrance as Much as a Help', New Scientist, Vol. 134, No. 1822, 23 May 1992, pp. 12-13. USA: The White House–Department of the Interior Panel on Waterlogging and Salinity in West Pakistan, *Report on Water Development in the Indus Plain*.

Chapter 5 Irrigation: a history of conflict
1. Clark, S.D., 'The River Murray Question: Part I', p. 27.
2. Powell, J.M., *Environmental Management in Australia*, pp. 132 — 3.
3. Davidson, B.R., 'Irrigation Economics', Ch. 13 in Frith, H.J., & Sawer, G., Eds., *The Murray Waters: Man, Nature and a River System*, p. 202.
4. *Ibid.*, p. 195.
5. Pope, D., 'Australia's Development Strategy in the Early Twentieth Century: Semantics and Politics', *Australian Journal of Politics and History*, Vol. 31, No. 2, 1985, pp. 218–29.

Chapter 6 Gridlock and Landslide
1. Clark, *op. cit.*, p. 36.
2. Kellow, A.J., *Saline Solutions*, p. 40.
3. Murray–Darling Basin Management Strategy Working Group, Framework for the Development of the Resource Management Strategy: Progress Report.
4. Kellow, *op.cit.*, pp. 64–5.
5. *Ibid.*, p. 3.

Chapter 7 Taking Stock
1. *Salt Action*, p. 13.

Chapter 9 Strategic Games
1. Haas, P.M., 'Do régimes matter? Epistemic communities and Mediterranean pollution control', *International Organization*, Vol. 43, No. 3, Summer 1989, p. 378.

Notes

2 Kellow, *op.cit.*, pp. 64—6.
3 *Ibid.*, p. 13.
4 Hobbes, T., *Leviathan,* Part I, Ch. 13, pp. 64-5.

Chapter 10 Social Choice
1 McLean, *Public Choice*, p. 27.
2 Coaldrake, P., *Working the System*, pp. 41,53.

Chapter 11 Making Decisions
1 Lindblom, C.E., 'The Science of "Muddling Through '," p. 157.
2 Condensed from Braybrooke, D., & Lindblom, C.E., A Strategy of Decision, pp. 48-54.
3 Lindblom, *op.cit.*, p. 157.
4 Dror, Y., 'Muddling Through — "Science" or Inertia?' pp. 167-8.

Chapter 13 The Modern State
1 Goudie, A., *The Human Impact on the Natural Environment,* pp. 8–9.
2 Giddens, A., *The Nation-State and Violence*, p. 5.
3 Rose, R., 'On the Priorities of Government: a Developmental Analysis of Public Policies', *European Journal of Political Research*, Vol. 4, 1976, p. 258.
4 Ibid., p. 267.

Chapter 14 Development and Environment in Australia
1 Gunn, J., *Along Parallel Lines: A History of the Railways of New South Wales,* Chs. 1 & 2.
2 Lines, W.J., *Taming the Great South Land: a History of the Conquest of Nature in Australia,* pp. 82–4.
3 Serventy, V., *Saving Australia,* p. 10.

Chapter 15 The Political Economy of Development in Australia
1 Butlin, N.G., Barnard, A., & Pincus, J.J., *Government and Capitalism,* p. 13.
2 Oliphant, J., & Morley, P., 'Goss backs zinc giant', *Courier-Mail* (Brisbane), 7 March 1991, p. 1.
3 Bell, S.R., 'Business, Government and the Challenge of Structural Economic Adjustment', Ch. 16 in Bell, S.R., & Wanna, J., *Business–Government Relations in Australia,* p. 172.

Chapter 16 Making Policy
1 Pusey, M., 'What's Wrong with Economic Rationalism?' in Horne, D., (ed.), *The Trouble with Economic Rationalism,* p. 64.
2 O'Neill, G., 'Rare possum's habitat destroyed by mistake', *The Age* (Melbourne), 12 May 1993, p. 5.
3 Wright, M., 'Policy Community, Policy Network and Comparative Industrial Strategies', p. 606.

Chapter 17 Evaluating Environmental Policy
1 Schindler, D.W., "The Impact Statement Boondoggle', *Science,* Vol. 192, No. 4239, 7 May 1976, p. 509.

Notes

2 Formby, J., 'Where has EIA Gone Wrong?: Lessons from the Tasmanian Woodchips Controversy', in Hay, P., Eckersley, R., & Holloway, G., *Environmental Politics in Australia and New Zealand*, pp. 3–17.
3 Economou, N., 'Problems in Environmental Policy Creation: Tasmania's Wesley Vale Pulp Mill Dispute', p. 49.
4 Goodin, R.E., *Political Theory and Public Policy*, p. 21.
6 *Ibid.*, p. 29.
5 *Ibid.*, p. 28.
7 *Ibid.*, p. 35.
8 Goodin, R.E., 'Ethical Principles for Environmental Protection', in Elliott, R., & Gare, A., *Environmental Philosophy*, pp. 3–20.
9 Walker, K.J., 'Methodologies for Social Aspects of Environmental Research,' p. 768.

Chapter 18 The Challenge of Environmental Policy
1 Joyce, C., 'US environmentalists oppose world trade deal', *New Scientist*, 15 February, 1992, p.6.

BIBLIOGRAPHY

Aitken, D., & Jinks, B., *Australian Political Institutions*. 3rd edition, Melbourne: Pitman, 1985; 4th edition, Melbourne: Longman Cheshire, 1989.

Arrow, K.J., *Social Choice and Individual Values*. New York: Wiley, 1951; 2nd edition 1963.

Ashworth, W., *The Late, Great Lakes: An Environmental History*. New York: Knopf, 1986.

Atkinson, M.M., & Coleman, W.D., *The State, Business, and Industrial Change in Canada*. Toronto: University of Toronto Press, 1989.

Atkinson, M.M., & Coleman, W.D., 'Strong States and Weak States: Sectoral Policy Networks in Advanced Capitalist Economies', *British Journal of Political Science*, Vol. 19, No. 1, January 1989, pp. 47-67.

Australia. Parliament. House of Representatives. Standing Committee on Environment and Conservation, *First Report: Environmental Protection, Adequacy of Legislative and Administrative Arrangements*. October 1979.

Axelrod, R., *The Evolution of Cooperation*. New York: Basic Books, 1984.

Bachrach, P., & Baratz, M, *Power and Poverty: Theory and Practice*. New York: Oxford University Press, 1970.

Barr, N., & Cary, J., *Greening a Brown Land: the Australian Search for Sustainable Land Use*. Melbourne: Macmillan, 1992.

Barry, B., *Political Argument*. London: Routledge & Kegan Paul, 1965.

Bartlett, R.V., 'Ecological Rationality — Reason and Environmental Policy', *Environmental Ethics*, Vol. 8, No. 3, September 1986, pp. 221–39.

Beder, S., 'Science and the Control of Information: An Australian Case Study', *The Ecologist*, Vol. 20, No. 4, July/August 1990, pp 136–40.

Bell, S.R., 'Business, Government and the Challenge of Structural Economic Adjustment', Ch. 16 in Bell, S.R., & Wanna, J., *Business—Government Relations in Australia* Sydney: Harcourt Brace Jovanovich, 1992, pp. 171–7.

Bell, S.R., & Head, B., *State Economy and Public Policy in Australia* Melbourne:

Bibliography

Oxford University Press 1994

Bilsky, L.J. (ed.), *Historical Ecology.* Port Washington, NY: Kennikat Press, 1980.

Birch, C., *Confronting the Future–Australia and the World: the Next Hundred Years.* Melbourne: Penguin, 1975, Chs. 1 & 2.

Birrell, R., Hill, D., & Stanley, J., *Quarry Australia? Social and Environmental Perspectives on Managing the Nation's Resources* Melbourne: Oxford University Press 1992

Blainey, G., *The Rush That Never Ended.* Melbourne: Melbourne University Press, 1963 (2nd edition 1969, 3rd edition 1978).

Blainey, G., *Triumph of the Nomads.* Melbourne: Sun Books, 1976.

Blainey, G., *The Tyranny of Distance: How Distance Shaped Australia's History.* Melbourne: Sun Books, 1966 (Revised edition 1983).

Bolton, G., *Spoils and Spoilers: Australians Make their Environment, 1788–1980.* Sydney: George Allen & Unwin, 1981; 2nd edition 1992.

Borgstrom, G., *Too Many: an Ecological Overview of Earth's Limitations.* New York: Collier, 1971.

Borgstrom, G., *The Hungry Planet: the Modern World at the Edge of Famine.* 2nd edition, New York: Macmillan, 1972.

Bottomore, T.B., *Elites and Society.* Harmondsworth: Penguin, 1966.

Boyden, S., 'Australia and the Environmental Crisis', in Dempsey, R., *The Politics of Finding Out: Environmental Problems in Australia,* (Melbourne, Cheshire, 1974) pp. 3–16.

Boyden, S., *Western Civilisation in Biological Perspective: Patterns in Biohistory.* Oxford: Clarendon Press (OUP), 1987.

Boyden, S., Dovers, S., & Shirlow, M., *Our Biosphere Under Threat: Ecological Realities and Australia's Opportunities.* Melbourne: Oxford University Press, 1990.

Bowman, M., *Australian Approaches to Environmental Management: the Response of State Planning.* Hobart: Environmental Law Reform Group, University of Tasmania, 1979.

Braybrooke, D., & Lindblom, C.E., *A Strategy of Decision.* New York: The Free Press of Glencoe, 1963.

Breckwoldt, R., *Wildlife in the Home Paddock.* Sydney: Angus & Robertson, 1983.

Breckwoldt, R., *The Last Stand: Managing Australia's Remnant Forests and Woodlands.* Canberra: AGPS, 1986.

Brown, L., *World Without Borders.* New York: Random House, 1972.

Brown, L., *The Twenty-Ninth Day.* New York: Norton, 1978.

Brownstein, R., 'The Toxic Tragedy', Ch. 1 in Nader, R., Brownstein, R., & Richard, J. (eds.), *Who's Poisoning America: Corporate Polluters and their Victims in the Chemical Age* San Francisco: Sierra Club Books, 1981, pp. 1–59.

Burton, J.R., 'Community Involvement in the Management of the Murray–Darling Basin,' *Canberra Bulletin of Public Administration,* No. 62, October 1990, pp. 76–80.

Butlin, N.G., Barnard, A., & Pincus, J.J., *Government and Capitalism: Public and Private Choice in Twentieth Century Australia.* Sydney: George Allen & Unwin, 1982.

Cannon, M., *The Land Boomers.* Melbourne: Melbourne University Press, 1966.

Carden, M., 'Land Degradation on the Darling Downs', Ch. 4 in Walker, K.J., (ed.), *Australian Environmental Policy,* Sydney: New South Wales University Press, 1992 pp. 58–83.

Carneiro, R., 'A Theory of the Origin of the State', *Science,* Vol. 169, 1970, pp. 733–8.

Catton, W.R. Jr., 'Why the Future Isn't What It Used to Be (And How it Could be Made Worse Than it Has to Be)' *Social Science Quarterly,* Vol. 57, No. 2, 1976, pp. 276–91.

Bibliography

Catton, W.R. Jr., *Overshoot: the Ecological Basis of Revolutionary Change.* Urbana: University of Illinois Press, 1980.

Chalmers, A.F., *What Is This Thing Called Science?* St. Lucia: University of Queensland Press, 1976.

Clark, S.D., 'The River Murray Question: Part I–Colonial Days', *Melbourne University Law Review,* Vol. 8, June 1971, pp. 11–40; 'Part II–Federation, Agreement and Future Alternatives', August 1971, pp. 215–53.

Clay, E.J., & Schaffer, B.B., (eds.) *Room for Manoeuvre.* London: Heinemann, 1984.

Coaldrake, P., *Working the System: Government in Queensland.* St. Lucia: University of Queensland Press, 1989.

Cocks, D., *Use With Care: Managing Australia's Natural Resources in the Twenty First Century.* Sydney: New South Wales University Press, 1992.

Cohen, R., & Service, E.R., (eds.), *Origins of the State: The Anthropology of Political Evolution.* Philadelphia: Institute for the Study of Human Issues, 1978.

Commoner, B., *The Closing Circle: Nature, Man, and Technology.* New York: Knopf, 1971.

Connell, D., & Miller, G.J., *Chemistry and Ecotoxicology of Pollution.* New York: Wiley, 1984.

Connell, R.W., *Ruling Class, Ruling Culture: Studies of Conflict, Power and Legitimacy in Australian Life.* Cambridge: Cambridge University Press, 1977.

Crabb, P., *The Murray–Darling Basin Agreement: An Examination in the Light of International Experience.* CRES Working Paper 1988/6. Canberra: Centre for Resource and Environmental Studies, Australian National University, 1988.

Crabb, P., 'Resolving Conflicts in the Murray–Darling Basin', Ch. 9 in Handmer, J.W., Dorcey, A.H.J., & Smith, D.I. (eds), *Negotiating Water,* Canberra: Centre for Resource and Environmental Studies, Australian National University, 1991, pp. 147–59.

Crouch, C., 'The State, Capital and Liberal Democracy', Ch. 1 in Crouch, C., (ed.), *State and Economy in Contemporary Capitalism,* London: CroomHelm, 1979, pp. 13–54.

Crough, G., Wheelwright, T., and Wilshire, T., *Australia and World Capitalism.* Melbourne: Penguin, 1980.

Crowley, K., 'Accommodating Industry in Tasmania: Eco-Political Factors Behind the Electrona Silicon Smelter Dispute', in Hay, P., Eckersley, R., & Holloway, G., (eds) *Environmental Politics in Australia and New Zealand,* Hobart: Board of Environmental Studies, University of Tasmania, 1989, pp. 45–58.

Crowley, K., 'Tasmania Greening: Ecopolitics from Pedder to Wesley Vale', paper presented to the Ecopolitics VII Conference, Griffith University, 2–4 July, 1993.

Davidson, B.R., *The Northern Myth.* Melbourne: Melbourne University Press, 1965.

Davidson, B.R., *Australia Wet or Dry?* Melbourne: Melbourne University Press, 1969.

Davidson, B.R., 'Irrigation Economics', Ch. 13 in Frith, H.J., & Sawer, G., (eds.), *The Murray Waters: Man, Nature and a River System,* Sydney: Angus and Robertson, 1974, pp. 193–211.

Davidson, K., 'The Failures of Financial Deregulation in Australia', Ch. 21 in Bell, S.R., & Wanna, J., (eds.), *Business–Government Relations in Australia,* Sydney: Harcourt Brace Jovanovich, 1992, pp. 221–30.

Davis, B.W., 'Water Resources', Ch. 8 in Forward, R., (ed.), *Public Policy in Australia,* Melbourne: Cheshire, 1974, pp. 249–76.

Davis, B.W., 'Federalism and Environmental Politics: An Australian Overview', *The Environmentalist,* Vol. 5, No. 4, Winter 1985, pp. 269–78.

Davis, G., Wanna, J., Warhurst, J., & Weller, P., *Public Policy in Australia.* Sydney:

Bibliography

Allen & Unwin, 1988. 2nd edition 1993.

Deegan, J., Jr., 'Looking back at Love Canal', *Environmental Science and Technology*, Vol. 21, No. 4, April 1987, pp. 328–31, & No. 5, May 1987, pp. 421–6.

Dickie, P., *The Road to Fitzgerald*. St. Lucia: University of Queensland Press, 1988.

Dickson, D., 'Love Canal Continues to Fester as Scientists Bicker Over the Evidence', *Ambio*, Vol. 9, No. 5, 1980.

Divale, W.T., & Harris, M., 'Population, Warfare, and the Male Supremacist Complex', *American Anthropologist*, Vol. 78, No. 3, September 1976, pp. 521–38.

Dregne, H.E., *Desertification of Arid Lands*, Chur: Harwood Academic, 1983.

Dror, Y., 'Muddling Through–"Science" or Inertia?' in Etzioni, A., (ed.), *Readings on Modern Organisations*, Englewood Cliffs, NJ: Prentice-Hall, 1969 pp. 166–171.

Dryzek, J., 'Ecological Rationality', *International Journal of Environmental Studies*, Vol. 21, 1983, pp. 5–10.

Dryzek, J., 'Present Choices, Future Consequences–A Case for Thinking Strategically', *World Futures*, Vol. 19, 1983, Nos. 1–2, pp. 1–19.

Dryzek, J., *Rational Ecology*. Oxford: Basil Blackwell, 1987.

Dudley, N., 'Acid Rain and Pollution Control Policy in the UK', *The Ecologist*, Vol. 16, No. 1, January/February 1986, pp. 18–23.

Economou, N., 'Problems in Environmental Policy Creation: Tasmania's Wesley Vale Pulp Mill Dispute', Ch. 3 in Walker, K.J., (ed.), *Australian Environmental Policy*, Sydney: New South Wales University Press, 1992, pp. 41–57.

Ehrlich, P., *The Machinery of Nature*. New York: Simon & Schuster, 1986.

Ehrlich, P., Ehrlich, A., & Holdren, J., *Ecoscience: Population, Resources, Environment*. San Francisco: W.H. Freeman, 1977.

Emy, H.V., & Hughes, O.E., *Australian Politics: Realities in Conflict*. 2nd edition. Melbourne: Macmillan, 1991.

Envirotest, Draft Report: Environmental Audit of Kingston Properties, Mt Taylor. Unpublished document prepared for Bureau of Emergency Services. Brisbane: Envirotest Health Environment and Workplace Pty Ltd. (Qld), 1990.

Etzioni, A., *The Active Society*. New York: The Free Press, 1968.

Falk, J., 'Economic Rationalism and Environmental Regulation', Ch. 17 in Vintila, P., Phillimore, J., & Newman, P., (eds.), *Markets Morals and Manifestos: Fightback! and the Politics of Economic Rationalism in the 1990s*, Murdoch, W.A.: Institute for Science and Technology Policy, Murdoch University, 1992, pp. 195–204.

Fenner, F., 'Population and Economic Growth', in Dempsey, R., (ed.), *The Politics of Finding Out*, (Melbourne: Cheshire, 1974) pp. 254–68.

Fitzgerald, R., *A History of Queensland, from 1915 to the 1980s*. St. Lucia: University of Queensland Press, 1984.

Formby, J., 'Environmental Policies in Australia–Climbing the Down Escalator', Ch. 6 in Park, C.C., (ed.), *Environmental Policies: an International Review*, London, Croom Helm, 1986, pp. 187–98.

Formby, J., 'Where has EIA Gone Wrong?: Lessons from the Tasmanian Woodchips Controversy', in Hay, P., Eckersley, R., & Holloway, G. (eds.), *Environmental Politics in Australia and New Zealand*, Occasional Paper 23, Centre for Environmental Studies, University of Tasmania; Hobart: Board of Environmental Studies, University of Tasmania, 1989, pp. 3–17.

Freedman, B., *Environmental Ecology: the Impacts of Pollution and Other Stresses on Ecosystem Structure and Function*. San Diego, Ca.: Academic Press, 1989.

Fried, M.H., *The Evolution of Political Society: an Essay in Political Anthropology*. New York: Random House, 1967.

Frith, H.J., & Sawer, G., (eds.), *The Murray Waters: Man, Nature and a River System*.

Bibliography

Sydney: Angus & Robertson, 1974.
Frohlich, N., & Oppenheimer, J.A., *Modern Political Economy*. Englewood Cliffs, NJ: Prentice-Hall, 1977.
Galligan, B., (ed.), *Australian State Politics*. Melbourne: Longman Cheshire, 1986.
Giddens, A., *The Nation–State and Violence*. Cambridge: Polity Press, 1985.
Gilbert, A., 'The State and Nature in Australia', *Australian CulturalHistory*, 1981, pp. 9–28.
Gilpin, A., *Environment Policy in Australia*, St. Lucia: University of Queensland Press, 1980, Ch. 4.
Goodin, R.E., 'Ethical Principles for Environmental Protection', in Elliott, R., & Gare, A. (eds.), *Environmental Philosophy*, St. Lucia: University of Queensland Press, 1983, pp. 3–20.
Goodin, R.E., *Political Theory and Public Policy*. Chicago: University of Chicago Press, 1982.
Goudie, A., *The Human Impact on the Natural Environment*. 3rd edition, Oxford: Basil Blackwell, 1990.
Gould, S.J., *Ever Since Darwin: Reflections in Natural History*, Harmondsworth: Penguin, 1980. (First publication Burnett Books/Andre Deutsch 1978.)
Gould, S.J., *Hen's Teeth and Horse's Toes*, Harmondsworth: Penguin, 1984. (First publication New York: Norton, 1983.)
Gould, S.J., *The Flamingo's Smile*. Harmondsworth: Penguin, 1987. (First publication New York: Norton, 1985.)
Gunn, J., *Along Parallel Lines: A History of the Railways of New South Wales*. Melbourne: Melbourne University Press, 1989.
Haas, P.M., 'Do regimes matter? Epistemic communities and Mediterranean pollution control', *International Organization*, Vol. 43, No. 3, Summer 1989, pp. 377–403.
Haas, P.M., *Saving The Mediterranean*. New York: Columbia University Press, 1990.
Haas, P.M., 'Banning chlorofluorocarbons: epistemic community efforts to protect stratospheric ozone', *International Organization*, Vol. 46, No. 1, Winter 1992, pp. 187–224.
Hall, R. H., 'Poisoning the Lower Great Lakes: the Failure of U.S. Environmental Legislation', *The Ecologist*, Vol. 16, No. 2/3, 1986, pp. 118–23.
Hamburger, H., *Games as Models of Social Phenomena*. San Francisco: W.H. Freeman, 1979.
Hardin, G., 'The Tragedy of the Commons', *Science*, Vol. 162, 13 December 1968, pp. 1243–8.
Hardin, G., *Exploring New Ethics for Survival: the Voyage of the Spaceship Beagle*. New York: Viking Press, 1972.
Hardy, F., *Power Without Glory*. London: Sphere Books, 1968. (First published Melbourne 1948).
Harris, M., *Culture, People, Nature: an Introduction to General Anthropology*, 2nd edition. New York: Crowell, 1975. (5th edition 1988).
Head, B.W., (ed.), *State and Economy in Australia*. Melbourne: Oxford University Press, 1983.
Head, B.W., (ed.), *The Politics of Development in Australia*. Sydney: Allen & Unwin, 1986.
Higgins, E., 'The Degradation of the Noble Murray', *National Times on Sunday*, 15 November, 1987, p. 15.
Hobbes, T., *Leviathan*. (First published 1651; numerous editions.) References in text are to the Everyman edition; London: J.M. Dent & Sons, 1914.
Hockett, C.F., *Man's Place in Nature*. New York: McGraw-Hill, 1973.
Houghton, R.A., & Woodwell, G.M., 'Global Climatic Change', *Scientific American*, Vol. 260, No. 4, April 1989, pp. 18–26.

Bibliography

Hughes, C.A., *The Government of Queensland*. St. Lucia: University of Queensland Press, 1980.

Hughes, J.D., *Ecology in Ancient Civilisation*. Albuquerque: University of New Mexico Press, 1975.

Hughes, O., 'Bauxite Mining and Jarrah Forests in Western Australia', Ch. 8 in Scott, R. (ed.), *Interest Groups and Public Policy* Melbourne: Macmillan, 1980, pp. 170–93.

Jakeman, A.J., Thomas, G.A., & Dietrich, C.R., *Water Resource Management in the River Murray: Models of Salinity Travel Time and Accession and their Application*, CRES Working Paper 1986/21. Canberra: Centre for Resource and Environmental Studies, Australian National University, 1986.

Johnsen, H., 'The Adequacy of the Current Response to the Problem of Contaminated Sites', *Environmental and Planning Law Journal*, Vol. 9, No. 4, August 1992, pp. 230–46.

Joyce, C., 'US environmentalists oppose world trade deal',*New Scientist*, 15 February, 1992, p. 6.

Kartawinata, K., & Vayda, A.P., 'Forest Conversion in East Kalimantan, Indonesia: The Activities and Impact of Timber Companies, Shifting Cultivators, Migrant Pepper-farmers, and Others', Ch. 7 in di Castri, F., et al., Eds, *Ecology in Practice: Establishing a Scientific Basis for Land Management*. Dublin: Tycooly International, 1984.

Keim, S., & Richards, C., 'Queensland Toxic Sludge', *Legal Service Bulletin*, Vol. 16, No. 2, April 1991, pp. 53–6.

Kellow, A., 'Managing an Ecological System 2: The Politics and Administration', *Australian Quarterly*, Vol. 57 Autumn/Winter 1985, pp. 107–27.

Kellow, A., 'Electricity Planning in Tasmania and New Zealand: Political Processes and the Technological Imperative', *Australian Journal of Public Administration*, Vol. XLV, No. 1, March 1986, pp. 2–17.

Kellow, A., 'The Environment, Federalism and Development: Overstated Conflicts?' Ch. 10 in Walker, K.J., (ed.), *Australian Environmental Policy*, (Kensington, NSW: New South Wales University Press, 1992) pp. 203–14.

Kellow, A., *Saline Solutions: Policy Dynamics in the Murray–Darling Basin*. Deakin Series in Public Policy and Administration No. 2. Geelong, Vic.: Centre for Applied Social Research, Deakin University, 1992.

Kennedy, K.H., *The Mungana Affair: State Mining and Political Corruption in the 1920s*. St. Lucia: University of Queensland Press, 1978.

Knight, K.W., 'The Reconstruction of the Mount Isa Railway', in B.B. Schaffer & D.C. Corbett, (eds.), *Decisions*, (Melbourne: Cheshire, 1965) pp. 45–68.

Laver, M., *The Politics of Private Desires*. Harmondsworth: Penguin, 1981.

Laver, M., *Invitation to Politics*. Oxford: Martin Robertson, 1983.

Laver, M., *Social Choice and Public Policy*. Oxford: Blackwell, 1986.

Leeper, G., 'Mammonism v. Luddism: Overpopulation, Ecology & Australia', in Dempsey, R., (ed.), *The Politics of Finding Out*, (Melbourne: Cheshire, 1974) pp. 205–9.

Levine, A.G., *Love Canal: Science, Politics, and People*. Lexington, Mass: Lexington Books, 1982.

Levine, A., 'Psychosocial Impact of Toxic Chemical Waste Dumps', *Environmental Health Perspectives*, Vol. 48, 1983, pp. 15–7.

Lindblom, C.E., 'The Science of "Muddling Through"', in Etzioni, A., (ed.), *Readings on Modern Organisations* Englewood Cliffs, NJ: Prentice-Hall, 1969, pp. 154–66.

Lindblom, C.E., 'Contexts for Change and Strategy: A Reply', in Etzioni, A., (ed.), *Readings on Modern Organisations* Englewood Cliffs, NJ: Prentice-Hall, 1969, pp. 171–3.

Lindzen, R.S., 'Some Remarks on Global Warming', *Environmental Science and*

Technology, Vol. 24, No. 4, 1990, pp. 432–435.
Lines, W.J., *Taming the Great South Land: a History of the Conquest of Nature in Australia.* Sydney: Allen & Unwin, 1991.
Lowe, I., 'Towards a Green Tasmania: Developing the 'Greenprint', *Habitat Australia,* August 1991, pp. 12–14.
Lowi, T., 'American Business, Public Policy, Case-Studies, and Political Theory', (review article) *World Politics,* Vol. XVI, 1964, pp. 677–715.
McConnell, G., 'Problems Surface: the Murray Valley', Ch. 14 in Birrell, R., Hill, D., & Stanley, J. (eds.), *Quarry Australia? Social and Environmental Perspectives on Managing the Nation's Resources* Melbourne: Oxford University Press, 1972, pp. 234–46.
McEachern, D., *Business Mates: the Power and Politics of the Hawke Era.* Sydney: Prentice Hall, 1991.
Mackay, N, Lawrence, B., & Eastburn, D., 'The Challenge of the River Murray', *Heritage Australia,* Vol. 4, No. 4, 1985, pp. 13–16.
Mackenzie, W.J.M., *Politics and Social Science.* Harmondsworth: Penguin, 1967.
McLean, I., 'Mechanisms for Democracy', Ch. 6 in Held, D., & Pollitt, C. (eds.), *New Forms of Democracy,* London: Sage/Open University, 1986, pp. 135–57.
McLean, I., *Public Choice: An Introduction.* Oxford: Blackwell, 1987.
Macpherson, C.B., *The Real World of Democracy.* New York: Oxford University Press, 1972.
McQueen, H., *A New Britannia.* Revised edition, Melbourne: Penguin, 1975. (1st edition 1970; 3rd edition 1986)
Makim, A., Possibilities for International Ecological Cooperation, Unpublished B.Sc. Honours Thesis, School of Australian Environmental Studies, Griffith University, 1991.
Marshall, J., *The Great Extermination.* Melbourne: Heinemann, 1966.
Martin, G., 'The Founding of Australia', Ch. 1 in Whitlock, G., & Reekie, G. (eds), *Uncertain Beginnings: Debates in Australian Studies,* St. Lucia: University of Queensland Press, 1993, pp. 7–13.
Mercer, D., 'A Question of Balance': Natural Resources Conflict Issues in Australia. Sydney: The Federation Press, 1991.
Mill, J.S., *On Liberty.* (Numerous editions are available).
Miller, S., 'Is This the Last Word on Love Canal?' *Environmental Science and Technology,* Vol. 16, No. 9, September 1982, pp. 500A–501A.
Misham, E.J., *Cost-benefit Analysis* Amsterdam: North Holland, 1969.
Mokhiber, R., & Shen L., 'Love Canal', Ch. 8 in Nader, R., Brownstein, R., & Richard, J., *Who's Poisoning America: Corporate Polluters and their Victims in the Chemical Age,* (San Francisco: Sierra Club Books, 1981) pp. 268–310.
Mudie, I., *Riverboats.* Melbourne: Sun Books, 1965.
Murray–Darling Basin Commission, *Annual Report.* Canberra: The Commission, annually from 1988.
Murray–Darling Basin Management Strategic Working Group, *Framework for the Development of the Resource Management Strategy: Progress Report.*
Murray–Darling Basin Ministerial Council, *Murray–Darling Basin Environmental Resources Study.* Sydney: State Pollution Control Commission for the Murray–Darling Basin Ministerial Council, 1987.
Niemi, R.G., & Riker, W.H., 'The Choice of Voting Systems', *Scientific American,* Vol 234, No. 6, June 1976, pp. 21–7.
Nosich, G.M., *Reasons and Arguments.* Belmont, Ca: Wadsworth, 1982.
Oliphant, J., & Morley, P., 'Goss backs zinc giant', *Courier-Mail* (Brisbane), 7 March 1991, p. 1.
Olson, M., *The Logic of Collective Action: Public Goods and the Theory of Groups.* Cambridge, Mass: Harvard University Press, 1965.
O'Neill, G., 'Land for food is the problem ahead', *The Age* (Melbourne), 10 June

Bibliography

1992, p. 6.
O'Neill, G., 'Rare possum's habitat destroyed by mistake', *The Age* (Melbourne), 12 May 1993, p. 5.
O'Neill, W.H., *The Pursuit of Power: Technology, Armed Force, and Society since AD 1000.* Oxford: Basil Blackwell, 1983.
Ostrom, E., *Governing the Commons.* Cambridge: Cambridge University Press, 1990.
O'Shaughnessy, T., 'Joh and Don: Capital and Politics in Two Peripheral States', *Intervention*, No.12, April 1979, pp. 3–28.
Parenti, M., *Democracy for the Few.* 4th edition, New York: St. Martin's Press, 1983.
Parkin, A., Summers, J., & Woodward, D., *Government, Politics and Power in Australia.* (Melbourne, Longman Cheshire, 1980; 4th edition 1990.)
Paterson, J., 'Managing an Ecological System 4: Reforming the Tariff', *Australian Quarterly*, Vol. 57, Autumn/Winter 1985, pp. 139–147.
Pearce, D.W., *Environmental Economics.* London: Longman, 1976.
Pearce, D., Markandya, A., & Barbier, E.B., *Blueprint for a Green Economy.* London: Earthscan, 1989.
Pearce, D., (ed.), *Blueprint 2: Greening the World Economy.* London: Earthscan, 1991.
Pearce, F., 'British Aid: a Hindrance as Much as a Help', *New Scientist*, Vol. 134, No. 1822, 23 May 1992, pp. 12–13.
Pearl, C., *Wild Men of Sydney.* London: W.H. Allen, 1958.
Perhac, R.M., 'A Critical Look at Global Climate and Greenhouse Gases', *Power Engineering*, September 1989, pp. 41–4.
Phillips, J., 'Sludge II–the sequel?' *Courier-Mail* (Brisbane) 15 February 1989, p. 9.
Pirages, D., *Global Ecopolitics.* North Scituate, Mass: Duxbury, 1978.
Ponting, C., *A Green History of the World.* Harmondsworth: Penguin, 1992.
Pope, D., 'Australia's Development Strategy in the Early Twentieth Century: Semantics and Politics', *Australian Journal of Politics and History*, Vol. 31, No. 2, 1985, pp. 218–29.
Powell, J.M., *Environmental Management in Australia, 1788–1914. Guardians, Improvers and Profit: an Introductory Survey.* Melbourne: Oxford University Press, 1976.
Powell, J.M., *Watering the Garden State.* Sydney: Allen & Unwin, 1989.
Prasser, S., Wear, R., & Nethercote, J., *Corruption and Reform: the Fitzgerald Vision.* St. Lucia: University of Queensland Press, 1990.
Public Interest Research Group (Brisbane), *Legalised Pollution.* St. Lucia: University of Queensland Press, 1973.
Pusey, M., *Economic Rationalism in Canberra: a Nation-Building State Changes its Mind.* Melbourne: Cambridge University Press, 1991.
Pusey, M., 'What's Wrong with Economic Rationalism?' in Horne, D. (ed.), *The Trouble with Economic Rationalism*, Newnham, Vic: Scribe Publications, 1992, pp. 63–9.
Pyne, S.J., *Burning Bush: A Fire History of Australia.* Sydney: Allen & Unwin, 1991.
Renard, I.A., 'The River Murray Question: Part III–New Doctrines for Old Problems', *Melbourne University Law Review*, Vol. 8, September 1972, pp. 625–84.
Rogers, P., 'Climate Change and Global Warming: A New Role for Science in Decision-Making', *Environmental Science and Technology*, Vol. 24, No. 4, 1990, pp. 432–5.
Rolls, E., *They All Ran Wild.* Sydney: Angus & Robertson, 1969.
Rose, R., 'On the Priorities of Governments: a Developmental Analysis of Public Policies', *European Journal of Political Research*, Vol. 4, 1976, pp. 247–89.
Rothman, H., *Murderous Providence: a Study of Pollution in Industrial Societies.* London: Rupert Hart-Davis, 1972.

Bibliography

Rudd, S. (Arthur Hoey Davis), *Dad in Politics and Other Stories.* St. Lucia: University of Queensland Press, 1968 (first published 1908).

Rudd, S. (Arthur Hoey Davis), *On Our Selection.* (first publication 1899; revised edition 1909; available in many editions.)

Rudd, S. (Arthur Hoey Davis), *Our New Selection.* (first publication 1903).

Runge, C.F., 'Institutions and the Free Rider: The Assurance Problem in Collective Action', *Journal of Politics,* Vol. 46, No. 1, 1984, pp. 154–81.

Russ, P., & Tanner, L., *The Politics of Pollution.* Camberwell, Vic: Visa, 1978.

Schindler, D.W., 'The Impact Statement Boondoggle', *Science,* Vol. 192, No. 4239, 7 May 1976, p. 509.

Schneider, S.H., 'The Global Warming Debate: Science or Politics?' *Environmental Science and Technology,* Vol. 24, No. 4, 1990, pp. 432–5.

Selinger, B., *Chemistry in the Market Place.* Canberra: Australian National University Press, 1981.

Sen, A.K., *Collective Choice and Social Welfare.* Amsterdam: North-Holland, 1979. Originally published San Francisco: Holden-Day, 1970.

Sen, A.K., *Choice, Welfare and Measurement.* Oxford: Basil Blackwell, 1982.

Serventy, V., *Saving Australia.* Sydney: Child & Associates, 1988.

Simmons, I.G., *Changing the Face of the Earth: Culture, Environment, History.* Oxford: Blackwell, 1989.

Smith, D., *Continent in Crisis: a Natural History of Australia.* Melbourne: Penguin, 1990.

Smith, G.J., *Toxic Cities and the Fight to Save the Kurnell Peninsula.* Sydney: New South Wales University Press, 1990.

Smith, J., (ed.), *The Unique Continent: An Introductory Reader in Australian Environmental Studies.* St. Lucia: University of Queensland Press, 1992.

Spann, R.N., *Government Administration in Australia,* Sydney: Allen & Unwin, 1979, Ch. 10.

Stewart, R.G., & Ward, I., *Politics One.* Melbourne: Macmillan, 1992.

Suzuki, D., *Inventing the Future: Reflections on Science, Technology and Nature,* Sydney: Allen & Unwin, 1990. (First publication Toronto: Stoddart, 1989).

Tamvakis, M., The Kingston Toxic Waste Problem: Government Control of Pollution and Land Development. Unpublished Honours thesis, School of Humanities, Griffith University, 1990.

Taplin, R.E., 'Adversary Procedures and Expertise: the Terania Creek Inquiry', Ch. 8 in Walker, K.J., (ed.), *Australian Environmental Policy* Sydney: New South Wales University Press, 1992, pp. 156–82.

Taylor, M., *Anarchy and Cooperation.* London: Wiley, 1976.

Taylor, M., *Community, Anarchy and Liberty.* Cambridge: Cambridge University Press, 1982.

Taylor, M., *The Possibility of Cooperation.* Cambridge: Cambridge University Press, 1987.

Thomas, W.L. (ed.), *Man's Role in Changing the Face of the Earth.* Chicago: Chicago University Press, 1956.

Thompson, J., Slessor, K., & Howarth, R.G. (eds.), *The Penguin Book of Australian Verse.* Harmondsworth: Penguin, 1958.

Tighe, P.J., 'Hydroindustrialisation and Conservation Policy in Tasmania', Ch. 7 in Walker, K.J., (ed.), *Australian Environmental Policy* Sydney: New South Wales University Press, 1992, pp. 161–95.

Tilly, C., *Coercion, Capital, and European States, AD 990–1992.* Revised paperback edition. Oxford: Blackwell, 1992.

Tsubaki, T., & Irukayami, K. (eds.), *Minamata Disease: Methylmercury Poisoning in Minamata and Niigata, Japan.* Tokyo: Kodansha, 1977.

USA. The White House–Department of the Interior Panel on Waterlogging and Salinity in West Pakistan, *Report on Water Development in the Indus Plain.*

Washington, DC: The White House, 1964.
Victoria. Natural Resources and Environment Committee of Cabinet, Department of The Premier and Cabinet, *Salt Action*. Melbourne: Government Printer, 1987.
Victoria. State Rivers and Water Supply Commission. *Salinity Control and Drainage: A Strategy for Northern Victorian Irrigation and River Murray Quality*. Melbourne: State Rivers and Water Supply Commission, 1975.
Vintila, P., Phillimore, J., & Newman, P. (eds.), *Markets Morals and Manifestos: Fightback! and the Politics of Economic Rationalism in the 1990s*. Murdoch, WA: Institute for Science and Technology Policy, Murdoch University, 1992.
Walker, K.J., Development Politics. Unpublished MA thesis, University of Melbourne, 1968.
Walker, K.J., 'The Politics of National Development: The Case of the Ord River Scheme', *Public Administration* (Sydney), Vol. 32, No. 1, March 1973, pp. 93–113.
Walker, K.J., 'Methodologies for Social Aspects of Environmental Research', *Social Science Information*, Vol. 26, No. 4, December 1987, pp. 759–82.
Walker, K.J., 'The State in Environmental Management: the Ecological Dimension', *Political Studies*, Vol. 37, March 1989, pp. 25–39.
Walker, K.J., (ed.), *Australian Environmental Policy*. Kensington, NSW: New South Wales University Press, 1992.
Ward, B., *Progress for a Small Planet*. Harmondsworth: Penguin, 1979.
Waterson, D., *Squatter, Storekeeper, and Selector*. Sydney: Sydney University Press, 1968.
Watson, I., *Fighting over the Forests*. Sydney: Allen & Unwin, 1990.
Webb, M.C., 'The Flag Follows Trade: An Essay on the Necessary Interaction of Military and Commercial Factors in State Formation', Ch. 4 in Sabloff, J.A., & Lamberg-Karlovsky, C.C., (eds.), *Ancient Civilisation and Trade*, (Albuquerque, University of New Mexico Press, 1975) pp. 155–209.
Webster, D., 'Warfare and the Evolution of the State: a Reconsideration', *American Antiquity*, Vol. 40, No. 4., 1975, pp. 464–70.
Weiskel, T.C., 'The Ecological Lessons of the Past: An Anthropology of Environmental Decline', *The Ecologist*, Vol 19, No. 3, May 1989, pp. 98–103.
White, L.T., *Mediæval Technology and Social Change*. Oxford, Clarendon Press, 1962.
Whitelock, D., *A Dirty Story: Pollution in Australia*. Melbourne: *Sun*, 1971.
Whiteside, T., *The Pendulum and the Toxic Cloud*. New Haven and London: Yale University Press, 1979.
Wildriver, S., 'Mining Mt. Etna: A Case Study in Statist Developmentalism and Environmental Irresponsibility', in Proceedings of the Ecopolitics VII Conference, Griffith University, 2–4 July, 1993.
Wilkinson, R.G., *Poverty and Progress: An Ecological Model of Development*. London: Methuen, 1973.
Williams, B.G., 'Salinity and Waterlogging in the Murray–Darling Basin', in Australian Science and Technology Council, *Environmental Research in Australia*, Canberra: Australian Government Publishing Service, 1991, pp. 87–120.
Winner, L., *Autonomous Technology: Technics-out-of-control as a Theme in Political Thought*. Cambridge, Mass: MIT Press, 1977.
Wright, M., 'Policy Community, Policy Network and Comparative Industrial Policies', *Political Studies*, Vol. 36, 1988, pp. 593–612.
Zakrzewski, S.F., *Principles of Environmental Toxicology*. Washington, DC: American Chemical Society, 1991.
Ziman, J.M., *An Introduction to Science Studies*. Cambridge: Cambridge University Press, 1984.
Ziman, J.M., *Knowing Everything About Nothing*. Cambridge: Cambridge University Press, 1987.

INDEX

102nd St. dump 27
2,3,7,8–tetra–
  chlorodibenzo–T–
  dioxin 28
2,4,5T 28
abattoirs 266
Aboriginal artefacts 234
Aboriginal environ-
  mental impacts 234
Aboriginal land rights
  244
Aborigines 182, 233,
  235, 236
absolutist states 217
abstraction 37
abundance 304
Acacia 75, 234
academic discourse 305
academics 199
accident rates 273
acclimatisation societies
  241
acid rain 54
acne 28
active society 197
ad hoc approach 301
adaptation 4
adaptation 48, 52, 68,
  75, 226, 233, 257, 311
adaptive capability 305
adaptive challenge 230
adaptive change 48
adaptive flexibility 310
adaptive radiation 38
Adelaide 78, 122
adhesives 13
adjudication 206
administration 111, 119,
  124, 128, 133, 134,
  209, 210, 225
administrative arrange-
  ments 131
administrative bureau-
  cracy 212
administrative capabili-
  ties 212, 213
administrative capabili-
  ty 272
administrative ineffi-
  ciency 215
administrative institu-
  tions 271
administrative reform
  115, 116
administrative routine
  120
administrative skills
  215
administrative struc-
  ture 117, 208
administrators 118, 137
adult franchise 88, 245
adventurers 232
adversarial methods
  276
advice 290, 298
advisory boards 277
aerosols 60
affection 285
Africa 38, 206, 240
Africans 219
agencies 267, 269
agenda 312
agenda control 181, 198,
  200, 226, 279
Agent Orange 28
aggregation 58
agistment 100, 286
agreement 108
agreement tendency
  300
agricultural 242
agricultural develop-
  ment 91, 123
agricultural history 238
agricultural land 79
agricultural machinery
  99
agricultural productiv-
  ity 92
agriculture 38, 43, 77,
  82, 133, 141, 155, 178,
  180, 206, 211, 222,
  227, 233–4, 241–2,
  244, 248, 258
Agriculture
  Departments 67
aid, to military 310
AIDS 307
air pollution 314
air quality 54
air transport 307
air travel 165
Albania 156
Albert Shire 13, 15
Albury–Wodonga 74
Albury 89, 91, 109–10
alg 45
algal blooms 78, 296
Algeria 156
algorithms 200
alien species 101
allegience 225
allocation, of resources
  3
ALP branch pre–selec-
  tion ballots 178
alternative policies 312
alternatives 181, 187,
  192, 194, 200, 279
altruism 164, 169, 186
Amazon forest 228, 307
amenities 289
amenity 3, 256, 259
Amethyst St, Kingston
  17
amphibians 77, 242
anaerobic bacteria 44
analysis 275, 290, 300,
  303
analytical capability 292
analytical parameters
  301
analytical techniques
  63, 288, 312
anarchic cooperation

331

Index

167
anarchists 59
anarchy 161, 165
ancient states 213
Anglo–American nations 269, 270
Angola 310
animal husbandry 206
animals 2, 101, 242, 309
anthropocentrism 298
anthropologists 208, 234
anti-welfare state ideas 277
anticipation 285
anticipatory policy making 56, 269, 271
antiquity 212, 215
antisepsis 218
aquifers 75, 81, 83, 85, 149
Arab political power 214
Arab states 155
Arabian Gulf 308
arable land 2
arbitrary 293
archological discoveries 233
archologists 41
archological evidence 208
Argentina 248
argument 57, 61–3, 66, 68, 275, 293, 297, 305
arid country areas 256
arid lands 233
arid zones 80
aridity 73, 81, 87, 238, 311
arithmetic 210, 215
armaments 224
armed force 263
arms races 221
army 160, 223
Arrow's theorem 170
Arrow, K.J. 180, 186, 285
Arrow impossibility result 301
arsenic 17
artificial fertiliser 43
asbestos 23
Asia 38, 232
Asia Minor 212
Asian empires 214
Asian nations 310
Asians 219
Assam 226
assumptions 65, 69, 180, 190, 192, 256, 296, 308
Assurance game 154–6, 165–6
sthetic costs 289
sthetics 3, 57, 241, 285
asymmetry 119, 128, 157, 160
atheoretical incremen-
talists 288, 293
atheoretical technique 298
Atkinson, M.M. 269
atmosphere 1, 75, 289
atomism 199, 292
attitudes 256, 312
auditing 278
Australia 5, 6, 31, 42, 67, 71, 80, 87, 96, 106, 142, 147, 159, 174, 176, 179–80, 182, 199, 205, 219, 222, 224, 232–3, 235, 238, 241, 243–4, 247–8, 256, 264, 268–70, 273–4, 280, 283, 286, 292, 307, 310–13, 315
Australia, aridity 80
Australia, Commonwealth 107, 110
Australia, Constitution 109
Australia, Federal Government 129
Australia, High Court 107, 280
Australia, Labor 51
Australia, Prime Minister 112
Australia, States 107
Australia's economy 242
Australia's history 247
Australia's potential 248, 255
Australian Aboriginals 206
Australian and New Zealand Environment and Conservation Council (ANZEC) 278
Australian Capital Territory (ACT) 120
Australian colonies 88, 90, 252, 270
Australian Conservation Foundation (ACF) 113, 273, 278
Australian Council of Trade Unions (ACTU) 278
Australian economy 258
Australian ecosystems 311
Australian electoral systems 178–9
Australian environment 236, 241
Australian environmental policy 312
Australian fauna 232
Australian flora 232
Australian government 256, 269, 270, 271, 278
Australian industry 244, 257
Australian Institute of Political Science (AIPS) 115, 118
Australian Labor Party 250, 268, 278
Australian Liberal Party 180
Australian manufactures 257
Australian manufacturing 253
Australian Medical Association (AMA) 180
Australian National (shipping) Line 251
Australian national politics 277
Australian policy making 269
Australian political parties 51
Australian political system 268
Australian politics 122, 233, 245, 256, 265
Australian products 249
Australian public policy making 197
Australian resources 313
Australian Senate 177
Australian species 242
Australian State lower Houses 175
Australian States 204, 253
Australian Workers' Union [AWU] 245
Australian workers 257
authority 106, 267
autocracy 269
autonomy 122, 198, 226
Axelrod, R 163

baby boom 21
Babylon 80
Bachrach, P 198–9
bacteria 45
bad seasons 207
bads 284
balance of payments 258
Balonne river 78
Baltimore 198
banditry 221
Bangkok 266
bank crashes 257
banking 146, 214
banks 240, 264, 266
Baratz, M 198–9
Barcelona Convention 156
bargaining 108, 172, 179, 186, 192, 196, 201
Barnard, A 249, 258
Barr Creek–Kerang area 77
Barr Creek 112, 119
barrages 110
Barron River electorate 179
base metals 243
Basques 226
Bass Strait 242, 275, 291
bauxite 244
beach–sand mining 244
beaches 288
beef cattle 67
behavioural rules 159
Bell, S.R. 257, 270
Bell Bay smelter 252
beneficiaries 160
benefit-to-cost ratio 286
benefits 285
Bentham, J 186, 283–4
benzene 21
Berri (SA) 78
beside 66
bias 6, 178, 200, 283, 290–1, 300–1
big men 208
big picture 276, 292
billabongs 75, 81
bioaccumulation 13
biocentric perspective 297
biodegradability 47
biodiversity 1, 304
biological oxygen demand (BOD) 45
biological production 219
biological resources 43
biological surplus 311
biological systems 299
biology 43
biomagnification 27, 44–5
biophysical regions 75, 77
biosocial systems 304
biosphere 43, 46, 56, 214, 304, 314
bird breeding 130
bird habitats 135
bird populations 45
bird species 241
birds 13, 77, 240, 242
birth defects 22, 24
birth deformities 28
birth rates 218
Bismarck 217
bison 141
bit decisions 194
Bjelke-Petersen, J 66, 270
Black, A 95–6
black box 294–5
black lung 23
blackberries 241
blackbirds 241
Blainey, G 235, 243
Bloody Run 27
blue-green alg 78, 129

Index

Blue Mountains 235–6
boom-and-bust pattern 243
boom, economic 225
boondoggle 283
border areas 221
border rivers 127
bores 243
Borgstrom, G 219
Borneo 307
Bosnia 3, 221, 306, 308
Bosnian Serb 'Republic' 226
Botany Bay 234
bottles and cans 47
Bougainville 226
boundaries 87, 122, 136, 148, 178, 221, 226
Bourke (NSW) 89, 91
brain-storming 194
branch-plant economy 249
branch plant 244
Braybrooke, D 189, 191
Brazil 1, 226–7, 248, 307
breeding 75, 78
Brewarrina (NSW) 89
bribe 158
bricks and mortar 250, 255
brigalow 75, 236
Brisbane (Qld) 9, 16–7, 289
Brisbane Courier-Mail 256
Britain 67, 80, 87–8, 92, 106, 147, 153, 173, 224, 226, 235, 242, 245, 249, 269, 288, 315
British Empire 284
British market 99
British Parliament 88–9
British settlement 236
Broken Hill 243
Broken River 115
brokerage 267
Brown, L 219
budget 212, 225, 250, 255, 293
budgeting 134, 190–1, 278
Buffalo 18
builders 266
building societies 272
bulk commodities 314
bullock–carts 89
Burdekin dam (Qld) 255, 294
bureaucracy 134, 220, 254, 265, 280, 304
bureaucratic conflict 269
bureaucratic goals 115
bureaucratic structure 117
bureaucratic system 271
bureaucratised routine 279

bureaucrats 226, 270, 277–9
bureaux 134, 310
Burma 227, 305–7
Burrinjuck dam 110
buses 54, 146
Bush, G 66, 305
business 180, 199, 224, 254, 258, 264, 268–9, 292, 311, 315
business as usual 6, 304
businessmen 236
Butlin, N 249, 258

Cain–Kirner Labor Government (Vic) 257
Cain Labor Government (Vic) 118
calendar-making 210
California 149
Cambodia 306
Canada 18, 26–7, 32, 222, 248, 269
Canadian citizens groups 27
Canadians 157
canals 223
Canberra 51, 118, 278, 312
cancer 22, 28, 43
candidates 175, 177, 181
cane–cutters 245
cannon 221
cannon fodder 160
capability, of government 268, 271, 280, 292
capacitors 13
capillary action 83
capital–intensive projects 252, 256
capital 97, 99–100, 119, 142, 225, 230, 242, 248, 255, 258, 271
capitalism 161, 170, 214, 217, 220, 222, 224, 247
capture thesis 266
car, private 54
carbon 233
carbon dioxide 44, 47
carcinogens 28
carrying capacity 4, 236
cartage 236
Carthage 213
case studies 5
cast iron 236
catchment 111, 131
catchment area 73
cats 63, 77, 241
cattle 71, 77, 90, 238, 244
cattle ranchers 307
Cattlemens' Union 268
Catton, W.R. 228, 308
cautionary principle 54, 56
cedar–getters 142, 242
censor 198

Central Intelligence Agency (CIA) 308
centralisation 105, 128, 134
centralised control of the means of violence 220
Centre for Resource and Environmental Studies (CRES) 126
Century zinc prospect (Qld) 253
cereals 100
Chaffey brothers 96–7
challenge 3, 6, 314
change 4, 267, 279, 310, 315
channels 96–7, 101
chemical curiosity 2
chemical industries 20, 26
chemical manufacturers 30
chemical waste dumps 20, 26
chemicals 13, 15, 20, 23, 29, 46, 53
Chicago 18
chicken 143, 154–6, 158, 160, 165, 167
chiefly power 208–9
child mortality 219
Chillagoe–Mungana 243
chilling, of meat 243
China 148, 162, 242, 257
Chinese empire 212, 214
chloracne 28
chloralkali plants 44
chlorates 27
chlorinated fluorocarbons (CFCs) 289, 294
chlorinated hydrocarbons 21, 45, 67
chlorinated organic substances 28
chlorine 28
chlorobenzenes 17
chloroform 21
choice, social 169–84
Chowilla dam scheme 123
churches 273
circumscription 208, 214
cities 222, 244, 247
citizen participation 51
citizens 9, 51, 172, 181, 209, 220–1, 309
civil rights 59
civil war 309
civilisation 42, 55, 211, 229, 304, 310
civility 148, 163
Clarence river (NSW) 242
class struggle 298
classical economics 283,

297, 308
clean air 2, 145, 147
clean water 2, 218
clearance controls 178
clearing, of land 77–8, 82–3, 85, 92, 141, 143, 236, 238, 240, 243, 245
Cleveland 18
client groups 279
clients 245, 291
climate 5, 237–8, 244, 311
climate change 2, 207
climatic modification 276
climatic restrictions 103
climatology 290
Cloncurry (Qld) 243
closed shop 160
closer settlement 89, 92, 98, 100, 102, 204, 236, 243–4, 248
clothing 4
coal 23, 54, 214, 218, 244
Coaldrake, P 179
coalition 148, 150, 172, 179, 182
coalition government 178
coast 243
coastal development 278
coastal forests 142
coastal fringe 235
coastal marshes 80
coastal zone 289
coercion 186
coercive powers 167
Cohuna (Vic) 78
Coleman, W.D. 269
collaboration 68, 131, 143, 149, 151, 265
collective action 144, 160
collective choice 4, 169
collective decision-making 206
collective goods 5, 141–50, 153, 156, 160, 164, 185, 200–1, 311
collective management 147
collective policy making 296
collective problems 167, 205
collusion 292
colonial affairs 89
colonial behaviour 307
colonial disputes 122
colonial economy 6
colonial era 307
colonial governments 249
colonial possessions 227
colonial socialism 249, 258
colonial system 232

333

Index

colonies 88–9, 93, 105, 228, 235, 246
colonisation 141, 248
colonised world 307
colonists 242
Comalco 252
combat 207
command economies 264
commensurability 284, 285
commerce 248
Commission of Inquiry into the Conservation Management and Use of Fraser Island and the Great Sandy Region 94
commodities 162
common-pool resources (CPRs) 148, 150
common goods 144, 146, 148–9, 163, 166, 205, 229
common land 161
common law 87, 89–90
common property 161–2
common sense 59, 63, 68, 297
Commoner, B 43–4, 288
Commonwealth Bank 251
Commonwealth of Australia 107, 123, 165, 253, 267
Commonwealth Parliament 107, 175
Commonwealth Scientific and Industrial Research Organisation (CSIRO) 113, 278
Commonwealth Treasury 99, 270
communal property 161
communication 36, 51, 181, 226, 248, 272, 274, 276, 304
communication technology 212
communications 213, 215, 220–1, 245, 247, 249–50, 259, 271
Communities of Common Concern 121, 130
community 80, 148, 159, 166, 180, 199, 206, 209, 225, 240, 254, 264, 273–4, 277, 286
community action 130
Community Advisory Committee 121
community groups 279
community leaders 178
commuter rail systems 253
comparative advantage 257
comparisons 191
compensation 158
competing states 314
competition 147, 244
competition 3, 5, 31–2, 77, 110, 135, 142–3, 149–51, 162, 208–9, 211, 213, 219, 221, 224, 229, 255, 257, 258, 269, 308
competitive multi-state system 215, 299
competitive redistribution 208
complete orderings 170
complex society 159, 165, 186
complexity 5, 44, 177, 298–9
compliance 134, 185
comprehensiveness 190, 196, 283, 285–8, 290, 293, 300–1
compromise 122, 179, 182, 187
compulsory taxes 160
computers 220
Comte, A 298
concensus 272, 315
conceptual framework 313
concertation 270
concessions 312
concurrent legislation 109
concurrent powers 109
Condamine river 78
conditional cooperation 164, 179
conditionally 172
Condorcet 170
confiscation 223
conflict 3, 43, 52, 59, 87, 89–90, 102–3, 122, 144, 155, 165, 187, 198, 205, 207, 209, 213, 226, 228–30, 244, 256, 279, 293, 306, 308, 310
confrontation 265, 282
congestion 3, 5, 32, 219
conquest 208–9, 211, 229
conscription 160
consensus 123, 126, 196
conservation 66, 107, 117, 142, 159, 224, 234, 244, 259
conservation councils 273
conservation lobby 274
Conservation Ministry (Vic) 116, 267
conservationists 275
conservatism 3, 6, 51, 62, 89, 192–3, 196, 244–5
conservative attitudes 315
conservative government 270
Conservatives 106, 174
consistency 57, 175
constitution 107, 166, 185
constraint 6, 258, 263, 296, 305
constraints on policy 299
constraints on state action 263
construction 255, 259
consultants 286, 291
consultation 272
consultative process 275
consumer demand 58
consumer electronics 218
consumer goods 220
consumption 39, 46, 58, 119, 146, 211, 219, 252
contamination 3, 18, 28, 31, 41, 149, 275
context–sensitivity 148
contextuating decisions 194
continental drift 2
contingency 187
contract zone 108, 179
contractors 255, 259, 266
contracts 163, 222, 255, 259
controversy 95, 102, 107, 163, 286, 303
conventions 165
Coode Island 31
cooperation 3, 36, 52, 86, 89, 93, 96, 103, 105, 109, 112, 115, 118–9, 122–3, 126–8, 130, 143–4, 146, 149, 153–5, 158–9, 161, 163–7, 169, 172, 182, 199, 207, 215, 229–30, 265, 296, 306, 309
cooperation 68
cooperative strategies 310
cooperatives 95
coordination 136, 155
coordination 4, 36, 52, 89, 102, 105, 109, 111–2, 120, 128, 131, 134, 136–7, 143–4, 146, 151, 155, 159, 165–7, 185–6, 200, 205–6, 268–9, 280, 285, 287, 293, 296, 301, 306
copper 17, 243
Corowa (NSW) 107
corporate interests 225, 230
corporations 47, 266
corruption 227, 250, 255, 258, 259
Corsica 226
cost–benefit analysis (CBA) 6, 99, 169, 189, 283, 285–6, 288, 290–1, 301
cost-effective analysis 300
cotton 275
counter–cyclical policies 225
counter-instance 63
counting heads 186, 284
country areas 89
Country Parties 176, 179, 180, 245, 268
Court, C 270
courts of law 166
covenant 163
cow-cockies 244
cow 161
Crabb, P 126
crackpots 273
crayfish 78, 101
creativity 200, 226, 312
credit 207, 220
creeks 238
Cretaceous 2
crime 79
criminals 148
crisis 147, 269
criteria for decision 193
critical intellectual stance 274
critique 279, 305
Croatia 3
crop failure 310
crop management 84
crops 78, 84–5, 132, 206, 225, 233, 237–8, 240–1, 243–4, 248, 313
Crown land 239, 245–6, 250
cuddly issues 293
cultivation 42, 85, 206, 235, 237–8, 275
cultural cringe 245, 311
culture 36, 38, 207, 211, 213, 225–6, 229, 232–3, 235, 245
Cunnamulla Nipple 179
Cup Day 244
currency 223
customs 88, 90, 250, 252
cyanide 10, 17
cyclicity 170, 173, 181
cylinders 221
Cyprus 155

dairy farmers 244
dairy produce 99, 243
damages 15, 90
dams 77–8, 82, 96, 101, 108–9, 124, 294
Darling Downs 74
Darling river 71–86, 89, 111, 129

334

Index

Dartmouth storage 123
data 188, 305
data analysis 300
databases 220
Davidson, B.R. 99–100, 135
Davis, G 195
DDT 13, 23, 45, 67
de Borda, J-C 170
Deakin, A 96
death of a thousand cuts 297
death rates 218
debt 99, 105, 227
deceit 291
decentralisation 102, 105, 116, 128, 314
decision-making process 148
decision analysis 279
decision calculus 283
decision context 197, 200
decision costs 196
decision maker 31, 186–7, 192, 200, 294
decision making 120, 301
decision processes 197, 200
decision space 200
decision strategy 188, 194, 196–7, 200, 279, 299
decision theory 5, 200
decisions 48, 52, 166, 181, 185–201, 268, 296, 300
decontamination 28
defection 153, 158, 163, 164, 208
defence 223, 245
deficits 225
defining functions 224, 269, 304, 306, 309, 314
defoliant 28
deforestation 40, 43, 46, 141, 212, 227–8, 307
degradation of pastures 238
demand 147, 149
democracy 51, 58–9, 66, 181–2, 196–7, 221, 308
democratic countries 225
Democratic Labor Party (DLP) 175
democratic political process 198
democratic political systems 181, 265
democratic social choice 182
democratic theory 169, 171
Democrats 177
demographic changes 175
demographic transition 219
Department of Conservation and Environment (Vic) 116
departmental battles 115
departments 134, 223, 267, 269, 277
dependence 178, 244–5, 248–9, 258, 267, 308, 310, 314
depletion, exponential 58
depletion 218, 242, 287
depletion rates, optimal 58
depression 110, 225, 245, 248
deregulation 227, 254, 277
desert 73, 208, 233
desertification 212, 311
destocking 239
detachment 292
detection 159
detergents 47
Detroit 18
developers 254, 288
developing countries 227
development 5, 6, 87, 102, 105, 117, 135, 141, 143, 149, 156, 180, 205, 221–2, 232–60, 264–5, 267, 270, 274, 292–3, 298, 312, 315
development strategy 251, 256
developmentalism 122, 250, 279
Diamond Street 13, 16
dictatorship 162–3, 181, 186, 196–7, 220, 265
diesel buses 253
dilution flows 119
dinosaurs 2
dioxin 27, 53
diplomat 226
diprotodon 234
disagreement 276
discipline 222, 300
discounting over time 58, 311
discrimination 307
discussion 275, 304
disease 42, 219, 233, 240, 296, 306, 308
diseconomies 248, 256, 298
disenfranchisement 196
disjointed incrementalism 190
displacement 53–5, 67, 226, 288
disruption 256

dissent 211
dissipative uses (of resources) 13
distortion 290–1
distribution 82, 110–1, 117, 122, 124, 129, 141, 255–6, 259, 268
distribution of power 299
distributive justice 182
distributive policies 254
distributive policy rgime 198
distributive politics 123
divide and rule 153
divisibility 145
divisible goods 145, 147
division of labour 206, 209
Division of Port Phillip 89
'do as you would be done by' 164
Doctors' Reform Society 180
doctors 180
doctrine 284
Dodgson, C.L. 170
domestic animals 28, 291
domestic market 244
domesticated animals 241
domestication 42
dominant strategy 153–4, 158
dominion capitalist 248
Doyle, C 54
drainage 73–4, 81–3, 96, 105, 121, 131, 132
drainage channels 130
drains 101
dried fruits 99, 243
Dror, Y 192–3, 279
drought 75, 91, 100, 103, 134, 233, 238, 240, 313
drought relief 240–1
drought resistance 238
drug Acts 222
drugs 266, 293
dry land agriculture 100–1, 243, 248
dry land produce 245
dryland salinity 79
dump 10
dumping 15, 155, 265
durables 307
dust–bowls 42
Dutch oil, embargoed 228
dynamic equilibrium 41

early–warning system 43
Earth Summit 1
East India Company 283
East Timor 226

Eastern Europe 147
echidna 233
Echuca (Vic) 91
eclecticism 300
ecological balance 314
ecological collapse 211
ecological condition of the world 1
ecological constraints 5, 237, 244, 313
ecological costs 307
ecological crisis 230, 304
ecological damage 215, 228, 232, 256, 307
ecological decline 228, 241
ecological destruction 240
ecological dictatorships 163
ecological disasters 212, 215
ecological disruption 56, 103
ecological diversity 207–8, 214
ecological epistemic community 280, 313
ecological failure 229
ecological gradient 208, 214
ecological imbalance 39, 205, 207–8, 314
ecological impact 107, 215, 236
ecological inputs 287
ecological issues 280
ecological knowledge 53
ecological laws 46
ecological management 68
ecological necessity 59
ecological network 44
ecological paradox 311
ecological pressure 207
ecological principles 32, 68
ecological problem 4, 40, 58, 68, 86, 126–7, 129, 228, 288
ecological processes 47, 53, 192
ecological rationality 57–8, 60, 68, 101, 143, 280, 303, 306–7, 309, 311–2
ecological relationships 291
ecological rules 56
ecological scarcity 256
ecological sensitivity 116
ecological stability 42, 57, 144, 211, 293
ecological stress 206, 209, 213, 214, 310
ecological sustainability 101, 278, 307

335

Index

ecological system 41, 53, 68, 285
ecological thresholds 146
ecological underpinning 306, 309
ecological variations 132
ecologically irrational behaviour 3, 311
ecologically irrational practices 312
ecologically rational management 257
ecologically rational policies 256, 299
ecologically sound practices 205
ecologically sustainable development (ESD) 57
ecology 2, 41, 43, 47, 56, 61, 75, 101, 235, 290, 297, 304, 310–1, 232
economic activity 205, 210, 223, 252, 254, 271
economic costs 283, 307
economic costs and benefits 128
economic criteria 135
economic cycle 225
economic 'development' 98, 208, 235, 315
economic environment 270, 309
economic evaluation 101, 103
economic forces 222, 258
economic growth 225, 229, 246, 248–9, 254, 265, 278, 304
economic hardship 80
economic 'imperatives' 314
economic irrationalists 307
economic 'laws' 309
economic losses 78
economic management 222–5, 229, 256, 259
economic marginality 241
economic orthodoxy 266, 273
economic payoffs 267
economic power 47
economic pressure 307
economic problem 107, 311
economic 'rationalism' 277, 301, 311
economic 'rationalist' policies 264
economic 'rationalist' policy makers 51, 309
economic 'rationalists' 58, 142, 147, 199, 270, 278, 312
economic rationality 58, 307
economic relationships 170
economic stringency 257
economic transformation 225, 264, 268, 271, 310
economic welfare 285
economics 3, 58, 119, 147, 150, 158, 180, 222, 309, 311
economics of irrigation 135
economists 51, 58, 135, 245, 276, 277, 285–6
Economou, N 292
economy 250, 258, 264, 272, 314
ecosystem 39, 40, 44, 313
ecosystem characteristics 4
ecosystem integrity 103
ecosystem management 271
ecosystem resources 149
ecosystem stability 147
ecosystemic integrity 5
ecosystems, overloading 45
ecosystems 4, 39–40, 42–4, 47–8, 57, 77, 101, 219, 309, 311, 313
education 130, 224, 247
efficiency 113, 120, 133, 147
efficient use of water 135
effluent 291
egalitarian ideology 245
Egypt 155, 208, 210–1
Ehrlich, P 309
eighteenth century 217, 235
eighteenth century thinkers 298
elaborately transformed products 314
elections 51, 88, 106, 173–4, 177–8, 180–1, 264, 293, 304
electoral advantage 179, 264
electoral backlash 178
electoral bias 178
electoral boundaries 175, 176
electoral considerations 250
electoral distortions 178
electoral division 175
electoral factors 264
electoral popularity 258
electoral rolls 178
electoral systems 89, 175, 178, 180
electorate 175, 198
electricity 251, 314
Electricity Commissions 251
electricity industry 275, 314
electricity supply 110, 222, 250, 253
Electrona carbide works 252
elites 199–200, 229
Elizabeth I 217
embryos 239
emergence of states 208
emperors 209
empire-building 304
empires 208, 212, 215, 217
empirical data 56
empirical evidence 54
employees 255
employment 256, 286
emus 313
endangered species 77, 242, 291
energy 39, 44, 46, 56, 214, 218, 227–8, 249, 310, 313
energy capture 314
energy wastage 312
enforcement 186, 293
enforcement costs 166
engineering 98, 100, 113, 117, 119, 123, 130, 132, 134, 143
engineering works 122
engineers 108–9, 122
England 161, 218, 235
English-speaking world 106
English ethical thought 284
enlightened self-interest 186
entrenched policy 252
entrepreneur 106, 148, 266
entropy 53, 55, 68
environment-related issues 280
environment 4, 75, 144, 147, 174, 200, 225, 232, 233, 256, 297, 309, 315
environment department 292
environment movement 274
environmental argument 60
environmental assessment 17
environmental battles 259
environmental catastrophe 228
environmental concerns 255
environmental conditions 210
environmental consultancy 283
environmental costs 134
environmental crisis 2, 3, 47, 282
environmental damage 53, 71, 230, 248, 305
environmental debate 55
environmental decisions 297
environmental degradation 2, 48, 65, 226, 304, 307
environmental destruction 60
environmental devastation 238
environmental disasters 31, 205
environmental evaluation 61
environmental exploitation 205, 214
environmental factors 124, 299
environmental goods 5, 59, 144, 149, 169, 185, 187, 311
environmental groups 273, 278
environmental impact 1, 5, 9, 17, 47, 214, 226, 236, 238, 243, 248, 254, 287, 314
environmental impact assessment (EIA) 6, 169, 189, 283, 287, 290–1, 293, 301
Environmental Impact Statement (EIS) 283, 290, 292
environmental insensitivity 204
environmental issues 6, 103, 135, 201, 230, 280, 293, 301
environmental law 266
environmental lobby 127
environmental management 6, 101, 205, 226, 230, 257, 259, 280, 306, 309–10
environmental modification 46, 245
environmental policy 1, 4–6, 61, 167, 169, 271, 278, 285, 290, 303, 312, 314–5
environmental policy evaluation 282, 301

336

Index

environmental policy problems 52
environmental preservation 215
environmental pressure 228
environmental programmes 117
Environmental Protection (Impact of Proposals) Act 1974–75 283
environmental protection 127, 129, 205
Environmental Protection Act 1972 (EPA) 283
Environmental Protection Authority (EPA) 26
environmental regulation 266, 308
environmental resources 120, 143
environmental standards 292, 306
environmental stewardship 309
environmental stress 103, 212, 230, 311
environmental values 276
environmentalists 127, 274, 278, 315
environmentally responsible policies 308
Envirotest 17
epistemic (knowledge-based) communities 274–7
equality 59, 106, 182
equity, in markets 147
equity 31, 120, 127–8
Erie, Lake 18
erosion 78, 82, 240
ethics 182, 284–5, 299
Ethiopia 228, 305
ethnic groups 199
ethnic minorities 197
Etzioni, A 194, 195, 197
Eucalyptus 75, 234
Euphrates river 80
Eureka 245
Europe 54, 213–5, 217, 223, 226, 228, 232, 235, 245, 247–8, 269, 306, 308, 311
European agriculture 103
European Australians 235
European colonial expansion 315
European crops 233, 235
European expansion 6, 214–5, 232

European impact 142
European males 307
European market 243
European nations 217, 306
European settlement 75, 77, 80–1, 101, 206, 232, 241
European shipping 307
European society 235
European species 241
European states 223
European techniques 235
European varieties 240
European villages 236
Europeanspecies 248
eutrophication 41, 45, 296
evacuation 28
evaluation 274
evaluation techniques 299, 301
evaluative capability 290, 292
evaporation 73, 81–3, 95, 103, 108, 132, 233
evaporation basins 113
evaporation pans 84
evapotranspiration 75
Everything is Connected to Everything Else 43–4, 54, 288
Everything Must Go Somewhere 43–4, 288
evolution 229, 233
ex–colonial nations 222, 226, 308
exchange 222
excise 250, 252
excludable goods 145
excluded middle 64
exotic carnivores 77
exotic flora 241
exotic organisms 77
exotic pests 307
exotic species 92, 241, 245
expansionism 211–2
expansionist states 215
expectations 52, 165, 166
expected-utility calculus 297
expenditure growth 254
experiment 61, 295, 305
expertise 290, 292
explaining the past 4
exploitation 2, 5, 6, 142, 149, 206, 208, 213, 215, 218, 222, 224, 226, 235, 248, 267, 268, 276, 304, 311, 312, 314–5
explorers 232–5
export 67, 92, 227, 242–4, 248–9, 254,

257, 311, 313–4
exportable surplus 310
expropriation 143, 246
extension services 134
extermination 226
externalities 58, 147
extinction 2, 77, 141, 192, 212, 241, 311
extortion 223
extractive industries 310–1
extractive power 210
extrasomatic transmission of adaptive information 37

factories 266
factors of production 58, 100
factory wastes 54
fair average quality(FAQ) 313
fair go 250
faith 236
families 218, 309
family farms 244
family size 219
famine 219, 227, 305, 308, 310
fanatics 198
Far East 214
farm bankruptcies 79, 84
farm buildings 240
farmers 84, 92, 99, 102, 113, 117, 119, 127, 129, 134, 136, 142, 148, 186, 238, 251, 255, 292, 313
farming 178, 238, 311
farming land 3, 241
farming practices 102
farming techniques 313
farms 101, 238, 243, 244, 311
fascism 186, 306
fast–tracking 187
fauna 134, 311
fauna 77, 101, 146, 233, 243, 248
faunal species management 137
favouritism 95
favours 250, 255
Federal Government 118, 164, 254, 270, 272, 278
Federal Parliament 176
Federation 87, 90, 96, 107, 253
feedback 3, 40, 166, 192, 195, 228, 273, 298, 300
feedback loop 42, 48
feedstocks 2, 44, 58
felicific calculus 284, 285
fencing 240, 243
feral animals 75, 241

feral species 101
fertile crescent 236
fertile land 235
fertiliser 78, 129, 255, 259, 289
Fifteenth Century 214–5
finance 121, 165, 223, 227
financial accountability 116
Financial Agreement 253
financial commitment 279
financial institutions 217
financial interest 273
financial stringency 254
financial system 239
fine grain 271
finite environment 298
fire-dependent flora 234
fire 13, 234
fire escapes 266
firepower 221
firms 224, 264, 268, 272
first-past-the-post 173
First Fleet 234
fiscal equalisation 267
fiscal ratchet 254
fish 13, 54, 77–8, 101, 242, 275
fish stocks 130
fisheries 111, 149, 275
fishermen 155, 292
fishways 101
Fitzgerald inquiry 94
Flemington (Vic) 244
flexibility 299
flood water 77
floods 1, 65, 75, 78, 91, 101, 111, 133, 210
flora 101, 146, 233, 238, 241, 248
flora conservation 137
flotation 243
flushing 81, 112
fogs 54
food 4, 88, 207, 214, 219, 227–8, 242, 293, 306
food chain 2, 27, 41, 44–5
food crops 92
foodstuffs 209
'footloose'industries 252
forbearance 311
forced labour 209
Ford, Bacon, and Davis 286
foreign affairs 196, 223
foreign capital 245, 249
foreign competition 244
foreign control 310
foreign currency 242
foreign exchange 228
foreign indebtedness 255
foreign interests 245

337

Index

foreign investment 244, 245
foreign ownership 244
foreign power 245
foresight 32, 116, 185
forest 'industry' 278
forest 57, 280
forest policy 275
forestry 111
Forestry Commission 276
forestry departments 266
forests 266
forests policy 280
formalisation 279, 299
formalised tools 286
formalism 312
Formby, J 289
fossicking 199
fossil fuels 222, 313
fossil resources 46, 214
fossils 233
fouling of streams 3
four laws of ecology 43
foxes 77, 241
fragmentation 6
France 156, 174, 217, 224, 226
franchise 88
free choice, among random fads 309
free good 162
free market 147, 277
free market economy 271
free market in ideas 198, 305, 310
free rider 154, 158, 160, 163, 165–6
free selectors 244
free trade 270, 273, 306
free transport 178
freedom 66
freedom of speech 305
friendly society 166
friendship 186
frontiers 141, 213, 221
fuel 46, 273, 313
fugitive resources 149
functionaries 215
functionary class 210
fundamental biological processes 2
fundamental challenge 314
fundamental issues 310
furniture 244
future–oriented thinking 309
future 32, 255
future social states 170
fuzziness 147, 149

game theory 5, 151–68, 169, 306
garbage 153
garbage can policies 199, 201, 269, 277, 279
gauging, river flows 110–1
Geelong 31, 257, 266
gems 211
General Agreement on Trade and Tariffs (GATT) 306
generalisation 63
genetic chain 45
genetic damage 23
genetic engineering 40
genocide 3, 226, 308
geographically 232
geography 238, 310
geologists 29
Germany 217, 226
gerrymander 178
get-rich-quick 236
ghost acreage 228, 314
ghost acreage 39
ghost resources 39, 228, 308, 314
ghost towns 243
Giddens, A 220, 223
gidgee 75
glaciation 74
global common 228
global environment 162
global financial system 264
global interdependence 229
global markets 249
global order 229
global system 219
gnomes 199
goal-direction 211
goats 241
gold 10, 92, 243
gold mining 204
gold rushes 92, 248
good society 309
Goodin, R 62, 193, 293, 296, 299
goods 145, 149, 169, 207, 220, 224, 284, 306–7
Gordon-below-Franklin dam 280
Gordon, G 95–6
Goss, W 253
Goss Government (Queensland) 18, 120, 293
Goulburn 115
Gould, R.J. 2
governemnt, British 67
Government-developer collusion 292
government, State 66–7
government 3, 6, 51, 59, 91, 94–5, 122, 129, 135, 143, 147–50, 153, 157, 160, 163, 166–7, 169, 171, 177, 185, 187, 199, 204–5, 209, 222, 224–5, 238, 240, 243, 246–7, 254–5, 258, 263–4, 268–70, 274, 278, 290–6, 298, 304, 313, 315
government agencies 134
government assistance 258
government debt 256
government decision-making 280
government departments 256, 259
government economic intervention 247
Government Entomologist (WA) 275
government infrastructure provision 251
government intervention 248
government investment 250
government policy 128, 240, 265–6, 277, 312
government services 254
governmental structure 115
Governor, States 88
Goyder's Line 237
grain–growing 235
grain elevators 251
grains 206, 227, 307
granary 213
grandiose schemes 255
grants 166, 223, 240, 254
grass trees 234
grasslands 75, 77, 241
graziers 142, 240, 243–4, 264, 286
grazing 161
Great Artesian Basin 74
Great Dividing Range 73, 236
Great Lakes 18, 30, 44, 156
great powers 229
greatest happiness of the greatest number 284
Greece 38, 155, 211, 226
greed 182
green belts 289
green democracy 305
greenhouse effect 313
greenhouse gases 1
Greenpeace 273, 278
Greenprint 256
Greens 174, 177, 268
gross national product (GNP) 230
ground parrots 233
groundwater 17, 74–5, 78, 80–2, 113, 121, 127, 130–2, 135, 137, 149
groundwater mound 82
groundwater recharge 82
groups 199, 225, 254, 264, 269, 276, 279, 292
growth, exponential 58
growth 55, 205, 210–1, 214, 225, 229, 236, 249, 274
Gulf 228
Gulf country 253
Gulf War 308
gullies 83
gums 130
gunpowder 213

Haas, P 274, 306
habitat 146, 187, 240, 243, 266, 280, 291, 309
habitat destruction 46, 244, 307, 311
Hall, R.H. 30
Hamburger, H 153
hamburger meat 307
handouts 225, 254
happiness 284
hard hooves 238
Hardin, G. 161–2
Hare-Clark system of PR 177
Harris 208
harvests 206–7
hauliers, road 273
Hawke, R 199
Hawker (SA) 238
hazardous substances 17
head-counting 5, 169
Head, B 265
health 20, 30, 252
health hazard 156
health insurance 145
health services 224
heavy metals 44, 275
Hegel, F 298
helmeted honeyeater 146
Henry VIII 217
herbicide 28
herring 27
herring gulls 27
hidden agenda 199
Hielscher, L 270
hierarchy 194
hierarchy of decisions 195
high-cost economies 257
high-value produce 99
high-wage 257
High Court, of Australia 107, 280
history 204, 298, 306
Hitler 59, 226, 308
Hobbes, Thomas 62, 162–3
Hockett 208
Hoffman–LaRoche 28
Holmes, S 54

Index

Home, Britain as 236
homeostasis 41
homestead Acts 236
Homo sapiens 2
honour among thieves 153
Hooker Chemical company 20–5
Hope, A.D. 235
Hopgood, D. 118
horse-collar 213
horse-drawn tramways 237
horses 213
horticulture 100
horticulture 98
House of Representatives 175–6
housing 251, 270
huerta 148
human–environment interface 5
human activity 3, 223, 294, 298
human adaptation 5, 305
human beings 309
human capital 224
human disturbance 291
human environment 283, 300
human impacts 214, 219
human needs 284
human numbers 4
human populations 308
human productive systems 301
human rights 106, 308
human society 3–4, 48, 53, 57, 68
human understanding 298
human values 285
human welfare 226
humanity 56, 215
humans 3, 235, 298
Hume reservoir 110
hunter-gatherer societies 206
Hunter Valley Water Board (NSW) 119
hunting 242
husbandry 313
hybrid wheats 99
Hyde Park dump 27, 29
Hydro-Electric Commission (HEC) 251, 259, 270
hydro–electric power 20, 251
hydrographic 96
hydroindustrialisation 251–2, 270, 280
hydrological cycle 81
hydrology 75, 85
hygienic conditions 266
hypotheses 2, 295, 298
ideas 4, 311

ideology 186, 197, 205, 255, 267, 271
ignorance 30, 32, 230
illiteracy 300
immunisation 218
impartiality 274, 276
imperfect information 172, 290
imperialism 227–8, 308, 315
implementation 4, 186, 195, 265, 271–2, 301
import-export trade 245
import competition 244
important issues 3
imports 245, 253, 306
imposition 186
impoverishment 227
inadequate data 287–9
incentive 157
incineration 13, 28, 55
income equality 250
income tax 253
incommensurability 297
incomplete data 5
incremental attrition 297
incrementalism 191–3, 196, 200, 279, 287–8, 283, 293, 295–6, 301
independence 88, 105
independence movement 226
India 80, 153, 219, 226, 242, 263
indigenes 249
indigenous development paths 248
indigenous plant species 313
indigenous population 246
indigenous species 101
individual rights 284
individuals 153, 272, 296
indivisibility 43, 145
Indonesia 214, 226, 257
industrial corporations 229
Industrial Court 276
industrial development 215
industrial economy 251
industrial nations 227
Industrial Revolution 6, 38–40, 106, 214–5, 217, 220–1, 304
industrial sector 265
industrial society 48, 222
industrialisation 39, 214, 222, 224, 251, 270
industrialisation of warfare 221, 224
industrialised nations 307

industrialised world 227
industrialist 145
Industrie Chimiche Meda Societa Anonima (ICMESA) 28
industries 249, 253, 256, 266, 279
industry 31, 54, 155, 160, 162, 218, 221, 244, 247, 249, 251, 253, 256, 258, 265–6, 268, 271, 278–9, 280, 292, 293, 314
industry policy 271
infant mortality 219
infiltration (of water) 121, 132
inflation 186, 223, 225, 254, 259
inflexibility 312, 315
influence 186, 263
information 3, 186, 190–2, 209, 220, 273–4, 291, 297, 300
information technology 119–221
informers 211
infrastructure 223, 247, 249, 254, 256, 258
injunctions 90
innovation 185, 194, 213, 229, 279, 290, 312
innovative farming techniques 313
ins and outs 273, 278
insecticides 45
insects 45, 64, 242, 307
insecurity 235
insensitivity 283
insiders 6
institutional arrangements 119, 126, 185, 259
institutional forms 220
institutional gridlock 123
institutional inadequacies 32, 116
institutional innovation 137, 208
institutional power 265
institutional rigidity 6
institutional structure 105
institutions 41, 87, 122, 124, 129, 148, 150, 166, 205, 214–5, 229–30, 247, 249, 265, 276, 298, 304–5
insulation 289, 307, 309, 314
insurgent minorities 226
integrity 291
intellectual orthodoxy 277

intentions 166
intercolonial conflicts 107
intercolonial disputes 5
interdepartmental conflicts 115
interdepartmental rivalry 115, 267
interdependence 42, 44
interest, on loans 97
interest groups 196, 199, 258, 264, 268, 279, 312, 315
intermarriage 244
internal colonial repression 226
internal order 223
international academic economics community 277
international agreements 307
International Bank for Reconstruction and Development (World Bank) 227
international compacts 274
international cooperation 229, 274, 306
international ecological epistemic community 274, 315
international environment 257
international law 32, 90
international markets 235
International Monetary Fund 264
international system 161
international trade rgime 308
internationalism 224
interpretation of data 276, 291, 305
intersocietal competition 209, 215
Interstate Commission 107–8
interstate competition 162, 205, 209, 211, 215, 221, 229–30
interstate conflict 123
interstate relations 108
interstate rivalry 306
Interstate Royal Commission 107
interstate system 221
intertribal strife 305
intervention, by government 271, 280
interventionist development 253
intimidating voters 178
intolerance 273
intransitivity 171, 182, 287

339

Index

introduced species 102, 238, 241, 245, 311
intuition 286
inundation 75, 101
inventions 213
invertebrates 242
investment 99, 100, 102, 119, 219, 224, 227, 244, 247, 249, 251, 253, 256, 258, 264, 267, 271, 276, 294, 314
investors 250, 266, 271
invisibility 43
Iran 66, 263
Iraq 80, 221, 308
Ireland 226
Irian Jaya 226
Irish Home Rule 245
Irish Republican Army (IRA) 226
iron ore 218, 228, 243
Iron Range 18
irrational outcome 169
irrational policies 256
irrationality, ecological 226
irreversibility 43, 61, 192, 228, 257, 296
irreversible threat 4
irrigated pastures 92
irrigation 5, 65, 75, 77, 80–1, 84, 87, 89, 91–100, 102–3, 105, 107–10, 113–4, 122–3, 128, 130, 132–5, 142–3, 148, 159, 162, 180, 204, 206, 210–11, 240, 243, 250–1, 258, 271
irrigation Trusts 96, 106
Islamic clergy 263
island ecologies 307
isolation, geographical 232
Israel 155
Italy 28
iteration 190–1, 300
Ivan, seventh son of Ivan 65

Japan 13, 44, 148
Java 210
Jews 59
jobs 250, 256, 259
Johnsen, H 31
joint ownership 148
joint resources 149
jointness of consumption 145, 149, 167
jointness of production 144, 149, 167
journalists 52, 272, 273
judgement 60–1
judges 94
Julia Creek 286
jurisdiction 87, 109, 123
justice 157, 284
justification 61

Kalimantan 307
kangaroos 234, 239, 243, 313
Karadzic, R 226
Keating, P 199
Kellow, A 116, 123, 128, 136, 164, 275, 290
Kennett, J 180, 275
Kerang (Vic) 113
key resources 2
Keynesian counter-cyclical policies 254
Keynesian economics 225
Keynesian ratchet 225
King, British 106
King Street, Kingston 15–17
Kingston 9, 15, 18, 30–1, 35, 60
Kirner 275
knockers 255
knowledge 4, 162, 194, 198, 230, 234, 282, 290, 292, 297, 300, 314
koalas 243, 289
!Kung bushmen 206
Kurdish minorities 308
Kuwait 308

Labor 177, 244
Labor government 106, 123, 164
Labor Party 17
laboratories 293
labour 100, 142, 174, 210, 222, 227, 245, 269
Labour movement 164, 245, 250
Labour Parties 106
labourers 244
lagoons 101
laissez faire 106, 308, 311
Lake Erie 18, 156
Lake Ontario 18, 26, 27
Lake Superior 18, 156
Lake Torrens 238
Lake Tyrell 113
lakes, Norway 54
lakes, Tasmania 251
land 2, 3, 59, 97, 100, 120, 134, 141–2, 144, 182, 210, 222, 229, 235–8, 240, 243, 249, 258, 306, 309–10
land administration 115
Land Banks 97
land booms 248
land clearance 81, 136, 186, 312
land degradation 40, 78, 111, 133, 141, 178, 180, 240, 311
land development 259
land management 117–8, 121, 313
land rents 250
land retirement 84, 133, 136–7, 178, 186
land sales 97, 252
land use 17, 31, 77, 84, 111, 141, 224, 244, 312
LandCare programme 117, 121, 134
landed gentry 106
landed interests 89
landfill 29
landholders 86, 90, 132
landless workers 222
landlords 161, 246, 250
landowners 91, 98, 143
landscapes 235
landslide 173, 175
language 5, 37, 305
large–scale problems 32
large dam schemes 80
laser levelling 132
law 59, 87–8, 90, 147, 162–3, 185–6, 217, 245
law and order 145, 159, 247
law courts 89
Law of the Sea 229
lawlessness 159
lawyers 30, 94, 166, 180
laypersons 275
leaching 82
lead 17, 146
Leadbeater's possum 146, 275
leadership 186, 206
leadership madness 198
learning 36
learning experience 291
leases 239, 252
legal decisions 90
legal framework 93, 217
legal rationality 59
legal reform 284
legal system 31–2, 59, 157, 223, 276
legends 209
legislation 89–90, 92, 109, 126, 177, 292–3
Legislative Assembly 88
Legislative Council 88, 179
legislators 166, 185
legislature 90
legitimacy 185, 273, 279
Lenin, V.I. 59
liability 15
Liberal 51, 59, 105, 177, 245
Liberalism 106, 180, 182
Liberals 268
libertarian 296
Libya 155
life expectancy 219
light industry 252
limiting factors 2

limits 192
limits to development 236
limits to growth 55
lindane 21
Lindblom, C.E. 188–9, 191, 193, 195, 264
lineages 233
linear programming 119
liquid fuels 314
literacy 210
litigation 32, 107
litmus test 302
litter 153
liver 45
livestock 90, 129, 233
living standards 4, 217, 298, 310, 313
loan funds 264
loans 96, 223, 240, 255
local action 131
local communities 79
local community 275
local conditions 85
local councils 17, 153, 252
local ecologies 307
local government 47
local industry 251
local initiatives 121
local needs 149
Locke, J 106
locks 82, 91, 109
Logan City 15, 16
logging 275–6
logic 5, 53, 65, 69, 187, 282
logrolling 181, 186, 196
logrolling paradox 172, 179
London 54, 88, 264
long-term issues 225, 299, 309
loot 213
looting 208
losers 157, 191
loss of production 79
Love, W 20
Love Canal 20–26, 30, 60
low-impact technologies 312
low-wage economies 257
lowest common denominator 306
Lowi, T.J. 117, 123
lubricants 13
lumpy goods 146, 149, 192
lung cancer 53
luxury goods 211
lynching 148, 159

macadamia nut 248
Macedonia 226
machine tools 221
Macquarie Island 242

340

Index

macro framework 271
Mad Cow disease 67
majority 170, 172–4, 176–7, 181
making choices 3
making decisions 206
making policy 192
malapportionment 179
malaria 45
Malaysia 306
maldistribution 227–8
maldistributions of wealth 182
male supremacy 207
mallee 74–5, 82, 84–5, 131, 133
malnutrition 227
mammals 13, 77, 242–3
management 120, 128, 149, 220, 223, 230, 304
management of interstate river basins 126
management problems 251
management programmes 115
management regime 137, 275
management techniques 85
mandate 180
manganese nodules 229
manpower 210
Mant, J 115, 118, 267
manufactured goods 227, 242, 245, 249, 257
manufacturing 28, 245, 257–8
manufacturing investment 249
manufacturing nations 238, 249
marine life 2
market economies 145
market failure 147, 150, 285
market fluctuations 248
market price 240
market society 161
marketing 248–9, 251
marketing boards 249
markets 99, 147, 149, 167, 181, 227, 252, 257, 309, 313
marsupials 233, 238
Marx, K 217, 222, 298
Marxism 62
mass education 222
mass extinction 2
mass marketing 220
mass media 51, 68, 145, 176, 181, 198, 304, 310
mass production 145, 221
materials 39, 46, 55, 209, 217, 227–8

mathematical puzzles 170
matter 44, 46
maximisation 191
maximising 195, 284
McEachern, D 292
McEwen, J 270
Mead, E 98, 113
means-end analysis 188, 191
means and ends 190
measurement 284
meat 99, 243, 313
meatworks 286
media 3, 199–200, 273, 277
medival Europe 161
medival history 222
medival society 161
medival states 217
medical care 220
medical technology 314
Medicare 180
medicine 218, 245
Mediterranean 156, 211, 213
Mediterranean Action Plan ('Med Plan') 155, 274
Mediterranean countries 274
Melbourne 18, 31, 97, 250
meliorative 65
mental experiments 305
mercantile capitalism 215, 217, 224
mercury 10, 17, 44
Mesopotamia 38, 80, 211
mesothelemia 23
meta-rules 185
metallurgy 221, 314
metals 218
meteor impact hypothesis 2
methodology 53, 299, 301
methylene chloride 21
Michigan 141
micro-economic reform 309
micro freedoms 271
Middle Ages 161, 214
middle class 245, 248
Middle East 42
migration 175, 232
Milan 28
Mildura 78, 96
military-industrial complex 224, 229, 265
military 205, 209, 211, 213, 221–2, 224, 227–9, 310, 314
military capability 215
military conflict 226, 305
military dictatorships 221

Military expansion 210
military force 257
military government 266
military mayhem 1
military nuclear activity 310
military policy 215
military rivalry 309
military service 160
military struggle 208
military technology 214, 221
Mill, J.S. 106, 283, 305
millionaire 284
Minamata (Japan) 44
Minamata disease 44
mineral development 270
mineral extraction 243
mineral industry 248
mineral production 227
mineral reserves 228
Mineral Reserves basin 113
mineral resources 228
minerals 144, 146, 243, 245, 249, 253, 257, 311, 313
miners 92, 142
mines 253–4
minimally processed products 249
mining 10, 66, 92, 178, 218, 243–4, 248, 255, 258
mining leases 66
Mining Warden (Qld) 66
Ministers 267, 272, 278
ministries 223
minorities, ethnic 226
minorities 158, 225, 230, 284, 310, 315
minority parties 173, 175–6, 179
misallocation, of resources 147
misappropriation of public funds 243
miscarriages 22, 24
misery 221, 266, 308
misplaced comprehensiveness 283, 288
mistrust 153, 155, 207
Mitchell grassland 75
mixed decision strategies 194, 200
'mixed scanning' 194–5, 197
Mobil oil company 10
mobilisation 210, 212
modern nations 304
modern societies 263
modern states 220, 224, 310
modern times 215
modernity 220
monarchy 217

monetarisation 222
monetary values 285
money 186, 249
monitoring 156, 290, 292, 307
monocultures 243
monopolies 142, 147, 221, 223
monopoly of force 221
monotremes 233
Montreal 18
morality 31, 57, 163–4
Moreton Bay 242
Morgan 91
Morocco 156
mortality 28, 57
mosquitoes 45
motor–racing 57
motor vehicles 47, 220, 253, 314
Mount Etna caves 276
Mount Isa railway 286
Mount Lyell 243
Mount Morgan 243
Mount Taylor 10, 16
Mozambique 310
muddling through 193
Mudie, I 91
mullahs 263
multi-member electorates 176
multi-state systems 211
multinational corporations 225, 229, 257, 264, 304
multiplier effect 173, 175, 176
multipliers, in CBA 286
municipal sewage 155
murder 182, 196, 226, 263
Murray–Darling basin 5, 64, 70–138, 141, 157–8, 160, 180, 198, 233, 251, 258, 271
Murray–Darling Basin Agreement 126
Murray–Darling Basin Commission, 121, 128, 130
Murray–Darling Basin Ministerial Council (MDBMC) 119–20, 130
Murray 112–3, 120
Murray 65, 74–5, 78, 87, 89, 91, 93, 107–8, 112, 115, 118, 122
Murray basin 73, 80, 250
Murray Cod 78, 101
Murray Groundwater Basin 74
Murray river, 74
Murray river, regulation 101
Murray River Main Canal League 107

341

Index

Murray trout 78
Murray valley 98, 131, 162, 164, 243
Murray waters 122
Murray Waters Agreement 123, 143
Murrumbidgee 89, 111
Murrumbidgee 99
Muslims 66, 263
mutual dependence 254
mutual pollution 153, 156
mutual self–interest 306, 315
mutual support 172, 267
mutual trust 162
myth 6, 209, 237, 248, 258, 276
myth of decisionality 197
mythologies 229
mythology 210, 244

n-person games 158, 160, 163, 167, 185
Namoi region 100
NASA moon programme 294
nation-building 248
nation-state 163, 205, 225, 229, 299, 304, 310, 312, 314
nation 204
national capital 223, 230
national defence 304
National Farmers Federation 268
National Health and Medical Research Council (NH&MRC) 275
national interest 58, 277
national parks 66, 143, 293
national policy debate 272
national survival 221
national vote 176
nationalist 153
nations 90, 114, 160, 219, 258, 264, 306, 308–9
native fauna 238, 241
native flora 132, 134, 240, 311
native forests 278
native grasses 241, 243
native species 130, 132–3, 236, 240–1, 243, 248, 313
native trees 84
native vegetation 77, 85, 135, 137, 241
natural disaster 240
natural ecological regions 136
natural ecosystems 205
natural environment 5, 127, 137, 206, 245, 298, 301
natural monopolies 253
natural phenomena 309
natural processes 84
natural resource base 215
natural resource conflict 149
natural resource extraction 253
natural resource management 121, 129
natural resources 58, 218–9, 224, 227, 253, 259, 268, 308, 310, 315
Natural Resources Management Strategy (MDBC) 121, 130
natural salination processes 81
natural selection 212, 229
natural systems 219
naturalists 235
nature 57, 235
nature reserves 275
navigation 75, 77, 91, 107–11, 122–3, 143, 162, 214, 247
navigators 89
negative feedback 57
neglect 30
negotiation 90, 226
neighbours 207
neo–colonies 228
neoclassical economics 273, 277, 297, 308
neocolonialism 227, 307–8
neolithic 206
nervous disorders 22
nested hierarchy 194
networks 6, 267, 304, 315
neutrality 258
New Caledonia 242
New Hebrides 242
New Right 223
New South Wales, Hunter Valley Water Board 119
New South Wales, Minister for Agriculture 275
New South Wales 75, 87–9, 91–3, 98, 107, 109–10, 112, 118–20, 122–3, 127–8, 157, 164, 236, 255, 271, 275
New South Wales Forestry Commission 267
New York State 22, 24, 26–7, 29
New York State Department of Environmental Conservation 21
New York State Department of Health (NYDOH) 22
New Zealand 248, 269, 275
newspapers 51, 272
Niagara County 20–2
Niagara Escarpment 18
Niagara Falls 18, 26–7, 29–31, 35, 156
Niagara Falls Board of Education (School Board) 20, 31
Niagara Falls City 21, 22
Niagara Gorge 27, 29
Niagara River 18, 27
Niigata 44
Nile 73, 210
nineteenth-century England 284
nineteenth century 88, 106, 122, 218, 223, 236, 241, 243, 245, 248, 283, 285, 298, 315
nitrogen 44–5, 78, 129, 233
noise, in culture 37
noise 252, 305
non-decisions 198
non-economic costs 285
non-economic values 285
non-excludability 145
non-governmental organisations (NGOs) 273
non-human species 298
non-Labour 245
non-renewable natural resources 218
non-return containers 47
nondictatorship 186
norms 164, 274
North Africa 43, 212, 213
North America 5, 18, 208, 228
North Atlantic Treaty Organisation (NATO) 160
Norwegian lakes 54
notables 304
novel solutions 279
novelty 5
noxious odours 29
noxious weeds 241
nuclear arms races 198
nuclear energy 40, 314
nuclear fusion 56
nuclear power 54
nuclear waste 221, 265
nutrient decline 78
nutrient loss 311
nutrient management 129
Nutrient Management Strategy (MDBC) 121, 130
nutrients 44, 228, 240, 289, 296

objectivity 274
obligation 148, 186
observations 61
Occidental Chemicals (see also Hooker) 29
oceanic pollution 274
oceans 2, 228
officials 210, 300
oil 80, 156, 214, 218, 308
oil companies 273
oil sludge 10
oil spills 308
old-boy networks 276
old-growth forests 280
oligarchy 182
oligopoly 142
Olson, M 159
omission 291
Ontario 18
open-cut mining 244
open-field systems 240
open language 36
open societies 191
open space 17
opinion 274, 303
opinion polls 174
opposition 134, 179, 187, 211, 272
optimal rate of depletion 58
optimality 296, 298
options 4, 258, 263
orchards 100
Ord River irrigation scheme 236, 275
ore 146, 253
organisation, economic 48
organisation, political 48
organisation 160, 205, 268, 304
organisational capacity 212, 214
organisations 134, 273
orthodoxy 181, 273
Ostrogorski's paradox 171, 180
Ostrom, E 148, 149, 161
over-consumption 307
over-free trade 309
over-representation 175
over-representation 89, 178, 180
overcapitalisation 240
overcentralisation 128
overcharging 266
overexploitation 127, 229
overloading 219
overpopulation 39, 227,

342

Index

304, 310
overpressures (of irrigation water) 82, 132
overpumping (of groundwater) 149
overseas borrowing 256
overseas capital 249
overseas markets 232
overseas money markets 258
overseas owners 249
overseas resources 232, 307
overshoot 304
overstocking 78, 240, 311, 312
ownership, joint 148
ownership 161
oxcart, transport costs 236
oxen 213
oxygen 2, 44
ozone layer 60, 289

Pacific islands 242
Pacific Northwest 208
Pacific Ocean 149
pack animals 89
packaging 289
pain 284
paints 13
pairwise comparison 170
Pakistan 80, 153
Palestine 38, 211
Palestinians 308
paper mill 159
paper production 28
Papua New Guinea 226
paradigm 61
paradox 5, 171, 186, 212, 215, 257, 270, 284, 304, 309
Paradox of Voting 170, 173, 180
paradoxes of social action 185
paradoxes of voting 182, 185
paramilitary police 221
Pareto, V 298
parks 17, 280
Parliament, Commonwealth 109
Parliament, Victorian 109
Parliament 88, 94, 106, 109, 175, 180, 292
parliamentary democracies 221
Parliamentary majorities 173
parliamentary majority 178
parochial politics 267
Parson's Nose at Caloundra 179
participation 121, 148, 287

parties, political 181
partisan mutual adjustment 196
Partition, India 153
party preference 171
passenger pigeon 141
pastoral exploitation 236
pastoral industry 248
pastoral settlement 246
pastoralism 100, 178, 238–240, 242, 258
pastures 82, 100, 132, 148, 161, 238, 243, 313
Paterson, J 118–9, 128, 135, 158, 160
patriotism 255
payoffs 119, 151, 153, 155, 156, 158, 160, 165
payroll tax 253
Pearce, D 286
peasant labour 210
peasant population 309
peasant revolts 222
peasants 210, 212, 215
peer pressure 296
pensions 195, 224
pentachlorophenol 28
Pental Island 89
periphery, in empires 213
persecution, of minorities 226, 308
personal loyalties, Tasmania 267
Peru 208
pest control 240
pesticide residues 67
pesticides 23, 46, 78, 275
pests 45, 233, 240–1, 275, 306–7
petrochemical industries 45
petrochemicals 47, 314
petroleum sludge 17
Pharaohs 209
Philadelphia 198
Philippines 148, 307
Phoenicia 38, 211
phosphates 129
phosphorous 27, 233
photography 220
physical systems 299
pianos 144, 244
piecemeal decision techniques 287–8, 293, 301
piecemeal social engineering 295
pigs 241
Pincus, J.J. 249, 258
piracy 217, 221
planning 17, 103, 120, 143, 147, 185, 189, 197, 199, 269–71, 278, 289
plants 2, 56, 101, 233,

242, 248, 309
plasticisers 13
plastics 47, 289
plastics industry 44
platypus 233
Playford, T 270
Playford Liberal government 251
pleasure, as 'sovereign master' 284
plunder 211, 213, 232, 236, 308
plural voting 88
pluralism 197, 199, 268, 279
pluralism 315
pluralist 201
pluralistic political systems 273
plurality 173
Point Wilson (Vic) 31
police 153, 166, 211, 223
policy–making, anticipatory 269
policy–making agencies 193
policy-making processes 4, 277
policy 65, 111, 120, 169, 181, 186, 190–1, 195, 197, 199, 211, 225, 240, 249–50, 252, 261–316
policy agenda 201, 269
policy analysis 282
policy biases 273
policy change 315
policy choices 56
policy community 123, 272–3, 276–8, 280, 312
policy coordination 200–1
policy dilemmas 256, 299
policy environment 129
policy evaluation 4, 64, 282–302
policy field 295
policy implementation 134
policy makers 191, 193, 256, 293–4, 309, 312
policy making 304, 312
policy networks 124, 277, 280
policy preferences 172
policy priorities 223, 229
policy problems 180, 219, 269, 272, 315
policy process 6, 198, 263, 268, 272–3, 279, 296, 303
policy settings 299
policy space 194, 279
policy trends 293
political action 150
political activity 51, 211

political actors 151
political agenda 181, 199, 264
political alliances 179
political behaviour 160, 165
political bias 186
political campaign 223
political climate 280
political convergence 120
political corruption 97
political culture 230
political debate 181, 199
political decision-making 142, 185, 301
political decisions 186, 190, 313
political determination 133
political discourse 63
political disputes 159
political ecology 32, 205
political economy 6, 68, 147, 198, 204, 263, 268, 298, 301
political enthusiasm 96
political entrepreneurs 150
political expediency 103
political framework 282
political history 86
political influence 250
political institutions 206, 250, 269
political intelligence 197
political interactions 151
political issues 31
political jobbery 97, 259
political leaders 148
political motives 290
political obligation 153
political organisations 51
political paralysis 136
Political parties 225
political parties 51, 171, 225, 244–5, 265, 268, 277
political philosophy 106
political power 207, 213–4, 240, 243, 312
political pressure 102, 127
political pressures 204
political programme 180
political rationality 59, 290, 293, 311
political scientists 190
political stratagem 309
political support 180
political survival 314
political system 3, 6, 9, 32, 43, 147, 171, 187, 196, 198–201, 263, 299, 303

343

Index

political theory 106, 180
political turmoil 148
political uncertainty 276
political will 32, 116, 128, 293
politically active scientists 276
politicians 52, 66, 118, 122, 136, 186, 190, 197, 236, 245, 254–5, 264, 267, 290, 298, 304, 315
politicisation 258
politico–economic system 311
politics 3, 186, 212, 222, 310
politics of distribution 225
pollutants 4, 15, 23
pollution 2–3, 5, 7–33, 39, 40, 45, 47, 54–55, 78, 111, 123, 145, 155, 186, 218, 244, 252, 287, 291, 304
pollution abatement 56
pollution control 147, 308
polyaromatic hydrocarbons (PAHs) 17
polychlorinated biphenyls (PCBs) 10, 13, 53
poor, the (in politics) 197, 227–8, 307
poor analysis 274
poor countries 1, 219, 227
Popper, K 295
popular vote 173
populate or perish 248
population 2, 4, 31, 38–9, 47–8, 55–6, 79–80, 92, 98, 141, 144, 146, 155, 175, 205, 207, 209–10, 212–14, 218–9, 227–9, 232, 239, 242, 244, 247, 253, 256, 297, 300, 308, 310
population growth 207, 210, 215, 219, 298
population policy 230
pork-barrel politics 245, 258, 267
Port Melbourne 250
ports 91, 236, 253
Portugal 214
positive–sum game 119
positive feedback 166
postal service 223
potlatch 208
poverty 227
power 43, 52, 56, 122, 162, 180–1, 195, 198–200, 201, 206, 208, 213, 220, 223–4, 229, 253, 259, 263, 265, 268, 271, 276, 293, 315
power bases 274
power struggles 185
powers of government 163, 269
pragmatism 300
pre-ecological social theory 298
pre-literate societies 210
pre-modern state 212, 214, 220, 223
pre–state societies 206–7, 209
precious metals 211
precipitation 95
predation 41, 77
predators 42, 44–5, 241
prediction 4, 286
preference 58, 169, 171–2, 174, 287
preference ordering 170, 285
preferential voting 175
prejudice 186, 256
Premiers'Conference 107–8
Premiers, State 270
premises, of arguments 63, 65
present value 58
press 1, 3, 200, 272, 277
pressure groups 268, 276
pressure pluralism 269, 279
price–makers 142, 181
price controls 251, 270
price mechanism 58
prices 244, 249, 257, 268, 307, 313
pricing 119
prickly pear 241
priesthood 210
primary benefit 286
primary goods 144, 147
primary produce 257, 307, 310
primary producers 248
primary production 257
primary states 209
primitive accumulation 217
primitive statecraft 215
principles 57
priorities 308
Prisoner's Dilemma (PD) 151–3, 155–67, 182
pristine states 209
private investment 178, 251, 254
private ownership 250
private property 161
privilege 159, 181, 213
Privy Council 88
probabilistic evidence 53
problem-orientation 301
problem displacement 68
produce 92, 98, 102, 185, 242, 311
production 56–8, 71, 79, 101, 103, 114, 133, 136, 146, 161, 208, 210, 220, 223, 238, 304, 311, 313
production 71
production costs 244
productive capacity 60, 68, 218
productive surplus 212
productive systems 227, 298
productive work 309
productivity 2, 115, 134, 210, 213–5, 233, 247, 257
professional ethics 291
professional middle classes 180
professional staff 292
professions 245
profit 43, 47, 99, 102, 217, 224, 249, 253, 258, 286
profitability 292
progress 38, 236, 293, 298
projectism 135
projects 134, 137, 187, 250, 253, 255, 259, 280, 285, 288, 290–1
property–owners 89
property 90, 143, 161, 240
proportional representation (PR) 177
prosperity 305
protection 57, 206, 297
protective capacity, of ecosystems 60, 68
protective capacity 60
protein swindle 227
proto-states 208, 214
public 148, 272, 275
public affairs 305, 310
public awareness 200
public concern 280, 293
public corporations 147
public cost 241
public debate 255
public discussion 198, 200
public enterprises 252
public funds 97, 245
public goods 144, 146–7, 161, 166, 255
public health 26, 156, 218
public hygiene 219
public interest 159, 187, 273
public investment 251
public opinion 181
public places 153
public policy 56, 135, 169, 191, 229–30, 251, 255, 269, 298, 312, 315
public policy failures 103
public policy making 186
public protest 276
public purse 47, 200, 247
public servants 186, 272
public transport 159
public utilities 248
public works 111, 115
publication 275, 305
pumping 84
punishment 159, 165
Punjab 80
puppet governments 227
Pusey, M 277, 312
Pyramid building society 266

Qantas 251
quantification 286
quantitative methods 300
quarantine 28, 307
Quebec 18
Queens Counsel 94
Queensland, Premier 253
Queensland 9, 17, 31, 66, 75, 88, 106, 118, 120, 127, 136, 142–3, 178–9, 243, 245, 255, 264, 270, 276, 288, 293–4
Queensland general election 1989 17
Queensland Government 16, 94
Queensland Legislative Assembly 179
Queensland Railways 253
quick and dirty analysis 286, 299
Quorn (SA) 238
quotas 119, 177

rabbit–proof fences 241
rabbits 77, 241
radiation 43
radical initiatives 279
radicalism, in small farmers 244
radicals 282
radio 51, 145
radioactive contamination 310
rail rates 251
rail transport 286
railway 146, 236
railways 91, 97, 107, 111, 146, 223, 236, 243, 249–50, 252–3,

344

Index

314
rain 233, 238
'rain follows the plough' 237
rainfall 73, 81, 91, 236
rainforest timber 306
rainforests 143, 276
rampant marketeering 309
random fads 309
Randwick 244
rangelands 311
rank 206
rapine 263
rare species 311
rates, of depletion 287
rates of growth 218
rational-comprehensive decision strategy 188, 200, 283, 286, 301
rational assessment 276
rational behaviour theory 57, 165
rational decisions 148,169
rational environmental management 311
rational intellectual discourse 305
rational self-interest 106
rational utility 58
rationalisation 61
rationalist 235
rationality, individual 60
rationality 57, 151, 153, 171, 193, 205, 211, 298
rats 307
raw materials 143, 147, 217, 227, 242, 252
reactive policy–making 199, 269, 280
reafforestation 130, 267
rebellion 222, 226
record-keeping 220
recreation 153, 244, 256, 266
red cedar 242
red gum 75
redistribution 175, 179, 209
reform 116, 123, 133
refrigerated ships 99
refrigeration 243, 289
regional boundaries 131
regional ecologies 309
regional plans 121
regulation 108, 111, 117, 124, 129, 141, 150, 185, 198, 222, 230, 256, 266, 268, 292
regulatory policies 272
regulatory responsibility 293
reinvestment 135, 211
relevance 301, 305
reliability 57, 160

religion 57, 153, 210, 225
religious groups 199
religious sanctions 210
relocation 24
remedial policy making 187
remedial policy strategies 65
renewable energy 312
Renmark 78, 96, 122
rent 239, 264
repetition 163, 191
replication 61, 305
representation 175, 177, 229
repression 211, 226
reproduction 27
reproductive system 45
reptiles 77, 242
research 9, 102, 120–1, 156, 291–2, 299–301, 312
research and development 147
research needs 291
Reserve Bank 272
reservoirs 81
residents 252
resilience 42, 57
resistance 45, 57
resource allocation 52, 230
resource availability 149
resource base 208, 213
resource conflict 314
resource depletion 242, 304, 310
resource exploitation 122
resource extraction 205
resource management 121
resource mobilisation 223, 264, 314
resource overuse 211
resource rgime 124, 141, 256
resource shortages 214, 310
resource substitution 55
resources 2, 4, 39, 56, 58, 101, 128, 130, 133, 137, 141–3, 147–50, 159, 161, 182, 186, 198, 206–11, 214, 217–9, 224, 228–9, 234, 255, 264, 267–8, 271, 274, 306–8, 310
responsibility 182, 274, 279, 312
restocking 239
restraint of trade 306
retiring land 84
revealed preference 174, 287
revegetation 84–5, 136
revenue 250, 252–3, 255,

267
reversibility 192, 296
revolt 209
rewards 165–6
rhyme 210
Ricardo 283
rights 59, 148, 284, 298
ringbarking 142, 235, 238
Rio de Janiero 1
riparian disputes 89
riparian issues 32, 90, 157
riparian rights 89, 90, 93, 96, 103, 128, 149
risk 54, 194, 289, 291, 309
rivalry 254
river basin management 161
river box 75
river flows 73, 108, 157
river maintenance 91
river management 124, 127
River Murray Commission (RMC) 108, 110–1, 118
River Murray Waters Agreement 108–9, 122
river navigation 91
river ports 91
river red gum 78, 101
river sediments 81
river structures 81–2
river systems 73
river trade 89, 91, 110–1
Riverina 99
riverine ecology 78, 130
riverine ecosystems 132
riverine environment 98, 108
riverine fauna and flora 137
riverine plains 74
riverine resources 127
riverine species 129
rivers 82, 155, 251
road haulage 277
road hauliers 272
road lobby 273
roads 145, 223, 247, 249, 314
robust 164
role of the state 6
Roman Catholic 245
Roman Empire 212–3
Rome 38, 211, 213
room for manoeuvre 200, 279
root zone 83
Rose, R 223, 269
Rousseau, J–J 106
Royal Commissions 93, 95
royalties 253
rubbish 153
rubbish bins 199

ruins 238
rule-of-thumb 299
rule of law 59, 106
rulers 198, 210, 212, 215
rules 52, 59, 162, 165–6, 169, 185
rules of the game 273
ruling class 211, 269
ruling clique 198
ruling elites 227, 229
Runge, C.F. 166
runoff (water) 121, 129
runoff election 174
rural-urban dichotomy 244
rural activities 313
rural crisis 257
rural development 100
rural economies 249
rural industries 258, 313
rural interests 178, 258, 268
rural misery 227
rural support 178
rural workers 245
rush that never ended 243
Russia 59, 226
rust 240
Ryan Labor Government (Qld) 264

'S' Area dump 27, 29
sabotage 186
Saddam Hussein 308
safe drinking water 2
safety 266
safety standards 222
salination 5, 64, 70–86, 97, 110–1, 113, 123, 130–2, 135–6, 141, 158, 165, 186, 207, 211
salination processes 116
saline agriculture 85, 132, 135
saline aquifers 82
saline dilution 110
saline discharge 112, 119, 158
saline groundwater 82–3, 135
saline seepage 80
salinity 5, 64–5, 70–86, 111–2, 116–9, 123, 130–1, 137, 162
Salinity and Drainage Strategy (MDBC) 121, 130, 158
salt 121
salt infiltration 113
salt pans 80
salt springs 83
saltbush 75
sanctions 159, 165–6
sandalwood 242
Sandridge (Port Melbourne) 250

345

Index

sanitary landfill 55
sanitation 218, 222
satellite remote sensing 194
satisficing 187, 190–1
scalding 80, 83
scale, problems of 159, 298
scandal 97, 243
scarce resources 2, 268, 310
scarcity 3, 141–2, 147, 149, 182, 206, 208, 217, 310
Schaffer, B.B. 200, 279, 296, 312
scholars 207, 219
schooling 244
schools 266
science-based industries 314
science 43, 210, 310
scientific access to government 276
scientific advice 274
scientific argument 2
scientific communication 274, 305
scientific community 274, 305
scientific data 53, 103, 204
scientific disagreement 276
scientific ecological epistemic community 313
scientific evidence 275
scientific information 186, 275–6
scientific integrity 283
scientific interest 233
scientific investigations 105
scientific issues 276
scientific knowledge 31–2, 61, 96, 156, 230, 276
scientific method 61
scientific opinion 274, 278
scientific principles 53, 198
scientific progress 61
scientific research 116, 276
Scientific testing 15
scientific uncertainty 4–5, 23, 26, 29, 53, 61, 68, 300
scientists 127, 234, 274–6, 293, 305, 313
scrub 240
scrutineers 178
sea 155, 214, 228, 288
sealants 13
sealers 242
sealing industry 142
second-best 186

second ballot 174
second preferences 175
Second World War 45, 253
secondary benefits 286
secondary industry 244, 249
secret ballot 178
secret compositions 97
sectional pressures 187
sectors 272
security 192, 219, 249
seed dispersal 75
seeds 307
seepage 27, 97
segmented societies 222
segmented states 211, 225
selective advantage 212
selective pressure 45, 205
selective process 229
selectors 240, 244
self-criticism 312
self-defence 148, 162
self-government 87–8, 178
self-help 105
self-interest 165, 169, 179, 187
self-preservation 57
self-regeneration 56
self-regulation 41, 57
self-selection 276
self-sufficiency 39
self-sufficient 308
semi-arid environments 73
Senate 177
Serbia 66
service sector 249
services 212, 222, 224, 249, 253, 267
settlement 6, 142, 235–6, 243–4
settlers 235–6, 238, 243–4
seventeenth century 106
Seveso 28
sewage 45, 78, 129, 153, 156, 288
sewage systems 218
sewage treatment 155
sex tours 307
shadow pricing 285
sharp practice 266
shearers 244–5
sheep 67, 71, 77, 90, 161, 235, 238, 244, 290
shellfish 44
Shi'ite 308
ships 243
shock clearance 143
shopkeeper 286
short-sightedness 3
short-term advantage 2
short-term market fluctuations 257

short-term rationality 309
shortages 2, 3, 182, 207, 209
shrubs 238
silicon smelter 252
silver-lead-zinc ores 243
silver 243
simple plurality voting 173–5, 179
simplicity 300
Sinclair Knight and Partners 16
single-interest groups 280
single-issue groups 278
single-issue problem 304
single-member electoral systems 175–6
skin diseases 22
slaves 212
sleeper effects 294, 311
sludge 16, 21, 129
slumps 225
small-scale military enterprise 221
small business 252
smallpox 296
Smith, A 106, 222, 283
smog 54
smoke abatement 54
smokestacks 288
smuggling 91
snagging 91
snowfields 73, 95
Snowy Mountains scheme 110, 123
soap 47
social behaviour 164
social belief systems 304
social causation 298
social change 48, 192, 214
social choice 5, 58, 182, 185, 287, 309
social choice processes 5, 287
social choice theory 169
social contract 106, 162
social control 163, 208, 222
social cost 153, 301
social decision processes 199
social decisions 148, 195
social disapprobation 166
social divisions 153
social grouping 269
social inequality 181, 229
social mobility 212
social order 213
social organisation 204, 209
social philosophy 195
social power 208

social preferences 182
social problems 3, 182, 196, 248
social reform 282
social sciences 4, 297–8
social scientists 4
social services 253
social stratification 106, 206, 209–10, 214
social structure 5, 166, 207, 220, 229, 257
'social sum' 169
social system 206, 282, 298, 309
social theory 170, 298
social values 263
social welfare 219, 223–4, 250, 310
social welfare function 285
socialism 66, 170, 250
socialist objectives 250
sociality 5
society 162, 186, 198, 222–3, 268, 305
socio-economic systems 299
socio-intellectual environment 297
socio-political system 208
sociological theories 297
sociology 298
soft-footed fauna 238
softwood packing cases 307
soil 1, 77, 161, 218, 240
soil acidity 79
soil conditions 85
soil degradation 65, 241, 293
soil erosion 240
soil exhaustion 78, 141
soil fertility 43, 85, 133, 207, 238, 240
soil permeability 85, 132
soil productivity 80
soil salination 82
soils 96, 102, 232, 238
solar hot water 314
solar radiation 289
soldier settlement 236
solidarity 164
Solomons 226
solvents 23
Somalia 221, 228, 284, 305, 309
song 210
South-East Queensland 17, 289
South Asia 42
South Australia 78, 87–9, 91–3, 98, 107, 109–10, 112, 115–6, 118–20, 122–3, 127–8, 136, 157, 164, 236, 257, 270

346

Index

South Australian Housing Trust 251
South Queensland Mines 10, 13, 15
Southeast Asia 313
Southern Right Whale 242
sovereign 163
sovereignty 127
Soviet Union 147
Spain 148, 156, 214, 226
sparrows 241
spatial displacement 54
specialisation 36, 143, 206, 215
specialisation of function 210
specialists 291
species 4, 5, 57, 77, 146, 187, 192, 235, 238, 242, 291, 298
species diversity 42
species extinction 2, 65, 228, 307
speculation 97, 98, 102
speculators 97
Spencer Gulf 236
Spice Islands 214
spices 211, 214
spies 211
spillages 10
spinoffs 213
spoilage 307
sponsors 290
sport 241
spray-painters 23
spray cans 289
squabbling 254
squatters 89, 92, 100, 106, 239, 244
St. Lawrence River 18
St. Vincent's Gulf 236
stabilisation 248
stability 212–3, 257, 309
stagnation 211, 312
stalemate 122
Stalin J 226
stalking–horses 290
standard of living 214–5, 218
Standard Oil of New Jersey (Exxon–Esso) 264
staples 207, 209
starlings 241
starvation 1, 284
state–subsidised profit–taking 247
state 211
state 269, 298, 309
state 68, 207
State banks 272
state control 107
State expansionism 215
state functions 220
State government 105, 239, 249, 275
state intervention 258, 271

State of Warre 162
state organisation 208
State Pollution Control Commission (NSW) 275
state power 220, 315
State revenue 275
State Rivers and Water Supply Commission (Vic) 98
state societies 6, 202–231, 304
state strength 269, 315
state system 205, 209, 224
statecraft 6, 218, 222, 224, 303, 314
states, defining functions 223
States (Australia) 88, 205–6, 221
states rights 59
statesmanship 186
static societies 212
stations 244
statism 271
statist developmentalism 250–6, 258–9, 311, 315
statistical analysis 28
status 206, 208
status quo 251
statutory corporation 251, 259, 277
statutory duties 267
steam locomotives 221
steamboat 89
steel industry 18, 264
sterility 299
stirrups 213
stock 75, 109, 238, 286
stock markets 222
stock pollutants 147, 287, 294
stock resources 147, 287
stockbrokers 180
stockmen 244
stocks 46, 209
Stone, J 199
storages 95, 109
Strategic games 160
strategic games 5
strategic positions 128
strategic situation 160
strategic voting 173, 179
stratosphere 289
stream flow 105, 112
streams 82
stress 160, 290, 299, 301
strong states 269
structural change 312
structural constraints 198
structural power 210
'stuffing'ballot boxes 178
subjectivity 60
suboptimality 147
subpoena 94

subsidies 117, 128, 135, 158, 166, 223–4, 240, 251–2, 312
subsidisation 107
subsidy 98, 100, 103, 159, 240, 255, 258, 267
subsistence economy 161
subsistence farming 212, 222
substantive issues 299
substitutability 58
subversion 265
successive limited comparisons 190
suicide 80
sulphur 233
sulphuric acid 17
Sumerian empire 80
supergames 163
Superior, Lake 18
superpowers 160
superstition 211, 298
suppression 275, 291
surveillance, heightened 220
Surveyor–General of South Australia 237
survival 2, 58, 146, 205, 209, 215, 221, 284, 309, 314
sustainability 57, 68, 120, 133, 297, 312–3
sustainable development 51
sustainable economic development 278
swales hypothesis 22, 24
swamps 81
Swiss 161
Switzerland 148
Sydney's notorious sewer outfalls 275
Sydney 18, 88, 242, 288
Sydney Cove 235
Sydney Water Board 275
syllogism 63, 65
symbolic politics 66, 264
symbolic rewards 267
symbols 66, 69
symmetry 157
synergism 24, 47
synoptic method 195
synoptic model 193
syntax 36
synthetic fibres 47
synthetic textiles 46

taboos 211
tailings 10, 13
Taiwan 257
tallies 209
Taplin, R 276
tariff barriers 227, 257
tariffs 88, 251, 253, 270

Tasmania 58, 89, 115, 118, 175, 179, 233, 251–2, 256, 259, 267, 270, 275, 280
Tasmanian Department of Deep Sea Fisheries 291
Tasmanian Fisheries Department 16
Tasmanian Forestry Commission 289
Tasmanian Government 252, 267, 272, 292
Tasmanian Parliament 177
Tasmanian Woodchip Export Study Group 289
tautology 64–5
tax 145, 210, 212, 253
tax burden 255
tax concessions 240, 251, 254
tax rgime 314
tax relief 249
tax revenues 249
taxation 215, 223
taxes 88, 209, 223, 249
taxi 146
taxing power 253
taxpayer 105, 109, 128, 252
taxpayers 102, 132, 137, 251
TCDD 28
technical assistance 117
technical criteria 290
technical data 210
technical decisions 275
technical experts 290
technical knowledge 98
technical measures 131
technical pitfalls 290
technique 279
techniques 6, 300
technological breakthroughs 99
technological capability 230
technological change 222
technological dependence 249
technological factors 275
technological innovation 217, 290
technological necessity 275, 279
technology 2, 5, 26, 36, 39–40, 42, 47–8, 52, 55, 113, 142, 146–7, 149, 210–1, 213–5, 220, 223–4, 229, 232, 235, 257, 304, 315
telegraph 223
telegraph lines 249
teleology 212

347

Index

television 1, 51
temperate rainforest 233
temperature 81, 233
Temple University 198
temptation 290
ten inch isohyet 237
tension 268, 309
Terania Creek Inquiry 276
teratogenic 28
Terms of Reference 94
terms of trade 227, 307
territorial conflict 305
territory 209–11, 213, 267, 304
terrorists 226
testability 305
testing 307
textile industries 242
textiles 211
Thailand 227, 266, 305–7
Thatcher Government (UK) 173
theft 80
theoretical understanding 296
theories 4, 299
theories of decision-making 5
theory building 305
thinkers 298
third world 218, 257, 310
threatened species 147
threshold 146–7, 149, 192, 294, 296, 311
throw–away economy 47
thrushes 241
Tierra del Fuego 290
Tigris river 80
tiled drains 84, 132
timber-cutting 275, 307
timber 'industry' 266
timber 141, 218, 242, 266
tips (rubbish dumps) 55
tit-for-tat 163
tobacco smoking 53
tokenism 66, 67
Toledo 18
toleration 226
tools 36
Toona australis 242
Tories 106
Toronto 18
torture 226
totalistic governments 263
totalitarian states 221
tourism 79
tourist 'industry' 278
tourist attractions 244
tourist resorts 254, 288
tourists 155
town planning 289
town water supply 78

Townsville 286
toxic chemicals 43
toxic contamination 1
toxic waste 9, 15, 18, 26–7, 30, 244, 293
toxic waste disposal 289
toxicity, long-term 23
toxicity 4, 21
toxins 44
trace elements 99, 233, 240
traction 213
trade-offs 191, 196
trade-offs 191, 196
trade-offs 191, 196
trade 91, 142, 208, 211, 214–5, 220, 223, 257, 277, 306, 311, 315
trade advantage 308
trade goods 217
Trade Minister 270
trades unions 106, 160, 164, 268
tradition 106
traffic 146
'Tragedy of the Commons' 161
tramways 253
transboundary issues 26, 32
transfer value 177
transformers 13
transpiration 82, 132
transplant 245
transport 213, 215, 220, 222, 236, 243, 248–9, 259, 271–2, 314
transport costs 235
transport systems 306
transport technology 212
travel 307
Treasury 199, 277
treaties 90, 172
treaty provisions 306
tree-planting 132
trees 240
trial and error 295
tribal territory 226
tributaries 81
trichloroethylene 21
trichlorophenol 21
trip wires 195
trivial decisions 289
troops 213
tropical 73
truck 54
truck manufacturers 273
trucking firms 277
trust 153
Trusts 251
Tudor England 217
Tunisia 155
tunnel vision 200, 276
turbidity 78
Turkey 155, 263
turmoil 305
twentieth–century economy 247
twentieth century 99,

218, 222, 243, 250, 257, 259, 306
twenty–first century 306
tyranny of distance 235, 247
tyranny of small decisions 289, 297
tyranny of the majority 284

Ulster 226
ultraviolet radiation 290
uncertainty 4, 61, 172, 174, 192, 276, 282, 284, 304
under-representation 178
underdevelopment 258
understanding the future 4
unemployment 65, 79, 227, 264
United Nations 218
United Nations Commission for Environment and Development (UNCED) 1
United Kingdom (UK) 242
United Kingdom (UK) market 243
universities 277, 305
unnatural selection 211, 215
unprocessed exports 249
unspoiled wilderness 3
upper classes 241
Upper House 89, 178
urban class structure 245
urban development 244
urban infrastructure 222
urban public transport 253
urban rich 244
urban transport 251
urbanisation 13, 222, 289
urgency 4, 43, 306, 310
urgent issues 298
urgent problems 312
urgent questions 299
Uruguay Round 306
US–Canadian relations 156
US Congressional Committees 190
US District Court 27
US Environmental Protection Authority (USEPA) 22, 25, 27, 29
US Government 283
US laws 306
US market 307

US protegs 308
USA 1, 18, 24, 27–8, 30–2, 41–2, 44, 48, 67, 80, 141, 147, 157, 160, 175, 221–2, 224, 227, 229, 265, 269, 308
USSR 160, 221, 224, 230, 264, 305
utilitarian values 57
Utilitarianism 3, 5, 193, 284, 285, 287, 296
utility 58, 166, 283, 284, 295, 300, 302, 305

value added 257
values 274, 297
variables 286, 295
varnishes 13
vegetation 75, 102, 237, 238, 309
vermin 116, 236
versatility 300
vertebrates 45
vested interests 159, 273, 276
veto 89, 175, 187
victims 145
Victoria, Grain Elevators Board 251
Victoria, Water Conservation Act 1881 96
Victoria 31, 77, 79, 85, 87–9, 91–3, 95–100, 105–6, 109–10, 112–9, 122–3, 127–8, 131, 134–6, 143, 146, 157, 164–5, 178–80, 257, 266–7, 271
Victorian attitudes 236
Victorian Environment Protection Authority (EPA) 267
Victorian Government 96, 98, 115, 271, 275
Vietnam 28
vigilante groups 159
villagers 161
villages 148, 206
vineyards 100
violence 222
voluntarism 134, 197
vote the cemetery 178
voter turnout 174
voters 119, 137, 172, 179, 181
votes 175, 186
voting 5, 88, 106, 169, 172, 174, 180, 182, 287
voting machines 178
voting procedures 181
voting process 180
vultures 55
WA Inc. 94
wages 257
Wagga Wagga (NSW) 89
Wainwright, J.W. 270

Index

Wales 218
Walgett (NSW) 89, 91
wallaby 243
Wanna, J 195, 286
war 90, 160, 211, 219, 221, 226, 229, 308
war crimes 306
war of all men against all men 162
Ward, B 306
warfare 207–9, 213, 215, 221
Warhurst, J. 195, 286
warlordism 148, 221
warlords 309
Warsaw Pact 160
waste–assimilation 57, 60, 68
waste 124, 149, 219, 310
waste disposal 20, 220
waste dumps 32
waste management 40
wastes 9, 44–5, 155
wastes, intractable 13
water 2, 59, 65, 71, 87, 95, 100, 123, 135, 142–3, 147–8, 155, 233, 255, 259
water, diversion 78
water, flow 108
water accounting 119
water administration 115
water allocation 107–9, 121, 122
water authorities 105, 116
water charges 96–8, 240
water costs 119
water distribution 107, 111
water management 113, 119, 137, 267
water policy 128
water pricing 128, 133
water purification 157
water quality 78, 83, 90, 122–4, 141
water quality monitoring 293
water resources 95, 102, 111, 143
water right 90, 97–8, 101
water rights 87, 93, 103, 149
water shortages 233
water storage 259
water stress 233
water supplies 129
water supply 71, 96, 103, 109, 122, 129, 141, 153, 222, 251
Water Trusts 90
water use 119
waterbirds 78
watercourses 238
waterfowl 101
waterlogging 82, 98, 121
watertable 74, 81–3, 132, 149
Waterworks Trusts 96
weak government 269
wealth 181, 198, 208, 210–2, 215, 217, 219, 223, 228, 244, 264
weapons 224
weathering 81
Webb, M.C. 208
Webster, D 208
weirs 77, 82, 101, 108
Weiskel, T 211
welfare 254
welfare economics 169, 285, 287
Weller, P 195, 286
Wesley Vale kraft pulp mill 272, 275, 291–3
Western Australia 88, 236, 257, 270
Western Australian Government 275
Western countries 228
Western Europe 308
Western New Guinea 226
Western societies 220, 308
Westminster 269
wetlands 75
whales 146, 242
whaling 142
wheat 244
wheat 79–80, 99, 236, 240, 242, 244, 248
Wheat Case 107
wheat farmers 96
Whigs 106
White Australia policy 248
Whitlam, G 112, 115, 117
Whitlam Labor Government 112, 283
wilderness 42, 57–9, 240, 256, 266
Wilderness Society 268, 273
wildfowl 75
wildlife 26, 75, 111–2
wildlife reserves 77
Wildriver, S 276
wind erosion (of soil) 78
Winner, L 290
wisdom 5
witness box 276
Wodonga (Vic) 91
woodchipping 255, 289
woodlands 75, 77
wool 92, 236, 242, 245, 248
wool cheque 244
work 220
workers 268, 286
working conditions 106, 160, 222, 266
Working Parties on Ecologically Sustainable Development 278
working people 250
World Bank 227, 264
world government 229
World Health Organisation (WHO) 218
World Heritage 143, 280
world markets 248, 249
world order 305
writing 209, 215
Wujal Wujal Aboriginal settlement 179

yabbies 97
Yellow Peril 248

zanjera 148
zero-sum 119
zinc 17

349